# Religious Narrative, Cognition and Culture

*Religion, Cognition and Culture*

**Series Editors**

**Jeppe Sinding Jensen and Armin W. Geertz**
**Aarhus University**

This series is based on a broadly conceived cognitive science of religion. It explores the role of religion and culture in cognitive formation and brings together methods, theories and approaches from the humanities, psychology, and the social, cognitive and neurosciences.

The series is associated with the Religion, Cognition and Culture Unit (RCC) at the Department of the Study of Religion, Aarhus University http://www.teo.au.dk/en/research/current/cognition.

Forthcoming

*Past Minds: Studies in Cognitive Historiography*
Edited by Luther H. Martin and Jesper Sørensen

*Origins of Religion, Cognition and Culture*
Edited by Armin W. Geertz

*The Burning Saints: Emotion and Motivation in the Fire-walking Rituals of the Anastenaria*
Dimitris Xygalatas

*Mental Culture: Towards a Cognitive Science of Religion*
Edited by Dimitris Xygalatas and William W. McCorkle, Jr.

# Religious Narrative, Cognition and Culture

## Image and Word in the Mind of Narrative

Edited by

Armin W. Geertz and Jeppe Sinding Jensen

Routledge
Taylor & Francis Group
LONDON AND NEW YORK

First published 2011 by Equinox, an imprint of Acumen

Published 2014 by Routledge
2 Park Square, Milton Park, Abingdon, Oxon OX14 4RN
711 Third Avenue, New York, NY 10017, USA

*Routledge is an imprint of the Taylor & Francis Group, an informa business*



British Library Cataloguing-in-Publication Data

A catalogue record for this book is available from the British Library.

ISBN 978-1-84553-294-9 (hardback)
ISBN 978-1-84553-295-6 (paperback)

Library of Congress Cataloging-in-Publication Data

Religious narrative, cognition, and culture : image and word in the mind of narrative / edited by Armin W. Geertz and Jeppe Sinding Jensen.
p. cm. — (Religion, cognition, and culture)
Includes bibliographical references and indexes.
ISBN 978-1-84553-294-9 (hb) — ISBN 978-1-84553-295-6 (pbk.)
1. Religion. 2. Narration (Rhetoric) 3. Culture. 4. Cognition.
I. Geertz, Armin W., 1948– II. Jensen, Jeppe Sinding, 1951–
BL51.R354 2010
210–dc22

2009035251

Typeset by Queenston Publishing, Hamilton, Ontario, Canada

Index by Richard Bartholomew

# Contents

# Contributors

**Terrence Deacon**, Ph.D., received his degree from Harvard University in 1984 and has served on the faculties of Harvard University, Harvard Medical School, Boston University, and the University of California, Berkeley, where he is currently Professor of Anthropology and Neuroscience. His research combines human evolutionary biology and neuroscience with a focus on the evolution of human language capacity and cognition. His work extends from laboratory-based cellular-molecular neurobiology to the study of semiotic processes and the special challenge of explaining emergent processes in biology and cognitive neuroscience.

**Merlin Donald**, Ph.D., FRSC, is Emeritus Professor in the Department of Psychology and Faculty of Education, Queen's University, Kingston, Ontario, Canada. A cognitive neuroscientist with a background in philosophy, he is the author of many scientific papers, and two influential books: *Origins of the Modern Mind: Three stages in the evolution of culture and cognition* (Harvard, 1991), and *A Mind So Rare: The evolution of human consciousness* (Norton, 2001).

**Laura Feldt**, Ph.D., is Assistant Professor at the Department for Cross-Cultural and Regional Studies, University of Copenhagen. She has published articles on ancient religious narratives from Mesopotamia and the Hebrew Bible, on monsters, miracles, and religion and fantasy literature. She is currently working on a project on conceptions of wilderness in ancient religions.

**Armin W. Geertz**, Dr.Phil., is Professor in the History of Religions at the Department of the Study of Religion, and Chair of the Religion, Cognition and Culture Research Unit (RCC), Aarhus University, Denmark. He is currently involved in coordinating the research coalition called MIND*Lab* at Aarhus University, financed by the Ministry of Science in Denmark. His publications in the cognitive science of religion range from articles and chapters on evolutionary theory, atheism, the neurobiology of prayer and introductions to the cognitive science of religion. He is currently writing a book on the cognitive science of religion and a book on mysticism. He is co-editor of the *Religion, Cognition and Culture* series at Equinox.

**Douglas L. Gragg**, Ph.D., is Associate Dean, Library Director, and Professor of Bibliography and Research at Louisville Presbyterian Theological Seminary. He is the author of several book chapters relating cognitive theories to the study of Graeco-Roman religions and the compiler of a bibliography of publications in the cognitive science of religion. He is currently revising the bibliography and writing a book with the working title, Forms and Modes of Religion in the Ancient Roman World.

**Kenneth Hansen**, Ph.D., is Associate Professor in Communication and Experience Design at the Department of Communication, Aalborg University, Campus Ballerup, Denmark. In his publications he has dealt with anthropological aspects of computer mediated communication and the design of mediated experiences. His first book is about the design of meaningful experiences and is due to be published soon. He is currently writing a new book on valuable experience management and international development.

**James M. Hegarty**, Ph.D., is Lecturer in Indian Religions at Cardiff University. This is his first publication on aspects of the cognitive study of religion. He has published numerous scholarly articles on Sanskrit narrative as well as on medieval Sikh hagiography. He is the author of *Religion, Narrative and Public Imagination in Early South Asia: Past and Place in the Sanskrit Mahābhārata* in the Routledge Hindu Studies Series. He is currently editing the *Oxford Handbook of Hindu Literature* (with Will Johnson) as well as working on a monograph on the role of genealogical discourse in the formation of early South Asian views of the past (as part of a three year research project supported by the Arts and Humanities Research Council U.K.).

**Hans J. Lundager Jensen**, Dr.Theol., is Professor in Religious Studies and in Old Testament, Aarhus University, Denmark. He teaches and publishes on structural studies of myth and ritual, classics in the study of religion, and Old Testament myths, legends, and cult.

**Jeppe Sinding Jensen**, Dr.Phil., is Associate Professor at the Department of the Study of Religion, and Coordinator of the Religion, Cognition and Culture Research Unit (RCC), Aarhus University, Denmark. His research interests include semantics and cognition in religious narrativity, myth and cosmology, and method, theory and the philosophy of science in the study of religion. He is the author of *The study of religion in a new key: Theoretical and philosophical soundings in the comparative and general study of religion* (2003) and the co-editor of *Rationality and the study of religion* (2003). He is editor-in-chief of the *Religion, Cognition and Culture* series at Equinox.

**Anita Maria Leopold**, Ph.D., received her degree at the Department of the Study of Religion, Aarhus University, Denmark. She is an independent scholar, lecturer and writer. Her publications on the cognitive science of religion range from articles and an anthology on a new theory of syncretism in religion explaining the nature of conceptual blending and religious categorizing in terms of the structure of human cognition. She is currently writing a novel based on cognitive theories of conscience and the “dreaming mind.”

**Anders Lisdorf**, Ph.D., is an independent scholar who earned his degree in the History of Religions from the University of Copenhagen, Denmark. His publications focus on how the cognitive science of religion can aid in understanding the historical reality of and the practice of divination in ancient Roman times.

**Guðmundur Ingi Markússon**, Cand.Mag., holds degrees in the study of religion and semiotics from Aarhus University, Denmark. His interests include the significance of religious information and signals as social markers (a pilot project with Human Behavior Laboratory, University of Iceland); Western esotericism; contemporary religiosity; atheism; evolutionary psychology and its limits; gene-culture co-evolution; biosemiotics, especially the "rehabilitation" of memetics through Peircian semiotics (in the work of T.W. Deacon) and the significance of semiotic ecologies for cultural evolution; Darwinism in the humanities and social sciences. He is a member of the Reykjavik Academy, association of independent scholars, in Reykjavik, Iceland, and a member of the International Association for the Cognitive Science of Religion (IACSR).

**Kirstine Munk**, Ph.D., is an independent scholar who holds a Ph.D. degree in Religious Studies and a Double Masters Degree in History of Religion and History of Philosophy and Ideas from the University of Copenhagen, Denmark. She has been working intensively with divination and is the author and editor of several publications on that topic.

**Rukmini Bhaya Nair**, Ph.D. (Cambridge) is Professor of Linguistics and English at the Indian Institute of Technology, Delhi. Widely recognized for her work in linguistics, cognition and literary theory, Nair received a second *honoris causa* doctoral degree from the University of Antwerp in 2006. Among her many books are *Lying on the Postcolonial Couch: The Idea of Indifference* (2002), *Narrative Gravity: Conversation, Cognition, Culture* (2003), *Yellow Hibiscus: New and Selected Poems* (2004) and *Poetry in a Time of Terror: Essays in the Postcolonial Preternatural* (2009). Nair, considered one of India's leading English language poets, has taught at universities ranging from Singapore to Stanford. She is currently working on a study of Salman Rushdie's fiction within the theoretical framework of "Literary Darwinism," a monograph on the evolutionary and pragmatic value of the human ability to deceive as well as a cycle of poems on the recent Mumbai terror attacks. Nair's latest research project, funded by the Indian Department of Science and Technology under their "Cognitive Science Initiative" is in the broad area of "Language, Emotion and Culture" (2010–2013).

**Anders Nielsen**, Cand.Mag., holds degrees in the History of Religions, Russian, and History from the University of Copenhagen, Denmark. He has published articles on Orthodox Christianity and the popular use of religious images.

**Ilkka Pyysiäinen**, Ph.D., is Academy Research Fellow at the Helsinki Collegium for Advanced Studies. He has published widely in international journals on a variety of topics in the cognitive science of religion. His most recent book is *Supernatural Agents: Why We Believe in Souls, Gods, and Buddhas* (Oxford University Press, 2009).

**Chris Sinha**, Ph.D., is Professor of Psychology of Language at the University of Portsmouth. He has taught in departments of Education, Psychology and Linguistics, in Britain, Brazil, the Netherlands, Denmark and India. He has published widely in anthropology, linguistics, education, evolutionary biology and connection science, as well as developmental and cultural psychology. His central research interest is in the relations between language, cognition and culture, and a main aim of his research is to integrate cognitive linguistic with socio-cultural approaches to language and communication, especially in relation to human development.

**Tom Sjöblom**, Dr. Theol, is Docent in the History of Religions at the Department of the Study of Religions, University of Helsinki, Finland. He is the author of the book *Early Irish Taboos—A Study in Cognitive History*. His interest areas in cognitive science include narrative cognition, the origins of religious cognition, emotional communication, and cognitive ethology.

# Introduction

Armin W. Geertz and Jeppe Sinding Jensen

This book is about religion as an expression of cognitive mechanisms that are both biological and social and which reveal themselves through culture. The cultural products of interest here are religion and narrative. Both, we believe, are fundamental human features and must be examined through diverse methods and theories that aid us in linking "the self" with "society" and "the mind." Assuming that cognition is neither solely nor perhaps even primarily located in individual brains, rather being distributed and situated in pragmatic social contexts, we are convinced that narrative is a key to discovering important insights into these complex relations.

Cognitive studies in relation to religion and culture have until recently mostly focused on what the mind "brings to the social world," but, here in line with our special point of departure, we focus on what the social world brings to the mind, that is, the formative effects that various kinds of cultural knowledge (in narratives, classificatory systems, rituals, institutions and everyday practice) may have on the ways in which minds are made and how they work. This project has become crucial because of recent breakthroughs in the neurocognitive and psychological sciences. Recent insights on the plasticity of the brain and the central role that social experience and culture plays in forming neural networks in the brain have persuaded neuroscientists and psychologists of the need to deal with areas of expertise in the human and social sciences. In other words, we need the help of colleagues from other fields, and they need help from us. It is exactly this interdisciplinarity which is so exciting about the study of cognition and religion more generally, and religious narrativity more specifically. The sense of excitement is enhanced by the fact that in one way or another and within the framework of one discipline or another, we are all trying to grasp the fundamental features of what it means to be human.

So, what is the mind in narrative, and of what does it consist? Is it the symbolic competence that pushed us along the evolutionary highway and provided us with a remarkably complex organ of self-assembly and self-organizing processes, as Terrence Deacon suggests? Or is it a key cognitive management technique, as Merlin Donald suggests, that creates our hybrid minds? Perhaps the mind is filled with iconology and imagination in a kind of coordinated sociogenetic economy, as Chris Sinha suggests. Perhaps it is a meme-vehicle that crosses cultural barriers conveying the emotions necessary for cultural survival, as Rukmini Nair suggests. Or does it consist of intuitive knowledge, agency, narratives, counter-intuitiveness and common knowledge, as Ilkka Pyysiäinen suggests? Our guess is that the mind in narrative is all of these things and much more.

Answering such rhetorical questions involves defining what we mean by "mind" and "narrative." We will not try to define the term "mind" in this introduction, but our basic assumption is to avoid being both "mind-blind" and "social-blind." Instead, drawing inspiration from scholars from a wide variety of disciplines and sciences, such as anthropologist Edwin Hutchins, neuroscientist Daniel Siegel, philosopher Andy Clark and psychologist Merlin Donald, we are following the middle road of what could loosely be called "distributed cognition," in other words, the idea that cognition is both inside and outside individual brains and, furthermore, that it is fundamentally cultural and social. As Daniel Siegel wrote, "human connections shape the neural connections from which the mind emerges" (1999, 4). Common to many psychologists, neurologists, anthropologists and linguists is the idea that narrative and communication are essential factors not only in shaping social relations but also in shaping selves. Thus narrative can be understood as the "glue" between individual and group cognition, in other words, between cognition and culture.

The topic of narrative is not new to scientific research. It has, however, been going on in mutually unintelligible circles. Especially since the 1950s, substantial work has been done on narrative, producing myriads of theories, interpretations and explanations that need to be integrated if they are to be of use to anyone outside their special disciplines. An array of disciplines have made their impact: 1) the text disciplines (such as folkloristics, semiotics and structuralism) with their interest in the formal aspects of written and oral narrative; 2) linguistics with its interdisciplinary interest in languages and communication; 3) anthropology, history and sociology with each their own particular take on narrative; 4) developmental psychologists and psychoanalysts with their interests in language and narrative for the developing child and adult; 5) social psychologists with their interest in what makes a good story and why; 6) neuroscientists with their interest in language and the brain as well as narrative and social cognition; and 7) philosophers with their interest in narrative as communicative act.

An integrative study of narrative involves three thematic areas: 1) theory, 2) text and 3) behavior.

1) A *theory of narrative* involves not only philosophical reflection on subjectivity, consciousness, representation and communication, but also more general reflections on the philosophy of science. An important aspect of the latter would be reflection on the strengths and weaknesses of the neo-positivist stance of cognitivism, and how it might contribute to our understanding of the human situation. A theory of *religious* narrative would necessarily require more general theories of religion and its relation to narrative. Thus, besides general topics like the interface between cognition and religion, theorizing is also needed to define and explore how myths, rituals, hagiographies, homilies and other sacred narratives relate to and function in relation to identity construction, memory talk and concepts of person.

2) *Religious expression* involves a whole series of media, but the bread and butter of most scholars of religion are sacred texts. Others work with oral literature that has been, or eventually becomes, reproduced in texts. But iconography, architecture, archaeological sites and remains, paintings and so on, constitute highly important expressions of religious narrative, in other words, non-linguistic media used to express religious sentiment and knowledge. The wide variety of approaches to the study of communication and language constitute the backbone of the history of religions. These approaches have dominated the

field for some 150 years, but they have been reluctant to view religious narrative in the wider perspectives of theory and behavior.

3) *Religious behavior* has been industriously studied from a variety of perspectives. Ethnographers and anthropologists have specialized in the narratives and cultures of indigenous peoples, whereas sociologists have specialized in the narratives and cultures of Western societies. The ethnosciences have used conversational analysis and narration to identify representations of self, forms of talk, and cultural models. Social psychology has focused on symbolic interaction and situated cognition. The psychology of religion has been interested in religious experience, visions, ecstatic behavior and so on. Recently, cognitive psychology and the neurosciences have made their impact on a variety of subjects. Here we find studies of memory, body and emotion, cognitive semantics and symbolization in connection with the neurophysiology and neuropsychology of mental processes.

Thus an integrative study of narrative draws on a wide variety of sciences and disciplines. It must also take into account the role that *religious* narrative has played in the history and current situation of humankind. Narrative demands attention because it is the single most important human enterprise.

The expressions of religious narrative, and the media used to communicate them, are myriad. Religious narrative is many things, ranging from interpersonal oral exchange to written texts, architecture and symbolic images. Each genre has its own limitations and potentialities, but all exhibit the fundamental interaction of self, society and mind. What we need to promote and orchestrate are systematic studies of the mechanisms involved. Describing and interpreting surface phenomena are not sufficient for us. We also need to find out how they originate and what they are made of. To do this, we need the help of colleagues from other fields.

## Themes and Issues Addressed

The chapters presented here address a variety of themes and issues relevant to religious narrative. The first two chapters by Armin W. Geertz and Jeppe Sinding Jensen explore various approaches, theoretical stances and definitions in an attempt to clarify and add precision to the issues at stake. Both chapters insist on a combination of top-down, bottom-up approaches.

The second section, which consists of five chapters, explores various explanatory interpretations. These chapters illustrate in more detail what a combined top-down, bottom-up approach implies. These keynote chapters are arranged in a bottom-up progression. Terrence W. Deacon's chapter begins with the neural correlates at play in the brain during the production of a speech-act. He shows that language is a neurologically emergent function produced through the complex synergy between multiple brain systems and levels of progressive neural differentiation and, he argues furthermore that a more sophisticated embedding of linguistic theory in semiotic theory is necessary in order to bridge the gap between linguistics and neurology. Merlin Donald's chapter moves on to phylogenetic and cultural levels where the origin of language is to be found in cognitive communities, understood to mean the "interconnected and distributed activity of many brains." Language evolved under the selective pressure of such communities, and as shared cognitive networks became more precise, language emerged. The drive was not to use words or develop grammars, but to "bond as a group, to learn to share attention and set up the social patterns that would sustain such sharing and bonding in the species." In Chris

Sinha's chapter, which attempts to combine literary with cognitive theory, is concerned with the links between cognitive processes and sociogenesis. Following philosopher Cornelius Castoriadis, Sinha argues that the specific site of sociogenesis is the "social imaginary," a term designating the order or pattern upon which any society grounds its symbolic order. Rukmini Bhaya Nair's chapter is concerned with the narratological and semantic levels, but analyzes the nature of narrative in terms of cognitive, neurobiological and speech-act theories. In challenging speech-act theories, Nair expands narratological analysis, which focuses on the sentence level, to a more encompassing conversation and story level. At these levels it becomes clear that narratives are most often co-created. Furthermore, they allow both speaker and participant listener to demonstrate their "implicultural competence." Nair applies these insights in an analysis of the religious elements in Salman Rushdie's work and why he has been condemned by Muslims seen from the perspective of evolutionary theory. With Ilkka Pyysiäinen's chapter, we come full round to religious behavior by showing how the psychology of ritual helps make narrative beliefs possible. Through the use of rich perceptual details that draw on our natural capacities of motor action, agency detection and mindreading, rituals transform imagined agents into real ones. Thus religious rituals are a means of "belief enhancement."

The third and final section, which consists of eleven chapters, addresses what is perhaps the most pressing issue in the comparative study of religions, namely, what and how can cognitive approaches contribute to the study of religion? Tom Sjöblom draws on a wide variety of cognitive scientists, philosophers, neuroscientists and developmental psychologists to show how narrative functions as emotional communication. These insights are then applied to the taboos in early Irish literature (600–1200 AD). He argues that emotional communication makes narratives relevant to the audience. Hans J. Lundager Jensen applies the theories of Émile Durkheim and Merlin Donald together with meme theory to an analysis of the book of Deuteronomy in the Hebrew Bible. In this perspective, Moses is seen as the great "meme-disseminator," and the Yahweh religion as having been in competition with other meme-generators wherever the Israelites wandered. It is a clear concern of the text that the many rules and norms of the Yahweh religion must be deliberately and repeatedly learned, "recollected" and embodied by each generation as a meme survival technique. In her analysis of the Hebrew Bible, Laura Feldt applies Maurice Halbwachs' concept of collective memory, by which individual memory grows through socialization processes, and Jan and Aleida Assmann's concept of cultural memory, by which shared memories are seen as the products of mediation, textualization and communication. Together with fantasy theory, Feldt analyzes the fantastic in the Exodus narrative and argues that it establishes specific mnemonic practices for the purpose of shaping the individual memories of each generation. Douglas L. Gragg focuses on the shock effect described by Harvey Whitehouse as a central characteristic of the imagistic mode of religiosity during which spontaneous exegetical reflection is claimed to arise. Gragg argues that the parables of Jesus, with their striking analogies, play a role similar to rituals performed in the imagistic mode. He argues that straightforward interpretations of the parables were not possible and that listeners received "cognitive shocks" because the parables not only violated social conventions but also expectations rooted in universal human cognition. These shocks encouraged spontaneous exegetical reflection, but, Gragg argues, the parables became "domesticated" by early Christian writers because they violated such universal expectations.

Anders Nielsen analyzes the cult of icons in contemporary Georgian Orthodox Christianity. He applies Alfred Gell's theory of art and illustrates how icons function in Georgian

Orthodox social exchange. Drawing on Pascal Boyer's work, Nielsen then explores the counter-intuitiveness of icon "personhood." He confirms that these superhuman agents are counter-intuitive, but he argues also that they are not counter-cultural. Thus even though they are felt by the practitioners to be quite natural, they are still "attention-demanding," as Boyer has argued. Nielsen then draws on the work of Edwin Hutchins and Lawrence Barsalou to show how icons function as material anchors and in embodiment. James M. Hegarty specifically addresses how cognitive and narrative theories can be applied to religious texts, in this case, the great Indian epic *Mahābhārata*. Thus Michael Tomasello's theory of the cultural origins of human cognition are combined with the work of Rukmini Nair and Mikhail Bakhtin who have analyzed the roles of narrative in human social life and cognition. Hegarty demonstrates that the *Mahābhārata* is structured along the lines of Tomasello's theory of socio-cognitive development, and it serves both as a repository of socially significant knowledge and "perspectivally based cognitive representation." The text itself provides techniques for dialogical thinking and ongoing public memory practice.

Anders Lisdorf takes a refreshing angle on how narrative, ritual and space fit together. Taking his departure in a systematic critique of both Mircea Eliade's and Jonathan Z. Smith's analyses of the Numbakulla myth and its concomitant Engwura ceremony among the Arrernte of Central Australia, Lisdorf demonstrates that both Eliade and Smith assume that narratives construct maps of the world and that these serve as maps for humans navigating in the real world. In drawing on the experimental evidence in neurology, robotics and cognitive sciences, Lisdorf shows that this is not the case. Narrative mapping is highly selective and restricts itself to symbolic or abstract landmarks with particular significance. Anita Maria Leopold draws on the work of neuropsychologist Louis Cozolino, cognitive linguists Gilles Fauconnier and Mark Turner and Rukmini Nair in an analysis of syncretism. Leopold argues that syncretism consists of three theoretical levels: the social, semiotic and cognitive. The social level involves issues of power, the semiotic issues of interpretation and the cognitive issues of conceptualization and categorization. Narrative, she argues, is a link between all three levels. She applies the resultant model on the Traditionalist Movement among the Hopi Indians in Arizona and their syncretistic interaction with New Agers as well as Rabbi Yanassan Gershom, an American Jew who employs New Age therapies on self-proclaimed American, reincarnated Holocaust victims.

Guðmundur Ingi Markússon examines the dynamics of the modern Icelandic Ásatrú neo-pagan movement. He argues that the Icelandic "national narrative" is one of the causes of the dramatic increase of membership in the movement, and he applies the cognitive concept of culture in Pascal Boyer's understanding as well as cognitive scientist of religion Jesper Sørensen's concept of the "immunology of cultural systems." Markússon argues that "narrative fit" makes the Ásatrú more salient and relevant among those disaffected with the national Christian Church. This salience, however, is culturally constructed and is not the result of innate cognitive mechanisms, as Boyer has argued. This critique is supported by Sørensen's immunology theory whereby societies develop cultural systems that select or resist new representations. Kirstine Munk analyzes the modern phenomenon of astrology in the U.S. and Scandinavia. She draws on developmental psychologists Katherine Nelson and Jerome Bruner, life story theorist Charlotte Linde, narratologist Manfred Jahn and others to show how people re-create themselves in new social contexts through the use of available cultural repertoires and actually form a new sense of self that has concrete influence on their lives. Through a biographical analysis of Danish astrologer Christian Borup, Munk argues that Borup's proclaimed "birth myth," which he uses as a tool in his astrological sessions and public talks, fulfills all the requirements of an effective and creative life story that uses drama, emotion and plot to define purpose in the individual lives.

This volume ends with a fascinating chapter by Kenneth Hansen. The topic of his analysis is 3D Cyberspace worlds. He shows how fictive agents in these worlds become real agents through virtual rituals. His point of departure is the "universal principles" which cognitive scientists of religion E. Thomas Lawson and Robert N. McCauley developed in their ritual competence theory. Through an analysis of virtual weddings in Alpha World, Hansen formulates a general theory of virtual rituals which supplement Lawson and McCauley's universal principles. Thus he introduces the Avatar Rule, the Epoché Rule, the Utopia Rule and the Meeting Rule as well as a typology of virtual rituals. The chapter ends with a cultural model of virtual rituals.

## Background

The papers in this volume are a result of a two-year research project called "Religious Narrative, Cognition and Culture" conducted at the Laboratory on Theories of Religion, Department of the Study of Religion, Aarhus University. The Laboratory is the culmination of two lines of research activity both of which were concerned with what we broadly call "theories of religion." We understand the phrase to mean theoretical reflections on religious symbolic systems and behavior conceived in terms of the philosophy of science. The first of the two lines of activities was initiated during the middle 1980s by Jeppe Sinding Jensen and Armin W. Geertz consisting of hosting theoretically oriented conferences, participation in the Groningen Working Group for the Study of Religious Symbols, which was founded in The Netherlands in 1983, and producing publications dealing with cognition, the philosophy of science and other theoretically interesting matters. The second of the two lines of activities which eventually led to the Laboratory on Theories of Religion was begun in 1999 by Hans J. Lundager Jensen and Anders Klostergaard Petersen, both at our Department of the Study of Religion, who established an interdisciplinary forum for the study of ritual—dubbed at the time Laboratory on Theories of Ritual—within which participants discussed a series of important books on theories of ritual, from Roy Rappoport's opus to Lawson and McCauley's and Harvey Whitehouse's groundbreaking work. Two seminars were held and publications were produced in a Danish journal. By 2001, we were all convinced that cognitive theory was here to stay and that we would most likely be reading and discussing publications on cognition in the near foreseeable future. We therefore changed the name of the laboratory to Laboratory on Theories of Religion and opened it to the research staff including doctoral students as well as students working on their master's theses. By this time a growing number of young scholars were writing master's theses and doctoral dissertations on cognition and religion, so it made good sense for their advisors to systematically study the literature and focus more research time on cognitive topics. In 2002 we were awarded a grant from the Faculty of Theology to pursue a two-year project on "Religious Narrative, Cognition and Culture." Our small group of 13 core participants, mostly in the study of religion, expanded during the next two years into a group of some 40-50 Danish and international colleagues, doctoral students and masters students from the study of religion, theology, anthropology, psychology, neuroscience, semiotics, modern and classical languages, archaeology and so on, all trying to grasp the roles, functions and processes of narrative, cognition and culture.

At the end of the two-year period, the Faculty of Theology awarded a grant for a four-year project with the broader focus of "Religion, Cognition and Culture." Our national and international contacts and partners increased to some 150 individuals as well as formal projects and agreements with a number of institutions such as the Center for Functional Integrative Neuroscience (CFIN) at the Aarhus University; the research focus area "Cognition, Communication and Culture" at the Faculty of the Humanities in Aarhus; the university hospitals in Aarhus and Skejby; the Institute of Cognition and Culture at Queen's University, Belfast; the Research Project "Mind and Society in the Transmission of Religion" at the Helsinki Collegium for Advanced Studies and the Department of Comparative Religion, University of Helsinki. We were also active in establishing the International Association for the Cognitive Science of Religion. As of 2008, the Religion, Cognition and Culture (RCC) group has become recognized as a research unit under the Department of the Study of Religion in Aarhus as well as a full partner in MIND*Lab*, a multimillion dollar project coordinating major centers and departments in Aarhus that produce research on brain, body and cognition.

The majority of the papers in this volume originate from two separate conferences. The papers by Rukmini Bhaya Nair and Ilkka Pyysiäinen were keynote addresses read at the conference entitled "Image and Word in the Mind of Narrative," October 3-4, 2003, hosted by our Laboratory and Department. That conference was one of the anniversary events held by the Faculty of Theology in celebration of the 75th anniversary of the Aarhus University. Terrence Deacon and Chris Sinha were the two other keynote speakers, but the papers in this volume are reprints from earlier publications dealing with the same topics. Merlin Donald gave a keynote address at the conference entitled "Religious Narrative, Cognition and Culture," December 2-4, 2004, which was hosted by our Laboratory, the Department and the Graduate School on "Religion, Identity, Culture" at our Faculty. The paper in this volume, however, is an abridged and revised version of two chapters in his book *A Mind So Rare*. Other papers presented at the December 2004 conference were read by Tom Sjöblom, Laura Feldt, Douglas L. Gragg, James M. Hegarty, Anders Lisdorf, Anita Maria Leopold, Guðmundur Ingi Markússon and Kirstine Munk. The papers by Hans J. Lundager Jensen, Anders Nielsen and Kenneth Hansen were contracted for this volume. Their origins are noted in each paper.

We extend heartfelt thanks to the Faculty of Theology, Aarhus University for continued financial support, conference facilities and secretarial assistance. We also thank the Aarhus University Research Foundation for financial support.

Thanks goes as well to David Alan Warburton, who played a key role in the two conferences and who has helped with the linguistic editing process. We also want to thank the discussants at the 2003 conference: Marianne Horsdal, Tine Biering, Anders Lisdorf, Jesper Sørensen, Guðmundur Ingi Markússon, Tinne Stubbe Østergaard, Lars Albinus and Jan Tønnesvang. Thanks is also extended to our student assistants at the two conferences: Anne Pedersen, Tina Nielsen, Ulla Winther and Jesper Østergaard.

Our special thanks goes to our publisher Janet Joyce who recognized our vision very early on and found it worthy of a special series at Equinox Publishers in London. Thank you, Janet, for your support and patience. Thanks as well to our editor Valerie Hall who helped guide us through the publication process. And special thanks to desk editor Dr. Russell Adams for his talented copyediting and typesetting and to Dr. Richard Bartholomew for his excellent indexing.

# — 1 —

# Religious Narrative, Cognition and Culture: Approaches and Definitions

Armin W. Geertz

Religion is about communication. It is a quintessential tool in harnessing our fundamental drives of sociality and meaning-seeking. We have evidently thrived as a species on realizing and materializing our virtual worlds. Some scholars, such as Steven Mithen (1996), even claim that the ability to imagine religious ideas was a result of our ability to materialize them in symbolic form. This may or may not be the case, however, it seems to be a truism that *Homo sapiens* excel in managing the world though the material and social extension of imaginary worlds (see Clark 1997, 2008; Rowlands 2003; Day 2004; Robbins and Aydede 2009). What is exciting for scholars today is trying to discover how we do this. Narrative is obviously one of the ways.

I defined religion in an earlier publication as "a cultural system and a social institution that governs and promotes ideal interpretations of existence and ideal praxis with reference to postulated transempirical powers or beings" (Geertz 1999, 471). Such beings populate religious worlds in the guise of cosmologies, theologies and anthropologies, and they are given real, concrete significance in social and cultural contexts. Thus religion and religious behavior consist of communicating these ideas and enforcing them in social contexts. More importantly these ideas are used in what Merlin Donald has called "cognitive governance," by which he means that the executive functions of consciousness are heavily dependent on cultural artefacts, symbols and models. Thus, culture serves as a governing factor in human cognition by providing top-down repertoires, frameworks and routines which are socially and collectively created and which shape the cognition of each individual. Further, these governing instruments—Donald also calls the process "enculturation"—involves emotional conditioning as well (Donald 2001, 59–70, 195–200, and Donald and Andreassen 2007). Thus any concept of cognition must include the brain, body and culture.

Religious narrative promulgates, extrapolates and investigates the significance of virtual worlds in real time contexts while at the same time refining them in all their majestic virtuality. Religious narrative provides paradigms for human identity thereby providing narrative governance of human cognition and emotion.

Religious narrative is more than written texts and/or oral traditions, although these are key expressions. The reasons for this are that we humans have many ways of communicating.

As Donald has shown, we still use our essential mimetic skills of the pre-linguistic phases of hominine evolution (Donald 2001, 259–274 and his essay in this volume). As he noted concerning intentional communication:

> We can say that human beings are inherently actors. They are truly theatrical in their actions, when *they know that they are communicating* with people by their expressions, body language, gestures, attitudes, postures, and tone of voice. There is a great tendency among human beings to play a role and to expect confirmation from others. We are a conforming species. This strong conformity of human beings is one of the best pieces of evidence for the importance of mimetic expression in human social life. The social domain is largely mimetic in the sense that it is a theater of action collectively encapsulating the events that define the society rather than an unfolding of events *per se*. (Donald and Andreassen 2007, 76)

One of the primary subject matters of mimesis is to communicate states of mind and emotion (feigned or otherwise) to others.[1]

Since our evolutionary mimetic phase, we have designed symbolic media that extend not only the types but also the reach of communication available to us. In fact, Donald claims that external symbols serve as external memory storage and thus allow "the human species to transcend time and space, and to escape the nervous system" (Donald and Andreassen 2007, 80). In terms of our interest in religious narrative, Donald claims that:

> External symbolic devices have enhanced our metacognitive understanding of human life. Indeed, one could argue that the primary function of the "text" in civilization is a metacognitive one. It provides the prototype or archetype against which individual lives are measured. (Donald and Andreassen 2007, 81)

Our stories dominate us. They provide identity, morality, class and authority, even models for everyday life. But "this collective leap into the narrative domain" brought us much more:

> Narratives, especially shared life narratives, are the basis of autobiographical memory itself. Stories and myths can completely reshape our semantic spaces, leading to a consensual definition of a shared virtual reality that is the core of oral culture. (Donald 2001, 296)

Narrative is a consequence of pre-linguistic mimetic communication, but when it became verbal in our evolutionary history, it took on a foundational role that changed our consciousness and the ways that we think. The advent of written media provided a further cognitive revolution, the consequences of which we as a species still have not witnessed to the end. Not only has literacy literally reworked our neuronal networks, it has introduced an unnatural, non-Darwinian instrument. "Literate culture," Donald argues, "has capitalized on untapped cerebral potential and reprogrammed the human brain in its own image" (2001, 304). Not only have individual brains been reorganized but also "the collective architecture of cognition...and how the larger human community thinks and remembers. They also enabled many new forms of mental representation" (2001, 304).

Faced with such immense complexity, what, then, is narrative? How can we define it in a way that captures its complexity? A few recent volumes have addressed narrative and cognition from different perspectives. In te Molder and Potter (2005) the topic is addressed from the perspectives of conversation analysis, discursive psychology and ethnomethodology. Hogan (2003) approaches narrative and cognition in terms of literary universals and emo-

1. For recent work on the neuroscience of dance, see Bachner-Melman *et al.* 2005; Brown *et al.* 2006a; and Brown and Parsons 2008.

tion. Herman (2003), on the other hand, promotes an interdisciplinary approach called "cognitive narratology." Gottschall and Wilson (2005) take on the burgeoning field of Darwinian literary studies in an attempt to present the evolutionary and cognitive propensities for literature and art (cf. Zushine 2006; Barash and Barash 2006; and Boyd 2009). Several of these publications contain good overviews of the literature but none have offered comprehensive definitions.[2] In the following, I will introduce an attempt to gain a conceptual clarity that encompasses the complexity of narrative.

## Approaching Narrative

It seems to me that there are three fundamental aspects of narrative at play.[3] These aspects are mechanisms and processes, media and expressions (Table 1.1). *The first aspect* includes neurobiological, psychological, social and cultural mechanisms and processes. The basis for narrative takes its departure in neurobiological and cognitive processes including consciousness, memory and sense of self. In fact, as already mentioned, narrative helps manage these processes, but there is a biological basis that makes such management possible. There are several ways that this management occurs, the primary one being the manipulation or stimulation of emotional drives. Another important process is that of cognitive mapping. This mapping process occurs, of course, at all levels of human existence: biological, psychological, social and cultural. But the point here is that already at the most fundamental biological level, the brain maps internal and external worlds on the basis of which it continually makes predictions about those worlds. Even in such basic and largely unconscious processes such as the senses, the brain predicts what it is sensing on the basis of its maps rather than just sensing raw data from the world (see Frith 2007; Blakeslee and Blakeslee 2007; Ramachandran and Blakeslee 1998). As Frith cogently put it, we do not experience the world itself, rather we experience our brain's map of the world. It is, in fact, surprising, Frith points out, "that our brain ever gets things right" (Frith 2007, 85):

> [O]ur brains discover what is out there in the world by constructing models and making predictions. Our models are built by combining information from our senses with our prior expectations. Both sensations and expectations are essential for this process. We are not aware of all the work our brain is doing. We are only aware of the models that result from this work. This makes our experience of the world seem effortless and direct.
>
> (Frith 2007, 138)

Furthermore, our brains create models of the minds of others:

> By making models of the minds of others (in the same way that it makes models of the physical world), my brain enables me to enter a shared mental world. By sharing my mental world with others, I can also learn from their experiences and adopt the models of others that are better than my own. From this process, truth and progress can emerge, but so can deception and mass delusions. (Frith 2007, 183).

It should be obvious from these neurological insights that narrative, so to speak, plugs

2. The very brief narratological definitions mentioned in Herman's introduction (2003, 2, including note 3) will not suffice for what I have in mind, as will become evident in the following.

3. The rest of this chapter is an expanded version of my essay "Religiøs narrativitet—Et definitionsforsøg," ("Religious Narrative: An Attempted Definition") in *Religiøs narrativitet, kognition og kultur* ("Religious Narrative, Cognition and Culture"), ed. Jeppe Sinding Jensen and Armin W. Geertz (Højbjerg: Forlaget Univers, 2008), 26–45.

| **Religious narrative** |
|---|
| *Mechanisms and processes* |
| 1. *Neurobiological aspects:* neurological and bodily processes, consciousness, memory, sense of self, emotions, cognitive maps<br>2. *Psychological aspects:* feelings, personality, dreams, etc.<br>3. *Socialization aspects:* nurture relations, language and communication, autobiographies<br>4. *Formalized institutions:* symbolic systems, universe maintenance techniques and institutions; lines of power and discourse; habitus and cultural capital<br>5. *Larger social and cultural aspects:* development, stagnation and destruction |
| *Media* |
| 1. *Bodily mimetic techniques:* dance, gesticulation, posture and movement<br>2. *Voice:* basic emotional sounds, song and oral traditions<br>3. *Symbols:* birchbark charts, stick bundles, palm leaves, rock art, beads<br>4. *Iconography (body):* body ornament and costumes, ritual and other symbolic behavior<br>*Iconography (objects):* natural objects like trees, mountains, caves; and human-made objects like statues, puppets and masks<br>*Iconography (architecture):* temples, houses, columns, poles<br>*Iconography (pictures):* paintings, films, photographs, videos, the internet<br>5. *Writing:* books, scrolls, tablets, letters, papyri |
| *Expressions* |
| 1. *Formal narratives:* myths, legends, songs, epics, sagas<br>2. *Personal narratives:* speeches, tales of travel, tales of conversion, autobiographies, dream accounts, vision accounts, diaries, counsel, jokes, letters<br>3. *Ritual texts:* sermons, hymns, mantras, homolies, litanies, magical formulas, oracles, glossolalia, prayers<br>4. *Traditional literature:* exegesis, apologetics, hagiographies, wisdom literature, laws, codes, decretals, informal didactic formulas (proverbs, riddles, epigrams)<br>5. *Handbooks:* magical books, ritual handbooks, meditation handbooks, prayer books<br>6. *Popular genres:* essays, novels, plays |

**Table 1.1** Religious narrative.

into this powerful process, thus taking advantage of it and extending it.

The differences between psychological and neurological processes that seemed so self-evident until about twenty years ago are quickly disappearing as neurologists gain more and more insight into the brain and the mind. A good example of this is the neurological studies of consciousness. Much of this research is concerned with neurochemical processes that seem to be associated with consciousness.[4] Neurologist Antonio Damasio's book title *The Feeling of What Happens* (1999) is a good indication of where neurologists are searching for consciousness: in emotion and senses of self. This exciting research is clearly within the domains of psychology, but it would seem to be counterproductive to maintain inviolable boundaries. On the other hand, however, neurologists are dealing with very basic and primitive senses of self, whereas what we usually associate with our feelings, personality or soul are much more complicated and, furthermore, integrated in social and cultural processes.

It should be clear that by psychological mechanisms and processes, I am not referring

4. For an good overview see Zalazo *et al.* 2007.

to private languages or representations. Language is a collective device, and it is used together with other socialization devices to influence, modulate and formulate our innermost feelings and to develop our personalities and selves. The latter is in plural form because research shows that we nurture and develop a multiplicity of selves and personalities in relation to contingencies and contexts (Rowan and Cooper 1999; Shweder *et al.* 2006, 750–764; Carter 2008). Autobiographical narratives are quite useful in representing such selves—most of them are virtual and are often specifically choreographed in relation to the persons and selves of grand narratives, or simply to the values of the group which ultimately are anchored in myths and legends. Others, however, can and often do contest our autobiographical narratives (family, friends, colleagues, opponents or even the general public) and thus, in many ways, our identities and selves are co-created in social contexts by socially significant others (Mead 1934; Ochs and Capps 1996, 2001).

Such mechanisms and processes are closely related to socialization, especially in the first nurture relations that a child meets. It is through attachment to significant others that the child acquires its first abilities as a social being. How this occurs—socially, neurologically and psychologically—is still a matter of intense research. Children are born with certain abilities and features that effectuate swift attachment to its parents. From that point, the intimate sharing between mother and child involve much more than vital fluids (although such are of course essential for the child's existence).[5] Neuroscientist and psychiatrist Daniel Siegel (1999, 2001) argues that nursing mothers are literally sharing their inner worlds, emotions and mental states with the child. Mothers and fathers and other significant people are constantly, even compulsively, speaking to the child, helping it to understand the names and nature of their own inner states and needs. Siegel has shown that parents who are adept at telling coherent autonoetic narratives to their children help the child to become balanced in mind and body as well as to learn to manage its emotions and inner states. The child "feels felt" and experiences an alignment of states of mind and primary emotions with its parents that lead to the resonance of minds necessary for the development of a socially competent person (Siegel 2001, 78–81). Such a person learns "mindsight" which allows an understanding of self and others. Siegel also calls the process "integrative" because it helps the child to integrate the left and right hemispheres thus enabling it to produce coherent autonoetic narratives. Coherent autonoetic narratives reflect "bihemispheric" coherence (Siegel 2001, 87). The child has, in fact, learned to make use of the neural maps, I mentioned earlier, thus allowing the child to create representations of the minds of self and others (Siegel 2001, 82).

When things go horribly wrong in a child's first years, narrative therapy, it is claimed, can even heal the damage. Thus unresolved states, disorganized attachment and impairments to integration can be repaired through narrative therapy by encouraging individuals to stimulate the mind's "innate tendency to move toward integration" (2001, 89).[6]

This integrative process must be in place so that the child is able to learn a language and to communicate effectively and meaningfully. Research has shown that by the age of twenty-two months, normal children already have an impressive vocabulary concerning the mental and emotional states of self and others in terms of past, present and future (Bretherton and Beeghly 1982; Nelson 1989), as any parent or grandparent can confirm.

5. See Relindis Yovsi's fascinating study (2003) on how nursing styles reflect basic normative patterns of behavior and belief.

6. See Onyut *et al.* 2005 on narrative therapy for child war survivors in Africa.

It is astounding, actually, how easily toddlers at that age move between real and imaginary states, senses of selves and sensitivity to the emotional and mental states of others.

But the development of individuals is never completely left to the individual and its parents and siblings alone. Collectives must maintain coherence and a sense of history. Its members must act in certain ways and must learn to communicate appropriately. Thus human societies develop institutions that maintain languages, ideas, concepts, rules, codes, symbolic systems and traditions. Social institutions maintain and pass such matters on from generation to generation, otherwise their languages, cultures and societies die out. Religions and religious institutions are often at the forefront of (symbolic) universe-maintenance, to use sociologists Peter Berger and Thomas Luckmann's phrase (1966, 110ff.). Maintenance mechanisms are both conceptual and social, according to Berger and Luckmann. Conceptual machineries of maintenance are mythologies, theologies, philosophies and science. Mythology is the most archaic form of universe-maintenance. They define mythology as:

> a conception of reality that posits the ongoing penetration of the world of everyday experience by sacred forces. Such a conception naturally entails a high degree of continuity between social and cosmic order, and between all their respective legitimations; all reality appears as made of one cloth. (1966, 128)

Social machineries of maintenance are elders, churches or temples, governing bodies, the military, educational and cultural institutions, etc. Their monopolistic activities can cause conflict in society, a fact which Berger and Luckmann dwell on (1966, 134–146), but they are essential to continuity. Activities such as education, public rituals, books, oral traditions and other media protect and develop the virtual worlds of a society. These institutions make themselves heavily present in public discourse and are incremental to the power of deciding who should speak, who should listen and what can and cannot be spoken, as has been criticized so forcefully by subaltern and post-colonial authors.[7] Finally, the mechanisms and processes of religious narrative also include larger social, cultural and even natural forces that contribute positively or negatively to social life such as social development, stagnation, revolution, war or natural catastrophes. These forces not only influence narrative, the arts and communication, but narratives are also often used to understand, interpret and sometimes control them (see Rukmini Bhaya Nair's chapter in this volume).

*The second aspect* of narrative is the many media and methods used in human communication. The point of departure is the body. As mentioned above, the use of the body for mimetic communication through posture, gesticulation, movement and facial expression is most likely our earliest type of communication before the advent of speech. Merlin Donald has argued that the mimetic phase was a necessary pre-condition for symbolic thought and subsequently for language. The fact that it is still with us, despite more sophisticated linguistic communication, is an indication of its ancient heritage, which, by the way, we also share with our sapient relatives. Furthermore, each newborn infant goes through a mimetic stage of communication long before learning a language.

Our mimetic abilities are further enhanced by the use of voice. The primary use of voice is to express basic emotional sounds which can, and often are, expressed through melody and song. Recent research indicates that inarticulate singing may also have been a pre-

7. See Asad 1973; Ashcroft *et al.* 1989, 2000; Foucault 1961, 1975, 1976–1984; Geertz 1994a, 1994b, 2000, 2004; Guha 1982ff.; Guha and Spivak 1988; Nencel and Pels, 1991; Ochs and Capps 1996; and Said 1978, 1989.

condition for language.[8]

As soon as *Homo sapiens* began to use external devices for symbolic communication, there were literally no limits to what could be used to convey symbolic information. The earliest devices most likely functioned as simple techniques for preserving significance and tradition. Thus symbols and marks on hides, birchbark, petroglyphs and rock art were effective means for long periods of time in conveying and containing traditions. The next step in symbolic communication is the use of the body, objects, architecture and pictures to convey iconographical information. Once again, the body is a primary vehicle through the use of body ornament, paint, tattoos and costumes. But also the use of the body in ritual dance, for instance, transforms it into an iconographic medium. The use of material objects for symbolic expression is very ancient, not only man-made objects such as statues, staves, puppets and masks, but also natural objects such as trees, mountains and caves, all serve to express and explore virtual, symbolic worlds. As soon as humans began building dwellings, architecture also became a medium of symbolic expression from simple poles to temples and monuments. When accompanied by frescoes, paintings and statues, religious worlds come alive through such materializations of narrative. Humans clothe themselves and their environments with symbolic worlds, thereby supporting and maintaining them. During the past century the types and impact of iconographical media has expanded in surprising variety. From movie film to the internet, there are hardly any restrictions on visualizing narratives and virtual worlds.

The creation of language was a singular revolution in human communication and thought. It significantly rewired the brain and expanded our cognitive abilities making it possible for us to develop new ways of thinking and new levels of abstraction. It allowed us to externalize our thoughts much more effectively and to submit them to public reception and analysis. Our abilities at mind-reading became much more refined, but also our abilities to deceive. Because written texts are the most publicly accessible, much research in linguistics, languages and culture is based on them. Subsequently, most theories of narrative are based on this medium of expression, which can, of course, skew the perspective.

*The third aspect* is the expressive one. We are immediately confronted by a chaotic jumble of genres. On the one hand you find formal narratives, such as myths, legends, songs, epics and sagas, in which traditions are passed on in colorful and dramatic stories. Personal narratives, on the other hand, provide the individual with a rich variety of avenues making one's life and thoughts the main subject of the text. Such narratives are exceptionally useful in historical and psychological research, from personal correspondence and diaries to more structured narratives on travel, conversion, dreams, visions and autobiographies. Ritual texts abound in the history of religions and give us unique insight into religious worlds and ritual logic. Sermons interpret and teach formal texts and doctrines. Hymns and liturgies express central dogmas and events. Oracles give substance to communication with the gods. And prayers express personal and public religious communication. As noted above, religious traditions survive through the use of maintenance devices, such as laws, codes and decretals. Exegesis is employed to reflect on and relate to religious doctrine. Religious traditions become more pedagogically available to congregations and adherents through hagiographies and informal didactic formulas. Some topics and traditions are so complicated that they are compiled in handbooks. Such handbooks are useful for practitioners,

8. See for instance Wallin *et al.* 2000; Brown 2001; Zatorre and Peretz 2001; Juslin and Laukka 2003; Hauser and McDermott 2003; Patel 2003; Janata and Grafton 2003; Brown *et al.* 2004; Mithen 2005; Brown *et al.* 2006b; and Dissanayake 2006.

especially religious specialists, and include books on ritual, magic, meditation and prayer. Religious worlds, as mentioned above, are spread throughout a population in many ways. One of them is through popular literature, plays, novels and essays. These genres not only rework religious ideas and symbols into new meanings, but they can also expose received traditions to critical and non-religious commentary.

Religious narrative is significant for the study of cognition because religion, as far as we know, is universal and has been with us since the beginning. Narrative is perhaps one of the single most important interfaces between intrapersonal and interpersonal meaning production. In religiously dominated societies, religion is crucial to primary socialization and identity development in individuals, and to the development of cultural and ethnic identity in groups of people. Thus, religion is an important factor in creating and regulating the genesis of individuals and groups. This process is primarily created through the articulation of meaning systems in narrative form.

It should be evident that narrative is a complex phenomenon that does different things at different levels of human existence. There is far too little communication and cooperation between specialists pursuing research at these different levels. Furthermore, an adequate definition of narrative in my opinion should pay heed to each level.

As intimated above, the levels I have in mind are the narratology of narrative, its psychological reception and pragmatic contexts, its neurobiological preconditions and its social psychological relations. Instead of beginning at the beginning with neurobiology, I will begin at the point of departure traditionally assumed by most scholars of narrative, namely, the narratological level.

## 1. The Narratology of Narrative

Best known to the humanities is the linguistic level. By linguistic I mean semiotic, narratological and literary. There are good reasons for this, of course. Narratives are after all, narratives, i.e. stories, written or told, with plots, characters, and various narrative ploys. Thus, a possible minimal definition of narrative is "a story based on an action caused and experienced by characters, and a narrator who tells it" (Jahn 2002).

This definition assumes that narrative communication (primarily in fiction) has a standard structure, which by the way, also matches that of religious literature. Professor of Narratology at the University of Cologne, Manfred Jahn, has developed a Chinese box model of narrative in which participants in the narrative each have their own level. Figure 1.1 is a slight modification of Jahn's model to accommodate the topic of *religious* narrative. There are three levels. The first level concerns informative religious communication between an author addressing a reader or participant. This is in principle a non-fictive level external to the actual narrative. The next level concerns religious mediation and discourse in which the narrator communicates with one or more addressees. Here communication occurs within the fictive world of narrator and reader. The third level concerns religious action between characters in the story. At this level, the actual story is in play. The second and third levels are internal to the narrative.

Manfred Jahn distinguishes between a narrative and a story:

> *narrative*: Anything that tells or presents a story, be it by text, picture, performance, or a combination of these. Hence novels, plays, films, comic strips, etc., are narratives.

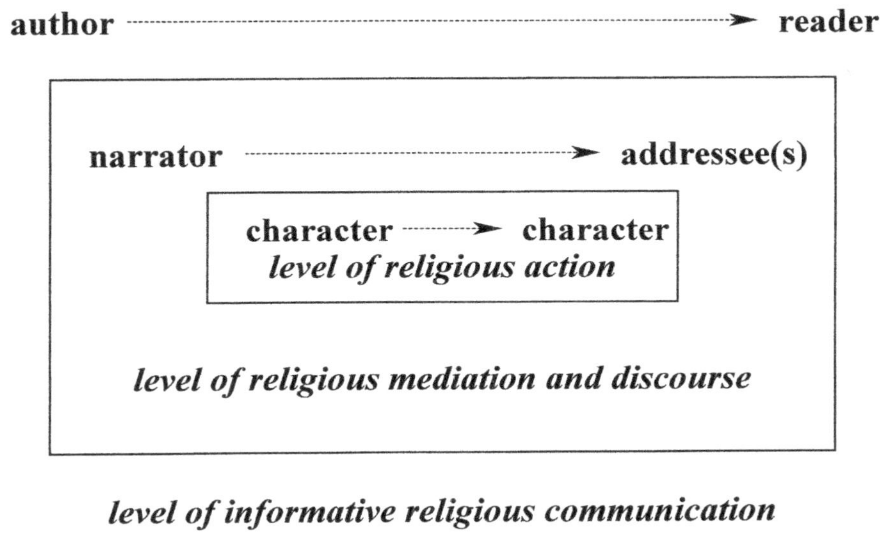


**Figure 1.1** Standard structure of religious narrative communication. Adapted from Manfred Jahn 2002.

> *story*: A sequence of events involving characters. "Events" include natural and nonnatural happenings like floods and car accidents. Characters get involved by being agents (causing an event), patients or beneficiaries (being affected by an event). Linguists further make a distinction between verbs which signal willful ('volitional') acts (*What does X do?*—jump from a bridge, watch a show) and verbs which signal nonvolitional acts or experiences (*What does X experience?*—falling from a bridge, seeing an accident). (Jahn 2002, N2.1.2)

Jahn defines discourse narratology as the study of stylistic techniques which determine the form of a narrative text or a narrative performance. He defines story narratology as the study of action units that "'emplot' and arrange a stream of events into a trajectory of themes, motives and plot lines" (Jahn 2002, N2.1.3). There are more complicated definitions in the semiotic and narratological literature, but Jahn's definitions will suffice for my purposes.

## 2a. The Psychology of Narrative Reception

Psychological linguistics and linguistic philosophers (especially speech-act theorists) have shown that there are other levels which are just as fundamental to narratives. Their main interest lies in the social and pragmatic contexts of narration and conversation. The social and especially psychological aspects of narrative and narration have been emphasized in social psychology, psychotherapy and, more recently, cognitive psychology. In these contexts, issues of text comprehension, identity and the construction of selves play central roles. There are many definitions possible in this regard. Psychologists Nancy L. Stein and Margaret Policastro have attempted to systematize the concept of story. In their analysis of twenty definitions formulated in the fields of linguistics, anthropology and psychology, they have identified four categories of definitions (Stein and Policastro 1984, 116ff.). The first category minimally defines stories in terms of "state-event-state change," in which there is no goal-directed behavior. An example of such a definition is that of Professor of Romance Languages, Gerald Prince:

> John was sad. Then John met a woman. As a result, John was happy.

Or:

> It was a quiet peaceful night in the forest. Suddenly, a hurricane descended from the sky and in a matter of minutes, chaos reigned. (Stein and Policastro 1984, 117; Prince 1973)

Even though the latter contains no protagonist, Prince nevertheless considers it to be a story.

The second category minimally defines stories as involving goal-directed behavior in relationship to personal social problems. Such stories involve a protagonist motivated to perform goal-directed behavior. Summarizing Stein and Glenn (1979), a story text must contain:

> 1. a specific protagonist capable of intentional behavior;
> 2. the motivation and/or goals of a protagonist;
> 3. overt actions carried out in the service of a goal; and
> 4. information concerning the attainment or nonattainment of the goal.
>
> (Stein and Policastro 1984, 119)

Psychologists Jean Mandler and Nancy Johnson also include nongoal-based sequences within this type of definition:

> Once there was a Czar (setting) who had three lovely young daughters (setting). One day, the three daughters went walking in the woods (initiating event). They were enjoying themselves so much (emotional reaction) that they forgot the time (cognition) and stayed too long (unplanned actions). They were then kidnapped by a dragon (ending).
>
> (Stein and Policastro 1984, 120; Mandler and Johnson 1977)

Even though they agree that such stories usually do not exist alone but are integrated in more complex narrative frames, they still maintain that such episodes contain the essential features of a story.

The final two categories of definitions criticize goal-based approaches and argue for more detailed models, on the one hand, and those that claim there are no set of features unique to stories. The former, more complex models are derived from "story grammar" approaches (Stein and Policastro 1984, 122).

Stein and Policastro are more interested in what constitutes a good story in real life rather than formalistic features. As a result, they conducted a fascinating study designed to identify those features necessary to a story. They asked a group of second-grade children (42 upper middle-class Chicago suburb, half male half female) and a group of elementary school teachers (38, from Chicago and suburban public school districts, more female than male) to judge the merits of and rank how good 31 different passages were as stories. The passages varied from single words and sentences to longer passages of several sentences. The passages were clustered to test each of the above-mentioned four categories of definitions, i.e. Prince's minimal definition, the difference between goal-based and nongoal-based stories, a more nuanced study of goal-based stories and the story grammar hypotheses of cognitivists John Black and Robert Wilensky (1979).

A detailed discussion is not warranted here, but the results showed significant consistency (with some variation) between what the children and the teachers felt constituted a good story. Prince's minimal definition was insufficient, and the more complex definitions were unnecessary. There was agreement on two things: "First, an animate protagonist and some type of causal relationship among events are almost always expected in any passage considered to be a story" (Stein and Policastro 1984, 147), and second, that both goal-based and nongoal-based behavior are sufficient as long as a protagonist is involved in some way. Most of the participants agreed that a good story should be goal-directed in which obstacles must be overcome by a protagonist (150).

One of the weaknesses in this otherwise excellent study was that the affective side of stories was not measured.[9] But the study did confirm their initial definition of a story:

> (1) a *setting* that includes or introduces an animate character; (2) either an *initiating event* or an *internal response* containing information from which the motives, goals, and emotional response of the protagonist can be inferred; (3) an overt *attempt* on the part of a protagonist or a *plan* outlining the overt attempt; and (4) a *consequence*, reflecting whether or not the goal has been attained. (Stein and Policastro 1984, 119)

## 2b. The Pragmatic Contexts of Narrative

Narratives use a series of linguistic and paralinguistic techniques that invite listeners to participate and co-create. Stories in such contexts are conversational, roughly hewn, and constantly interrupted, questioned, evaluated and *emergent*. Thus for literary theorist Rukmini Bhaya Nair, who has developed a modified version of speech-act theory, a narrative has as its function:

> to make some parts of a communicative loop or chain—in our particular case, casual conversation—both potentially detachable and iterable. A "good story" is one that can be "taken away" by listeners and/or tellers and repeated in other conversations, other contexts, other cultures. (Nair 2002, 5)

At the same time, it allows listeners to participate in an already incipient, potential "cross-examination of the story and story-teller by a community of listeners, as well as the self-reflexive examination of her own story by a teller" (Nair 2002, 10). Thus, narrative allows the individual to participate existentially:

> Listening to a narrative, like witnessing a magician's act, is a relatively inexpensive means of trying out patterns of living, situations and emotional reactions that one may never experience, or want to experience, in one's own life. The feelings, among others, of curiosity, awe, pity, sheer fun, and final reassurance that a narrative evokes—those indirect "perlocutionary effects" of speech act theory—must occur in our minds, not just or even necessarily, within the story-teller's. (Nair 2002, 15)

One of Nair's important points is that narratives have cultural aspects. Some stories are so universal in scope that they are capable of reaching across cultural boundaries, such as the ancient Babylonian Gilgamesh Epic, the Indian *Mahābhārata* or Shakespearean plays. Narratives, however, are products of particular cultures. They are not only spoken or written in particular languages and draw on culturally significant characters and scenographies,

9. See Murphy *et al.* 2009 for an interesting experiment on the moral dimensions of narrative comprehension.

but they are also informed by specific cultural assumptions. A good term for this, coined by Nair, is *impliculture*. A take-off on Grice's theory of *conversational implicature* (Grice 1975, 1978, 1981), the term impliculture designates cultural inferencing in narrative and how cultural conventions govern story telling. This significant aspect allows us to identify cultural specificities, things left unsaid in a cultural community as well as to gain insight into cultural frames and the "tone" of particular conversational or narrative contexts (Nair 2002, 187). These inferences are key instruments for listeners and readers:

> Every narrative has, so to speak, a gnomon coiled within it. As the narrative unwinds in its structural linearity, this gnomon of time lets loose various sparkling implicatures which listeners must infer in order to derive pleasure and profit from the story. These inferences alter the structure of the world for us. If we are indeed 'cogs' genetically programmed to make our communicative moves through the interlocking machinery of conversation, we are constrained then to enquire: exactly what mental events take place after we make a correct narrative inference? (Nair 2002, 203–204)

The answer is that the story and conversation create the illusion that we are individuals with each our own individual self:

> What do we see with that enormous illusion-making internal eye? Well, flowing, rushing, gushing into the human cognitive machine at all times are the variety of visual, linguistic, aural and somatic sensations. The task of the mind's eye is to read narrative coherence into this jumble and to make such inferences as render this chaotic material relevant to the mind's "I"—in Dennett's words. Shiva's third eye, we might say, organizes the wild dance of perception into a rationalizing story-grammar.... Narrativity as a conversational mechanism for information exchange is crucial within human cultures because it underwrites our collective illusions of being unified selves with reliable social identities.
>
> (Nair 2002, 206, 211–212)

Nair further developed and refined the analytical tools of story-grammarians especially in terms of what she calls inferential constraints or strategies which provide a framework for the listeners' or readers' active participation in and understanding of the narrative: "Can there be a *system* of inferential strategies which listeners adopt in order to recover the information required to render a narrative coherent/contextually relevant?" (212). Even though Nair wishes to retain the terminology and insights in speech-act theory, she essentially abandons speech-act interests in segmental units of narrative and highlights the inferences themselves "as constituting the central nodes of an abstract, possibly universal narrative tree" (213). She therefore provides a set of inferential rewrite-rules for narrative which allow an analysis of the continuous processes of inferencing (214):

Inferring an evaluative judgement from the story, when it is not made explicit by the teller, as in minimal narratives, enables the listener to

a. actively participate in the construction of the narrative.
b. display her cultural knowledge and membership-status, her meme-awareness.
c. cooperate with the teller in resolving inherent textual enigmas. (Nair 2002, 214)

The point is that narratives and conversations carry a logic that is equivalent to that of language grammar. Furthermore, this logic has significant cognitive value because it binds people together in linguistic and cultural unity, common inductive premises and common deductive rationales. To Nair, "narrative and conversation are just the most advanced routines we possess for 'social grooming'" (Nair 2002, 208).

### 3. The Neurobiology of Narrative

Which brings us to the next level of analysis. I have already discussed some aspects of the neurobiology of narrative. In this section, I will first address interesting evolutionary issues.

Narrative is central to our evolutionary history. Although much of what can be said about that history is speculative, it still warrants consideration. One theme that pervades a number of studies is the parallel growth of our brain size and capacities with the development of language. Liverpool psychologist Robin Dunbar argues that language and gossip developed out of primate grooming: language evolved among women with their concern about other people and relationships, thus strengthening social cohesion. Like grooming, gossip and everyday talk stimulate endorphin production inducing states of relaxation and mild euphoria. It also induces social bonding:

> If being human is all about talking, it's the tittle-tattle of life that makes the world go round, not the pearls of wisdom that fall from the lips of the Aristotles and the Einsteins. We are social beings, and our world—no less than that of the monkeys and apes—is cocooned in the interests and minutiae of everyday social life. They fascinate us beyond measure. (Dunbar 1996, 4)

The neurobiology of our social intelligence concerns the hemispheric specialization of the brain and the expansion of the neo-cortex which, according to psychologist Louis J. Cozolino, "created neural space for the development of more complex forms of social communication" (Cozolino 2002, 155). It is especially the development of the left hemisphere, which normally helps individuals navigate their social worlds, that may have been crucial to evolving language systems. Furthermore, the use of words probably led to or at least arose parallel to the inner dialogue, thus bringing to conscious awareness not only inner thoughts, but also coordinated group behavior. Thus the interpersonal self, the private self and the conceptual self (using part of Ulric Neisser's typology, 1988) consist of "a matrix of conscious and unconscious memories organized into episodes, stories, or narratives" (Cozolino 2002, 154). One of their functions is to perpetuate *the internal illusion* of a single, unified, controlling agent on the one hand, and *the external illusion* of an identifiable, predictable and accountable individual member of the group on the other. This is, to use Chris Frith's term, one of the final illusions produced by the brain (Frith 2007, 184). In truth, we are none of these. Our personal memories are fragmented and contain gaps (Neisser and Fivush 1994; Rubin 1996), our behavior is seldom rational or controlled, and we are driven by thoughts, feelings and emotions without conscious knowledge (Damasio 1999; Fine 2006). Thus, a lot of social construction is happening all the time (Mead's "generalized other," 1934, 154; Miller 1994), and the majority of our autobiographical narratives are tales spun by family and culture (Fivush 1994; Barclay 1996; Hirst and Manier 1996). As Cozolino formulated it, "We come into consciousness as a story that, in many ways, has already been written" (Cozolino 2002, 157).

Cozolino views narrative storytelling as "the primary model humans possess for the sequential and meaningful integration of human action." Furthermore, he claims, "Narratives are emotionally meaningful, causally linked sequences of actions and consequences that aid in the organization, maintenance, and evaluation of behavior" (Cozolino 2002, 166). Narratives thus regulate experience and help "to anchor us in our bodies through time." Neurobiologically, the cerebral cortex allows mental representations to occur which promote "the internationalization of the group and significant others," primarily, of course, the child's mother. Following Daniel Siegel, Cozolino argues that the nurturing mother and resultant patterns of attachment determine "the child's affective core and eventual self-image" (Cozolino 2002, 167). Parents "organize memory and create the story of the self with their children" (Nelson 1993; Cozolino 2002, 167), and these narratives have significance and value embedded in them, thus expanding the competence of the child's evolving social brain. As Cozolino notes, "Children who have been taught to tell stories that include mental states demonstrate a greater frequency of secure attachment" (2002, 168).

What is at stake, then, is the co-construction of narratives in the context of family and culture that help the child discern own and other mental states. This process continues throughout life, and one of the important maintenance techniques is gossip (see among others Rosnow and Fine 1976; Bergmann 1993; and Goodman and Ben-Ze'ev 1994). This process also continues on more institutional levels through the process of universe maintenance mentioned above.

## 4. The Social Psychology of Narrative

At this level, narrative plays a crucial role in the construction of selves and in monitoring social relationships. Narrative shapes and socializes identities and selves. As psychologist Peggy J. Miller and colleagues wrote:

> Of all the intersections of narrative and self, perhaps none is more common than that which occurs in ordinary talk when people relate to one another their personal experiences.... Stories of this sort...provide one widely available means by which people create, interpret, and publicly project culturally constituted images of self in face-to-face interaction.... There may be a special affinity between narrative and self such that narrative can be said to play a privileged role in the process of self-construction. (Miller *et al.* 1990, 292)

This understanding of narrative is supported by a large body of research (see Shweder *et al.* 2006, 742–749 for an overview; see also Linde 1993 and Eakin 1999). Anthropologist Elinor Ochs and psychologist Lisa Capps claim further that "narratives situate narrators, protagonists, and listener/readers at the nexus of morally organized, past, present, and possible experiences" (Ochs and Capps 1996, 22). Narratives are versions of reality where protagonistic selves are co-authored. Such narratives, primarily in small-talk and gossip, consist of identity-in-the-making. In fact, narratives can generate a multiplicity of partial selves depending on the contexts and what is at stake. Ochs and Capps draw attention to extensive research on the dimensions along which selves are multiplied: subject and object; male and female; id, ego, and superego; good and evil; normal and aberrant; and public and private (1996, 22) They claim that "as narratives are apprehended, they give rise to the selves that apprehend them" (1996, 22).

These processes are not necessarily positive. Narratives are versions of reality that often raise uncomfortable and challenging questions. Narratives are points of view and are therefore selective. Some experiences are left out as others are brought to the fore. Thus narratives make remembering a kind of forgetting (Ochs and Capps 1996, 21). Sometimes it is a technique of silencing and suppression (Ochs and Capps 1996, 32). Thus experience and memory become fragmented. These are powerful instruments in monitoring social relations between individuals and groups. They are linguistic tools for portraying individuals or groups and absolutely essential to socialization processes. They fuse single identities to collective identities, but they can also isolate, alienate and oppress individuals and groups. The power to decide what tales are to be told, which are to be silenced or ignored, who the narrator should be, what may or may not be said, who should be allowed to listen and when it should be told is frightening and often used. Parents use this power all the time on their children, and they attempt to do it to each other. Leaders and politicians use this power and manipulate others with it in their professions, as do religious leaders and specialists. They are all incarnated narrative specialists. Narratives told by competent speakers can both destroy and heal.

Minorities and marginalized individuals are not, however, always passive victims. They can also make effective use of narratives to undermine dominant narratives. Comedians, clowns, jesters and even academics play with forbidden narratives by directly or indirectly retelling them, satirizing them or inverting them (Ochs and Capps 1996, 35–37). Such narratives can also have more serious consequences through counterversions and outright rejections of dominant narratives such as those of democracy, equality and freedom by fundamentalist religious groups. All narratives, religious or not, are not necessarily good. They can lead to violence, unrest, persecution and destruction.

## A Definition

A lot more can be said about narrative. I have only provided a selective sketch of problems and themes in this chapter. The question is whether it is possible to provide a comprehensive definition that does justice to one of the defining elements of the human condition. My proposal is that narrative can be defined as follows:

> Narrative is a speech act between narrator and listener or author and reader. Narrative is ideally a story about a protagonist causally involved in a sequence of events which typically consist of obstacles to goal-directed behavior which the protagonist attempts to overcome and the results of which complete the story. Narratives can be formal or informal, and they can be co-created by teller and listener. Narratives assume cultural values and implications thereby serving fundamental cognitive and social functions for participants. The primary function of narrative is to maintain the illusion of an individual, controlling agent on the one hand and a member of the group who can be identified, held responsible and made predictable on the other. Narrative is essential to constructing selves and monitoring social relationships. Narratives draw on and stimulate our neurochemistry and make optimal use of the brain's map-making capacities.

What can be defined in this manner as narrative also applies to religious narrative if the narrative in question is in reference to postulated transempirical powers or beings. As Figure 1.2 indicates, narratives are at the center of a matrix of relations between individuals, groups, cultural repertoires and social institutions. Narrative integrates the four levels of interaction through time and space. In referring to Merlin Donald, it can be concluded that the evolution of language was primarily a response to a consciousness immersed in culture and residing "in the temporary state of a semantic network honed through a public process":

> The difference between us and other primates is only that our conscious-beast-within is much better equipped than theirs to understand and navigate the complex environment needed for public idea laundering.... Symbolic idea laundering takes a lot of management. A creative, interactive cultural process increases the sheer quantity of knowledge available to a person, and that knowledge has to be filtered, sorted, stored, and made retrievable.... Our real-time, on-line mental lives become much busier because in addition to having to

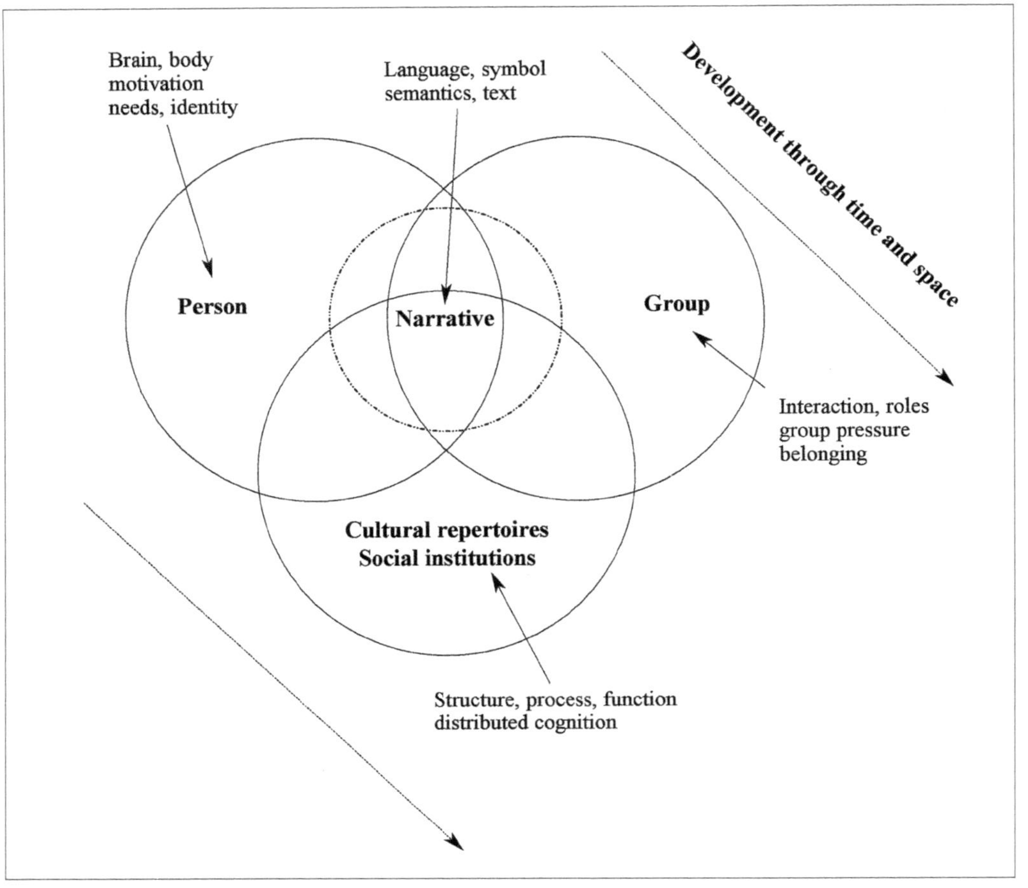


**Figure 1.2** The integral role of narrative.

> deal with the cognitive challenges that confront other mammals, we have to deal with a parallel cultural context every moment of our waking lives.... This takes executive management. It also takes plasticity because cultures constantly reorganize the inner universes of their members and multiply their mental models. This loads us down with thousands of ideas and facts, and above all, it demands our close attention to the rules and codes that allow symbolic cultures to operate. (Donald 2001, 288, 289)

Ultimately this hybrid process of cognition and culture has liberated our consciousness from the solipsism that characterizes other mammals. With the dramatic expansion of external memory banks, symbolic systems and a multiplicity of communication types, we as a species are now in a position to see ourselves as we really are: the result of a collective process (Donald 2001, 326).

## References

Asad, Talal, ed. 1973. *Anthropology and the Colonial Encounter*. Atlantic Highlands: Humanities Press, rpr. 1990.

Ashcroft, Bill, Gareth Griffiths and Helen Tiffin. 1989. *The Empire Writes Back: Theory and Practice in Post-Colonial Literatures*. London and New York: Routledge.

———. 2000. *Post-Colonial Studies: The Key Concepts*. London and New York: Routledge.

Bachner-Melman, Rachel, Christian Dina, Ada H. Zohar, Naama Constantini, Elad Lerer, Sarah Hoch, Sarah Sella, Lubov Nemanov, Inga Gritsenko, Pesach Lichtenberg, Roni Granot and Richard P. Ebstein. 2005. "*AVPR1a* and *SLC6A4* Gene Polymorphisms Are Associated with Creative Dance Performance," *PLOS Genetics* 1(3): 394–403.

Barash, David P. and Nanelle R. Barash. 2006. *Madame Bovary's Ovaries: A Darwinian Look at Literature*. New York: Delacorte Press.

Barclay, Craig R. 1996. "Autobiographical Remembering: Narrative Constraints on Objectified Selves." In *Remembering Our Past: Studies in Autobiographical Memory*, edited by David C. Rubin, 94–125. Cambridge: Cambridge University Press.

Berger, Peter L. and Thomas Luckmann. 1966. *The Social Construction of Reality: A Treatise in the Sociology of Knowledge*. Garden City: Doubleday, 1966, rpr. Harmondsworth: Penguin Press, 1967, 1987.

Bergmann, Jörg R. 1993. *Discreet Indiscretions: The Social Organization of Gossip*. Translated by John Bednarz, Jr. and Eva Kafka Barron. New York: Aldine de Gruyter. Originallly published as *Klatsch: Zur Sozialform der diskreten Indiscretion*. Berlin: Walter de Gruyter, 1987.

Black, John B. and Robert Wilensky. 1979. "An Evaluation of Story Grammars." *Cognitive Science* 3: 213–230.

Blakeslee, Sandra and Matthew Blakeslee. 2007. *The Body Has a Mind of Its Own: How Body Maps in Your Brain Help You Do (Almost) Everything Better*. New York: Random House. Reprint New York: Random House Trade Paperbacks, 2008.

Boyd, Brian. 2009. *On the Origin of Stories: Evolution, Cognition, and Fiction*. Cambridge: Belknap Press.

Bretherton, Inge and Marjorie Beeghly. 1982. "Talking About Internal States: The Acquisition of an Explicit Theory of Mind." *Developmental Psychology* 18(6): 906–921.

Brown, Steven. 2001. "Are Music and Language Homologues?" In *The Biological Foundations of Music*, edited by Robert J. Zatorre and Isabelle Peretz, 372–374. New York: New York Academy of Sciences.

Brown, Steven, Michael J. Martinez, Donald A. Hodges, Peter T. Fox and Lawrence M. Parsons. 2004. "The Song System of the Human Brain." *Cognitive Brain Research* 20: 363–375.

Brown, Steven, Michael J. Martinez and Lawrence M. Parsons. 2006a. "The Neural Basis of Human Dance," *Cerebral Cortex* 16: 1157–1167.

———. 2006b. "Music and Language Side by Side in the Brain: A PET Study of Generating Melodies and Sentences." *European Journal of Neuroscience* 23: 2791–2803.

Brown, Steven and Lawrence M. Parsons. 2008. "The Neuroscience of Dance." *Scientific American* 299(1): 78–83.

Carter, Rita. 2008. *Multiplicity: The New Science of Personality, Identity, and the Self*. New York: Little, Brown and Company.

Clark, Andy. 1997. *Being There: Putting Brain, Body, and World Together Again*. Cambridge, MA: MIT Press.

———. 2008. *Supersizing the Mind: Embodiment, Action, and Cognitive Extension*. Oxford: Oxford University Press.

Cozolino, Louis J. 2002. *The Neuroscience of Psychotherapy: Building and Rebuilding the Human Brain*. New York: W.W. Norton.

Damasio, Antonio. 1999. *The Feeling of What Happens: Body, Emotion and the Making of Consciousness*. London: Heinemann. Reprint, London: Vintage, 2000.

Day, Matthew. 2004. "Religion, Off-Line Cognition and the Extended Mind." *Journal of Cognition and Culture* 4(1): 101–121.

Dissanayake, Ellen. 2006. "Ritual and Ritualization: Muscial Means of Conveying and Shaping Emotion in Humans and Other Animals." In *Music and Manipulation: On the Social Uses and Social Control of Music*, edited by Steven Brown and Ulrich Voglsten, 31–56. Oxford: Berghahn Books.

Donald, Merlin. 2001. *A Mind So Rare: The Evolution of Human Consciousness*. New York: W.W.Norton.

Donald, Merlin and Lars Andreassen. 2007. "Consciousness and Governance: From Embodiment to Enculturation—An Interview." *Cognitive Semiotics* 0: 68–83.

Dunbar, Robin. 1996. *Grooming, Gossip and the Evolution of Language*. London: Faber and Faber.

Eakin, Paul John. 1999. *How Our Lives Become Stories: Making Selves*. Ithaca, NY: Cornell University Press.

Fine, Cornelia. 2006. *A Mind of Its Own: How Your Brain Distorts and Deceives*. New York: W.W. Norton.

Fivush, Robyn. 1994. "Constructing Narrative, Emotion, and Self in Parent-Child Conversations about the Past." In *The Remembering Self: Construction and Accuracy in the Self-Narrative*, edited by Ulric Neisser and Robyn Fivush, 136–157. Cambridge: Cambridge University Press.

Foucault, Michel. 1961. *Histoire de la folie à l'âge classique—Folie et déraison*. Paris: Plon.

———. 1975. *Surveiller et punir*. Paris: Gallimard.

———. 1976-1984. *Histoire de la sexualité*. Paris: Gallimard.

Frith, Chris, 2007. *Making Up the Mind: How the Brain Creates Our Mental World*. Oxford: Blackwell Publishing.

Geertz, Armin W., 1994a. "On Reciprocity and Mutual Reflection in the Study of Native American Religions." *Religion* 24: 1–7.

———. 1994b. "Critical Reflections on the Postmodern Study of Religion." *Religion* 24: 16–22.

———. 1999. "Definition As Analytical Strategy in the Study of Religion." *Historical Reflections/Reflexions Historiques* 25(3): 445–475.

———. 2000. "Global Perspectives on Methodology in the Study of Religion." In *Perspectives on Method and Theory in the Study of Religion: Adjunct Proceedings of the XVIIth Congress of the International Association for the History of Religions, Mexico City 1995*, edited by Armin W. Geertz and Russell T. McCutcheon, assisted by Scott S. Elliott, 49–73. Leiden: E.J. Brill.

———. 2004. "Can We Move Beyond Primitivism? On Recovering the Indigenes of Indigenous Religions in the Academic Study of Religion." In *Beyond "Primitivism": Indigenous Religious Traditions and Modernity*, edited by Jacob Olupona, 37–70. New York and London: Routledge.

Goodman, Robert F. and Aaron Ben-Ze'ev. 1994. *Good Gossip*. Lawrence: University Press of Kansas.

Gottschall, Jonathan and David Sloan Wilson. eds. 2005. *The Literary Animal: Evolution and the Nature of Narrative*. Evanston, IL: Northwestern University Press.

Grice, H.P. 1975. "Logic and Conversation." In *Syntax and Semantics, 3: Speech Acts*, edited by Peter Cole and Jerry L. Morgan, 41–58. New York: Academic Press.

———. 1978. "Further Notes on Logic and Conversation." In *Syntax and Semantics 9: Pragmatics*, edited by Peter Cole, 113–127. New York: Academic Press.

———. 1981. "Presupposition and Conversational Implicature." In *Radical Pragmatics*, edited by Peter Cole, 183–198. New York: Academic Press.

Guha, Ranajit, ed. 1982ff. *Subaltern Studies: Writings on South Asian History and Society*. Delhi: Oxford University Press.

Guha, Ranajit and Gayatri Chakravorty Spivak, eds. 1988. *Selected Subaltern Studies*. New York: Oxford University Press.

Hauser, Marc D. and Josh McDermott. 2003. "The Evolution of the Music Faculty: A Comparative Perspective." *Nature Neuroscience* 6(7): 663–668.

Herman, David, ed. 2003. *Narrative Theory and the Cognitive Sciences*. Stanford, CA: CSLI Publications, Center for the Study of Language and Information.

Hirst, William and David Manier. 1996. "Remembering as Communication: A Family Recounts Its Past," in *Remembering Our Past: Studies in Autobiographical Memory*, edited by David C. Rubin, 271–290. Cambridge: Cambridge University Press.

Hogan, Patrick Colm. 2003. *The Mind and Its Stories: Narrative Universals and Human Emotion*. Cambridge: Cambridge University Press.

Jahn, Manfred. 2002. "Narratology: A Guide to the Theory of Narrative," Part III of Poems, Plays, and Prose: A Guide to the Theory of Literary Genres. English Department, University of Cologne. Digital excerpts at http://www.uni-koeln.de/~ame02/pppn.htm.

Janata, Petr and Scott T. Grafton. 2003. "Swinging in the Brain: Shared Neural Substrates for Behaviors Related to Sequencing and Music." *Nature Neuroscience* 6(7): 682–687.

Juslin, Patrik N. and Petri Laukka. 2003. "Emotional Expression in Speech and Music: Evidence of Cross-Modal Similarities." *Annals of the New York Academy of Sciences* 1000: 279–282.

Linde, Charlotte. 1993. *Life Stories: The Creation of Coherence*. Oxford: Oxford University Press.

Mandler, Jean M. and Nancy S. Johnson. 1977. "Remembrance of Things Parsed: Story Structure and Recall." *Cognitive Psychology* 9: 111–151.

Mead, George Herbert. 1934. *Mind, Self, and Society: From the Standpoint of a Social Behaviorist*. Chicago, IL: University of Chicago Press. Reprint 1962.

Miller, Peggy J. 1994. "Narrative Practices: Their Role in Socialization and Self-Construction," in *The Remembering Self: Construction and Accuracy in the Self-Narrative*, edited by Ulric Neisser and Robyn Fivush, 158–179. Cambridge: Cambridge University Press.

Miller, Peggy J., Randolph Potts, Heidi Fung, Lisa Hoogstra and Judy Mintz. 1990. "Narrative Practices and Social Construction of Self in Childhood." *American Ethnologist* 17(2): 292–311.

Mithen, Steven. 1996. *The Prehistory of the Mind: A Search for the Origins of Art, Religion and Science*. London: Thames and Hudson. Reprint, London: Phoenix, 1998.

———. 2005. *The Singing Neanderthals: The Origins of Music, Language, Mind and Body*. London: Weidenfeld and Nicolson. Reprint, Cambridge, MA: Harvard University Press, 2006.

Murphy, Fionnuala C., Gemma Wilde, Neil Ogden, Philip J. Barnard and Andrew J. Calder. 2009. "Assessing the Automaticity of Moral Processing: Efficient Coding of Moral Information during Narrative Comprehension." *The Quarterly Journal of Experimental Psychology* 62 (1): 41–49.

Nair, Rukmini Bhaya. 2002. *Narrative Gravity: Conversation, Cognition, Culture*. Oxford: Oxford University Press.

Neisser, Ulric. 1988. "Five Kinds of Self-knowledge," *Philosophical Psychology* 1(1): 35–59.

Neisser, Ulric and Robyn Fivush, eds. 1994. *The Remembering Self: Construction and Accuracy in the Self-Narrative*. Cambridge: Cambridge University Press.

Nelson, Katherine, ed. 1989. *Narratives from the Crib*. Cambridge and London: Harvard University Press. Reprint 2006.

Nelson, Katherine. 1993. "The Psychological and Social Origins of Autobiographical Memory." *Psychological Science* 4(1): 7–14.

Nencel, Lorraine and Peter Pels, eds. 1991. *Constructing Knowledge: Authority and Critique in Social Science*. London: SAGE Publications.

Ochs, Elinor and Lisa Capps. 1996. "Narrating the Self," *Annual Review of Anthropology* 25: 19–43.

———. 2001. *Living Narrative: Creating Lives in Everyday Storytelling*. Cambridge, MA: Harvard University Press.

Onyut, Lamaro P., Frank Neuner, Elisabeth Schauer, Verena Ertle, Michael Odenwald, Maggie Schauer and Thomas Elbert. 2005. "Narrative Exposure Therapy as a Treatment for Child War Survivors with Posttraumatic Stress Disorder: Two Case Reports and a Pilot Study in an African Refugee Settlement." *BMC Psychiatry* 5(7): 1–9.

Patel, Aniruddh D. 2003. "Language, Music, Syntax and the Brain." *Nature Neuroscience* 6(7): 674–681.

Prince, Gerald. 1973. *A Grammar of Stories: An Introduction*. The Hague: Mouton.

Ramachandran, V.S. and Sandra Blakeslee. 1998. *Phantoms in the Brain: Human Nature and the Architecture of the Mind*. London: Fourth Estate. Reprint, 1999.

Robbins, Philip and Murat Aydede, eds. 2009. *The Cambridge Handbook of Situated Cognition*. Cambridge: Cambridge University Press.

Rosnow, Ralph C. and Gary Alan Fine. 1976. *Rumor and Gossip: The Social Psychology of Hearsay*. Amsterdam: Elsevier.

Rowan, John and Mick Cooper, eds. 1999. *The Plural Self: Multiplicity in Everyday Life*. London: SAGE Publications.

Rowlands, Mark. 2003. *Externalism: Putting Mind and World Back Together Again*. Montreal and Kingston: McGill-Queen's University Press.

Rubin, David C., ed. 1996. *Remembering Our Past: Studies in Autobiographical Memory*. Cambridge: Cambridge University Press.

Said, Edward. 1978. *Orientalism: Western Conceptions of the Orient*. London: Routledge and Kegan Paul 1978. Penguin Books 1991.

———. 1989. "Representing the Colonized: Anthropology's Interlocutors." *Critical Inquiry* 15 (2): 205–225.

Shweder, Richard A., Jacqueline J. Goodnow, Giyoo Hatano, Robert A. LeVine, Hazel R. Markus and Peggy J. Miller. 2006. "The Cultural Psychology of Development: One Mind, Many Mentalities." In *Handbook of Child Psychology Sixth Edition. Volume One: Theoretical Models of Human Development*, edited by Richard M. Lerner, 716–792. Hoboken, NJ: John Wiley and Sons.

Siegel, Daniel J. 1999. *The Developing Mind: How Relationships and the Brain Interact to Shape Who We Are*. London: The Guilford Press.

———. 2001. "Toward an Interpersonal Neurobiology of the Developing Mind: Attachment Relationships, 'Mindsight,' and Neural Integration." *Infant Mental Health Journal* 22 (1–2): 67–94.

Stein, Nancy L. and Christine G. Glenn. 1979. "An Analysis of Story Comprehension in Elementary School Children." In *New Directions in Discourse Processing*, edited by Roy O. Freedle, 53–120. Norwood: Ablex.

Stein, Nancy L. and Margaret Policastro. 1984. "The Concept of a Story: A Comparison Between Children's and Teachers' Viewpoints." In *Learning and Comprehension of Text*, edited by Heinz Mandl, Nancy L. Stein and Tom Trabasso, 113–155. Hillsdale: Lawrence Erlbaum Associates.

te Molder, Hedwig and Jonathan Potter, eds. 2005. *Conversation and Cognition*. Cambridge: Cambridge University Press.

Wallin, Nils L., Björn Merker and Steven Brown, eds. 2000. *The Origins of Music*. Cambridge, MA: MIT Press.

Yovsi, Relindis Dzeaye. 2003. *Ethnotheories about Breastfeeding and Mother-Infant Interaction: The Case of Sedentary Nso Farmers and Nomadic Fulani Pastorals with Their Infants 3-6 Months of Age in Mbven Sub Division of the Northwest Province of Cameroon*. Münster: Hamburg: LIT Verlag.

Zatorre, Robert J. and Isabelle Peretz, eds. 2001. *The Biological Foundations of Music*. New York: New York Academy of Sciences.

Zelazo, Philip David, Morris Moscovitch and Evan Thompson, eds. 2007. *The Cambridge Handbook of Consciousness*. Cambridge: Cambridge University Press.

Zushine, Lisa. 2006. *Why We Read Fiction: Theory of Mind and the Novel*. Columbus: Ohio State University Press.

# — 2 —

## Framing Religious Narrative, Cognition and Culture Theoretically[1]

Jeppe Sinding Jensen

The combination of "religious narrativity," "cognition" and "culture" is both humanly and scientifically interesting because these phenomena tell us something very essential both about human beings and about being human. As with other higher species in the animal world, especially primates, humans have cognition: they can see, hear, act on their perceptions, remember, express mental states and so forth. But more than all other animals, humans may also use their cognitive abilities to communicate in symbolic format.[2] They can paint, draw, dance, lie, promise, tell crazy stories, they can share norms and values, ideals and not least: they can imagine superhuman beings and imaginary worlds. In brief: they can have culture and religion. And it has been that way for a long time. Archaeologists and others who work on evolution and development may be in disagreement over exact dates but a reasonable estimate tells us that we have been what we are now, that is "anatomically modern" for about the last 150,000 years. A lot has happened during that time. Not so much with our inherited anatomy as with that which we have created ourselves, that is: culture and society, language and religion. In the cross field of religious narrativity, cognition and culture we not only see the results of interactions between these fields as they are now, but also how and why the cross field has become what it is now. It is a long and great history of how humankind has created itself as a hybrid being consisting of both nature and culture.[3]

### On Method and Disciplinary Backgrounds

The issue, or topics, of religion, narrative, cognition and culture are each extremely comprehensive and complex in their own right, and the resulting mixture is no less so. It is a mixture with countless aspects but thus also scholarly challenging. The contributors to this volume come from various disciplines and fields, and they convene in a common

1. This chapter is a translated and revised version of an essay in Danish entitled "Introduktion: Religiøs narrativitet, kognition og kultur," in *Religiøs narrativitet, kognition og kultur*, edited by Jeppe Sinding Jensen and Armin W. Geertz (Højbjerg: Forlaget Univers, 2008), 5–25.
2. Classical studies on this topic and worth a re-visit are Cassirer 1955 and Langer 1951.
3. Clifford Geertz (1973) anticipated this history, now also stressed in Donald 2001.

interest in just this combination of issues—one that it would be impossible for any one discipline to handle alone. In the process of scholarly interaction they have gained much from each other's competencies and criticisms and experienced that just as inspiring cross-disciplinary work may be, it is just as problematic approaching the fields of others without the traditionally required skills. You must simultaneously be intrepid and humble. This is not news to scholars of religion because this field is—almost by tradition—oriented towards pluralism in method, if not simply eclectic. Once upon a time, the study of religion was primarily historical, philological and partly inspired by advances in anthropology. The study of religion has no separate method or theory, and at times it seems that its practitioners are in doubt as to whether it really has an object. What is "religion" anyway? It is, at least, a denomination for a broad range of human activities that cannot be explained in the light of any single theory or treated by one lone method. In a department of the study of religion, we usually find faculty with very diverse background: historians of religion, of course, but also theologians, philologists, sociologists, anthropologists, philosophers and so on. Thus, it is common that we borrow from one another in the study of religion in terms of practical and relevant solutions to our problems. But such an approach has pitfalls and misunderstandings that may render specialists in other fields somewhat suspicious because you cannot know the full history of their theoretical traditions and debates. It is therefore of invaluable benefit when colleagues from other fields wish to take part in a collective effort such as the one presented in this volume.

As I was originally trained in Semitic philology and then became a historian of religions, the Muslim pilgrimage to Mecca, the *Hajj*, is a favorite example of mine when presenting the multifariousness of religious phenomena to students. Of course it is (if you accept the terminology, that is) a "pilgrimage," as we normally understand and interpret this action complex, with rites, cosmology and sacred history, but it is also just as much a social, cultural and psychological fact, and it is often politicized. There is a lot of economy involved, and it is an immense complex occurrence of infrastructural requirement when more than a million participants must eat and drink, wash and go to the loo. It is also an instance of sexual harassment when female participants (according to various testimonies) get pinched on the bottom along the way. In our context of theories here, it is obviously important to see the *Hajj* as an individual, psychological and cognitive phenomenon where intentionality, perception, emotion and body are involved in a ritual as part of the activation of a total narrative cosmology. The participants become actors in the story, the program and the schemata of which the total ritual consists. There is, accordingly, enough to take into consideration when you aim at a "thick description" of anything that has to do with religion. To do so, we need to use all kinds of tools from various theoretical and methodological repertoires. Such a course of action, as already mentioned, is not without its problems, and I shall, therefore, attempt to place some of the issues that we work with in a relevant theoretical setting.

In a cross-disciplinary discussion, it may turn out that a theory that is current in one subject has been deemed out-of-date in another. But, then again, the first subject may have kept and developed the theory in new and important ways so that things are not as bad as they seem in other subject area. In this volume, there are several examples of this, as for instance structuralist theory and method which has proven to be very useful in the study of religion and yield good results in analyses of traditional cultural matters. Structuralism was once new and contested, later it was considered a "French academic fad" from the 1970s, but now it has become very much of a normal paradigm in the study of religion

(Jensen 2005). Another and related example is semiotics, the science of the production and organization of meaning and meaning-systems, which is a vexed issue in many disciplines, whereas in other places (such as Aarhus) it has developed into cognitive semiotics. The concept or idea of evolution was much contested in the study of religion in the twentieth century. Biology is not possible without evolutionary theory, but in the human and social sciences many associate evolutionary theory with the triumphalism of nineteenth century evolutionism with its normative scales and distinctions between "civilized," "savage" and "primitives as the children of the human race." However, all things point to cultural evolution being a fact. Otherwise I would have had to write this chapter on stone tablets. The work presented in this volume would have been impossible without taking cultural evolution into account.

Furthermore, there is the concept of "religion" which has proven difficult to handle in academic discourse. If any concept tends to be misunderstood in the academy, it is "religion." For reasons which it would be opportune to investigate more closely, it seems that there is a tendency among many to associate "religion" with a childhood experience of it as it was propagated (or not) in socialization, primary school and so forth. In my part of the world, it would be Christianity in one form or another that would serve as the prototype example. And, no matter how much my colleagues in other fields may dissociate themselves from religion, their dissociations are driven by those Christo- and Eurocentric prototypes. When taking in a new batch of students, a great deal of my teaching consists in un-learning and dismantling such prototypes before we can get to the subject matter and start theorizing about it in a more advantageous manner. Perhaps there is something about religion—again to be analyzed—that makes childhood experiences more salient than those that we have of art, politics and economics, where very few cling to their childhood conceptions of these matters. So in the case of religion, there seem to be very persistent prejudices that are based on childhood familiarity understandings of "it." It is an attempt to counter these ethnocentric conceptions that we as scholars of religion tend to introduce examples from all over the world and from the depths of history. That does tend, however, to produce more misunderstanding than we wish, so we are basically challenged with the task of explaining very carefully what it is that we want to do. And one of those things is to make matters more clear and explainable.

Added to these problems of concepts, theories and methods are the more or less hidden agendas and discourses in both public opinion and the academic world. There are many things about which it is better to remain silent not least because you run the risk of being grossly misunderstood or suspected of harboring suspicious motives. On the other hand, there are of course limits to what you should be expected to master before you bring a new idea, theory, method or subject area into your work. This you will readily realize as soon as you begin to get involved in a new area and then delve into its deeper teachings. My own realization of this happened in the 1990s upon entering the field of cognitive science as a new approach to the study of religion. As it soon turned out, there are several groups or camps of scholars who seemed to ignore one another or who were in obvious disagreement over the range and properties of cognition in the first place. They also disputed on levels of description, explanation and understanding. Naturalism, mentalism, physicalism, functionalism, identity theory, emergence theory, holism and other "-isms" are the philosophical, metaphysical, ontological and epistemological foundations of the impressive range of cognitive science research, reaching from brain surgery to learning theory. No wonder that a certain amount of cognitive dissonance appears to be inevitable.

## Definitions and Theories

The easiest thing you can say about definitions is that they are theories in their shortest possible version. For example, Marx's famous dictum that "religion is the opium of the people" rests on certain aspects of a definition of religion as projection and simultaneously as both psychological comfort and social protest. A definition of religion tells us what religion is or how it functions, and this will always depend on a larger theory as the explanatory backdrop (Jensen 2003, 61–65). A common problem in the definition of religion concerns the possibility of a "real" definition of what religion *is* or whether we shall content ourselves with a functional definition. In a sense this discussion belongs to a positivistic or empiricist past as it presumes that something as general as the concept of religion should have a direct empirical reference to a factual entity. Many concepts do not have such direct references to specific matters. On the other hand, empirical objects and entities do not contribute to our understanding of them by offering their own labels, a horse has neither "Pferd" nor "cheval" written on its side. The concept of religion is thus an abstraction, an analytic construct, but it is a useful abstract construction since it may refer to a multitude of human activities, individual and collective, in thought and in action. Thus, there may be many different aspects of a subject—as in the example of *Hajj* above—and it should therefore be easy to convince oneself that the idea of a singular, exhaustive and exclusive definition is illusory (at least concerning religion). This does not imply, however, that "anything goes" or is equally rational. It does not mean that we may just as well give up on empirical research or that religion "does not exist." What it does mean is that we are obliged to specify our methods and theories and locate them in the topology of scientific and scholarly endeavours. Definitions then take on a stipulative function as they inform us prescriptively about our subject matters, how to look at them and how to treat them. One could also view definition as "generalized interpretation" as when anthropologists speak of "sorcery" or "marriage" or when historians of religions talk about "sacrifice" or "myth." In that way we may escape the definitorial muddle, but precise and strict it certainly is not. All things point to the fact that we must live with this situation. Some have suggested that we should also give up our everyday psychological notions such as feeling, personality, grief and so on, because these concepts do not indicate anything scientifically valid. Instead of "love" we could talk about gland secretion, dopamine and oxytocine production, synaptic activity etc., most of which is more scientifically "accurate" but also a bit unwieldy. So we shall have to live with a flexible and hence problematic conceptual apparatus. Had the furniture of the world come with its own labeling it would have been easier. It did not, so it is up to us humans to figure these things out. Science is thus provisional.

## Cognition

Cognition is, as a generalizing interpretation, a rather loose notion covering all that which has to do human perception and conceptualization of the world in memory, consciousness, language etc.[4] All of this has been the topic of philosophical reflection for hundreds (and thousands) of years. In the middle of the twentieth century, however, the idea of cognitive science was introduced along with the invention of computers. It was a characteristic of the early development of cognitive science that it focused rather narrowly on human cognition

4. A good publication on what is traditionally considered to fall within the range of the cognitive sciences is Wilson and Keil (1999).

as having to do with the individual only and then primarily concerning the perception of the physical environment or "Umwelt" and descriptions of the cognitive mechanisms (i.e, brain activities) responsible for how this perception worked. It resembled very closely what we have since come to know as information processing in general, and this development was strongly inspired by the nascent computer technology. This conception is now largely abandoned because it proved to be too simple. A part of the more traditional cognitive science referred to in this volume continues to focus on cognition primarily as physical perception, mostly visual and tactile. When cognitive science took physical perception as its point of departure, it was also in order to reduce the range of variables, but the price paid was that matters beyond such perception were ignored. The human environment, the "Umwelt," is much more than the physical, as it includes mental, social, symbolic, and linguistic aspects. The perspectives offered in many contributions to this volume underline that cognition is much more than perception and they direct attention to cognitive activities in social relations to other humans and to symbolic information. Cognition takes place in brains and between brains. In addition, we include bodily and emotional aspects and dimensions which have all too often been excluded from cognitive analysis. We find it justified to work with a modified and expanded idea of cognition. A broader definition of the cognitive field as we understand it in this context runs like this:

Cognition is the designation for the mental capacities and processes located and working in human bodies and brains in interaction with material, social and symbolic environments.

## Culture

The standard cognitive science of religion argues that "culture" is either non-existent or simply a metaphysical concept without causal value. "Culture" is admittedly a somewhat imprecise and unwieldy concept. But that is no reason for despair, for this only means that it is also a flexible and open-ended concept. Not so long ago it became a fashion in anthropology to see "process" everywhere instead of "culture," because most things in human life move, are changing and exchanging. It is difficult to disagree, but for some it meant that we could no longer use the concept of culture because such a concept would be static, useless, erroneous and old-fashioned. This scenario is similar to that of the notion of religion, but there criticism mostly came from historically oriented and particularist scholars who disliked talking about religion in the singular because they felt it was theological and normative. In all its simplicity, the problem often was that critics had overlooked the distinction between concepts and empirical facts—as the distinction between map and territory. It goes without saying that we cannot distinguish unambiguously, firmly and finally between phenomena in a world of object-language and concepts in a world of meta-language, for the one is nothing in terms of human knowledge without the other. This argument is not as circular as it appears. It is rather an acknowledgement of the dialectics that underlie the production of most of our knowledge. The point is simply that in order to understand anything in human behaviour we do need "static" concepts in order to define such realms of activity as culture and religion. Defining culture as the "production of meaning" or "symbolic behavior" is not to use violence against our lived reality. It is no more than necessary, theoretically validated attempts to comprehend the flows of impressions and thoughts. Our thoughts about things depend on the language at our disposal, and this makes narrative so important not least for social life. Considered in their cultural aspects, societies are stories. We understand them as narratives about how society ought to be based on ideals, on stories about humanity, liberty, democracy, equality and so forth. It is therefore imperative that the members of a society subscribe to the same narratives, the same stories or the same myths. For these reasons I maintain a concept of culture that is oriented towards narrative, in which culture can be understood as the "sum of its texts." This brings us to the textual nature of human worlds and thus their narrative foundations.

## Narrativity

In the research project "Religious Narrativity, Cognition and Culture" at Aarhus University, we are specifically interested in religious narrativity (the term "religious" will be explained below) and to develop and further narratology as a field in the study of religion. Narrativity is in itself quite a problematic term as there are several competing and complementary schools: the "older" theory of narrativity in the Parisian school, e.g. Roland Barthes, and further into structural semantics with butterfly and actantial models as represented in the work of Claude Bremond and Julien A. Greimas (Barthes 1994; Bremond 1966; Greimas and Courtès 1979).[5] Most of these theories have since become "mainstream" approaches, and they gradually turned into literary methods applied in secondary teaching systems. In the work of some linguists, e.g. Rukmini Nair, narrativity is primarily an aspect of conversational analysis (Nair 2002). Well known philosophers and theoreticians, such as Paul Ricoeur, Roland Barthes and Tzvetan Todorov have contributed not only to our understanding of narrative in general, but also to the development of literary theory where, however, narrativity and narratology are seen primarily as relating to genres and techniques of authors.

An additional theme, also in the context of literature, is the fundamental question about fiction: What is it? What is it good for? Why is it that we humans apparently cannot do without it? Why do we find it in all cultures and in all religions? Undoubtedly, there are good evolutionary reasons for this, the ability to imagine things is an invaluable cognitive resource albeit it sometimes runs wild (Byrne 2007). That could be the explanation in the "case of religion" (e.g. Boyer 2003). There is the additional epistemological issue, which I find extremely interesting, namely the impression (or could it be a testable hypothesis?) that we humans as a species seem to be able to understand matters only when they have been turned into narrative in one form or another. There seems to be a fundamental, deep cognitive demand which requires that perception and cognition (and probably also emotion) can become knowledge only when we can talk about it. And, further, for anything to be a subject of discourse, it must be placed in a certain grammar and syntax, and our perception and recognition of "things" appears to take place in consecutive spaces so that our inner mental representations and conceptions are ordered into some kind of "grammar," that is, be situated in specific frames, etc. before they may become re-cognition and knowledge at all (Polkinghorne 1988). The formal languages used in the natural sciences are in principle describing matters in a similar way: "Given such and such, and then under certain conditions, such and such will follow...." The knowledge produced in the natural sciences becomes available to us when it has been put into stories, so to say, or as the philosopher W.v.O.Quine once termed it: "cultural posits" (1953).

5. It was the ambition of the Parisian school of semiotics to establish a science of the production of meaning, modeled on the success of linguistics. An idea of the scope and approach can be seen in, e.g. Greimas and Courtès 1979.

Analogous conditions appear to hold also for our inner, seemingly private representations and knowledge as they are not nearly as private as we intuitively believe, and so our "I," or "the myth of the subjective," is a lot more dependent upon language and shared knowledge than we are aware of (Davidson 1989). This is not to say that there are no bodily sensations that exist independently of language, such as, e.g. pain, but, as Wittgenstein once pointed out, for that sensation to become known as "pain" or "toothache" then more is required, a kind of "talk-about-ability," so that it can be communicated, and this requires shared language and intersubjectivity (1973; McDowell 1994). Our subjectivity is to a very large extent intersubjectivity that is constituted in communication and mainly through language (Harré and Gillett 1994). This may have been different before the "symbolic revolution" and the advent of language some 50,000 to 100,000 years ago (Donald 2001).[6] But in whichever way the matter is put, narrativity is a central part of the perspective.

## Religion and Religious Narrativity, Transcendence and "Worlds"

The psychologist Louis Cozolino once wonderfully remarked that "We come into consciousness as a story that, in many ways, has already been written" (2002, 157). A very prominent and important aspect of religion is the importance of narrative; religions are made of strongly narrative "materials." All religious traditions, oral or literate, have narratives; some more than others, but most have narratives in abundance. The larger literate and scriptural traditions have organized collections of canonized narrative, such as the Jewish Torah, Buddhist Tripitaka, the Christian Bible and the Muslim Qur'an. Attached to such "foundational" scriptures are copious collections of other story materials: stories about founders, reformers, saints, and all other accounts that may serve to foster propriety and morals, freedom or social control, insight or superstition. In Chapter 1, my colleague Armin Geertz presented a table with an instructive diagram of various kinds of narrativity and the many levels to be implemented in description, explanation and understanding of them. The complexity is evident because religious narrativity concerns all that which we, the human species, are able to imagine. That is a lot, to say the least, and therefore there is much to engage in from many angles.

In the modern Western and predominantly secularized world, religion is often considered to be either something special and positive or something strange and negative. As "special," religion is considered to be related to matters metaphysical or transcendent and the deeper meaning of life, whereas, on the other hand, as "strange," religion is viewed as, e.g. bizarre superstition. The attitudes towards religion are countless, and thus it is apt that I place the present exposition in a "spectrum of attitudes." Once, when lecturing on cognitive theorizing on religion, I was asked the question: "Is there transcendence in a cognitive theory of religion?" The answer is an uncompromising "no," if by transcendence one thinks of gods, the holy, the spiritual as something factual as, e.g. the cement industry is. But in that respect, cognitive theories of religion are not special, they are like all other critical theories of religion. Perhaps some of these theories appear harsher than previous ones not least in the ways in which cognitive theories have been applied by some of the materialist "hard-liners," that is, scholars who demand that everything must be explainable in empirical, material terms.[7] In their theoretical universe, everything metaphysical and

6. The dating of the "symbolic revolution" is a controversial issue which I will not dwell on here. In this context, the most important issue is that it has taken place and that we are products of it.

7. Examples with public fame (that is, known from the media) in recent years are Daniel Dennett and

transcendent is best avoided, that is, anything that cannot be explained with the methods of natural science such as validation through the replication of experiments. The problem is that if religion as a subject matter was to be studied by such methods, then many things would fall outside the scope of investigation. How does one conduct controlled, double-blind experiments with the sources of the Hebrew Bible, the enlightenment of the Buddha or the revelations of the prophet Muhammad? The aspects that we are able to investigate, however, we certainly *shall* investigate, and that is precisely what has been undertaken in cognitive theory-based studies on religion where the widespread conceptions of magic, superhuman agents, ghosts and life after death have been analyzed and explained. As it turns out, these issues can be exposed in controlled experiments, but many other matters cannot (yet) be subjected to such methods.

So, to the question "Is religion transcendent?" one is forced to answer "no." On the other hand, however, religion *is* transcendent, but only in the sense that language is also transcendent. That is, transcendent in the sense that it reaches and extends beyond the individual, not least when and as religion is full of narratives about who we are and how the world should be conceived. Plunging into the heart of the matter, we may claim that from the narrative perspective, which is our main subject here, religion is a human project which aims at making the world, the cosmos, speak so that everything known to us may become an object of discourse through the use of narrative. This is even more interesting, because this activity of "making the cosmos speak" concerns not only the discursivity of the material and physical world that has been the subject of philosophers' epistemological search for valid and verifiable knowledge and the criteria for truth, that is, for "what exists" in a more conventional sense. This is, of course, of absolute concern, but perhaps less so when studying religion, since religious discursive modes express and articulate very different ideas about "what there is" and with very few restrictions in the way of ontological commitments. Religious forms and modes of language are highly inclusive and holistic because they comprise both physical and metaphysical worlds; they speak about the mental realms, the social worlds and not least about superhuman or supernatural entities and agents. There is, as we know, room for gods of all sorts, for parrots as ancestors and so much else—for such is religion. Religious "world-making" is thus not only about "what there is" in a physical and material sense but about whatever humans may imagine. As such imagination could, however, become rampant or run amok in idiosyncratic behavior, solipsistic pseudo-knowledge or wild hallucinations, some kind of regulation and control is required. This is where narrativity comes in again, since the existence of sociality demands intentionality adjustment and conceptual convergence. Language is the prime medium for that, as we see in something as banal as gossip. Ritual, in the world of religion, is another medium for these intersubjective transactions. It is also remarkable how children in their play activities create social as well as fictive universes using language as their tool (Sinha 2005).

## On Methodology: In and Out of the Brain

We discover two general methodologies when exploring the connections between language and cognition. They may briefly be characterized as an "inside-out" and an "outside-in" methodology, respectively. They resemble a more classic distinction between language as

Richard Dawkins. They emphasize in each their own way that religion is a figment of the brain and so "nothing but" a human fabrication. That is however, no new insight; see, e.g. the critique of them in Geertz (2008, 2009).

external and psychology as internal, and the discussion about what is the cause and the effect of what. The distinction is crucial in deciding where it is most important to focus in research. Should we study the brain and what it produces, or should we study the influences on the brain from outside? In my opinion we should do both—and, fortunately, we can. There is no point in denying the importance of the division of labor in the academic world. If the brain did not produce anything, it would be pointless to study it, and brains probably would not be if not for the "materials" they deal with. We need both sides. The "inside-out" programs and hypotheses explain religion on the basis and as the result of innate cognitive capacities and mechanisms inherited from our evolutionary past. These capacities and mechanisms produce religious representations because of our universal human propensity for producing illusions and projecting human properties and intentions to real or imagined entities in our environments. Thus, humans have invented ghosts or spirits for as long as we know. Religion has been created through the creation of counter-intuitive representations in the mind and brain, and these were then brought out into the world, and over time, these representations solidified into religions as we have come to know them. This is the essence of the best known paradigm of cognitive theory in the study of religion.[8] It is an interesting paradigm which has initiated and spurred a host of interesting studies, but it also has its limitations (for critique, see Day 2004; Laidlaw 2007). Because these scholars primarily restrict their research to hypotheses and investigations that can be tested in controlled experiments in the lab, the possible scope of research is considerably restricted. The spectrum of phenomena under investigation is subsequently limited mostly to the perception of the physical environment through the senses and how this is dealt with in human cognitive mechanisms. A decisive problem that often leads to disagreement in this kind of research is the determining of what we may actually set up as proper objects of investigation for empirical testing. In the "inside-out" program, the main concern is what happens in the brain and the cognitive properties and mechanisms of individuals—all that which goes on without our conscious knowledge of it. A fundamental premise is that these cognitive functions are outside the reach of introspection—we do not perceive their operations. This type of research is strongly oriented towards the individual and so is based on an individualistic methodology.

It is therefore necessary to attempt to clarify some important ontological and epistemological considerations on the subjects of religious narrativity, cognition and culture. Many cognitive science hard-liners are not just individually oriented materialists, but also (self-confessed) reductionists and eliminativists. Briefly characterized, the reductionist would say that religion is "nothing but," e.g. an illusion, brain-control or social projection. The eliminativist would say, for instance, that meaning in language does "not exist." The catchwords "nothing but" and "not exist" easily disclose the ideals and objects of research programs, their ideas about the world and what we may know about it. There are eliminativists in the cognitive science of religion who tell us that consciousness does not exist, that culture does not exist, that meaning is "nothing but" metaphysical gibberish and that only material objects exist. Most cognitive oriented scholars in the study of religion have declared themselves "true materialists" (Sperber 1996). Consequently they espouse some form of "semantic eliminativism," that is, they denounce the idea that "meaning" is a scientific notion and that it consequently does not belong in cognitive science terminology.

8. Represented by, e.g. Dan Sperber, Pascal Boyer, Scott Atran, Ilkka Pyysiäinen, E.Thomas Lawson and Robert N. McCauley. The volume edited by Jason Slone (2006) focuses on that project.

Rather, it belongs on the heap of other discarded notions such as "phlogiston" or "soul." Before one begins to worry too much, it should be kept in mind that part of this line of argumentation is rooted in what can be characterized as "narrow theorizing," research policy maneuvers and—not least—limited epistemological aptitudes. Some of this discourse, or jargon, does not make sense itself; one might just as well say that literature does not exist, only books, or that traffic does not exist, only moving vehicles. And what, then is a vehicle, other than a heap of metal? Or, what do you call it when traffic does not move? Is it still traffic?

Here again the problem lies in lacking the recognition of the difference between map and territory, concept and observable fact, between words and objects.[9] It also refers to the problem of levels of adequate description. For this contribution, the main issue is the existence of the semantic levels, that is, all that which has to do with (primarily) linguistic and symbolically mediated meaning. We must take for granted that meaning exists, in one way or another, or it would be meaningless to talk about these matters.... To declare that meaning does not exist is a statement that refutes itself.

The second major theoretical program and methodology is the "outside-in." It is very well known and has been for a long time, for it simply assumes that we are all somehow products of our social and cultural surroundings, of our socialization, and that during our entire lives we acquire all that which makes us into what we are. On this account, human brains and cognitive systems are conditioned by that which comes to us from the outside. In the history of theories, this has become known as "social constructionism" or "social constructivism." A relevant example is how religious narratives condition our consciousness about the cosmos, morality and our own personality systems. One serious problem with this "outside-in" theory is that many of its former proponents have been guilty of what the cognitive hard-liners call "culturalism," that is, the assumption that human minds and brains are blank slates upon which anything can be written equally well and with the same ease. But mind or brain is not a "tabula rasa," and the "inside-out" theorist are right in disclosing how the properties and functions of mind and brain are constituted and configured in particular ways that are essential for the creation and reproduction of socio-cultural patterns. Experiments demonstrate that we humans have certain basic and universal patterns of behavior and mental operations, and that, for example, some things are better and more easily remembered than others. On the other hand, the ambassadors of the "outside-in" approach also rightly maintain that our experiences mold us for better or worse and that we are not and have not only what is innate. The degree of brain plasticity discovered in recent research, turns out to be much more pronounced that was earlier imagined. It has been customary to think of the infant brain as plastic and malleable, but that the adult brain becomes set and inflexible with no further development possible. Contemporary brain scanning techniques have made possible experiments that demonstrate how the architecture and functions of the brain are not predetermined once and for all, but that they are susceptible to continual stimuli and change much more than we assumed (LeDoux 2002; Cozolino 2002; Frith 2007).

This development provides us with some neurological grounding for the "culturalist," outside-in view of how the conditions of our environmental contexts contribute to the conditioning of behavior and thought in perception and cognition. Our understanding

9. This is a very complex, not to say vexed, problem in philosophical metaphysics and the philosophy of language, and I shall refrain from going into the matter. See, e.g. Jensen (2003) for a discussion of these problems in relation to the vocabulary of religious studies.

of ourselves and our relations to others and to the world(s) we operate in are governed, regulated and controlled by norms, prescriptions and ideas. We may thus speak about the formation of "normative cognition" when we learn to handle such concepts as justice, mercy, equality, democracy and so forth and apply them in our everyday action. These higher-order cognitive (and emotional) skills are learned in social interaction, not least through participation in narrative activities, from the reading aloud of fairytales, across gossip to academic studies of ethics and moral philosophy.[10]

It is imperative that we remember how all that which we learn and experience becomes "installed" in us as the theories, norms and values, that is, as the effects by which we understand ourselves and our world(s). These effects become our filters. In that sense, the narratively experienced and symbolically mediated world becomes primary and dominant. It determines how we conceptualize perceived reality. In this way, humans are able to invert the normal order of animal cognition: our imagined scenarios in semantic memory may be there before the actual experience of something, and so they come to govern the making of our episodic memories, that is to say, of how we perceive, conceptualize and remember a situation. And the filters change all the time: if today, we see a child being physically punished, we will be upset by it, but a hundred years ago, we would have thought that the child must have misbehaved and only had what was "coming to it" as by some unquestionable necessity. Similarly with our conceptions of nature, gender, and race—they change all the time. That is because our normative stories about life and death, violence, childhood and so on also have changed. The stories and representations that we find in them are also closely connected to our understanding of motivations and authority. In social reality, we may re-arrange things because the social world is created through conventions of our own making so that we may jointly decide that "X counts as Y," as for example when physically punishing children changes status from necessary nurture to unacceptable violence. Other quite obvious results of new narratives are the redefinitions and re-descriptions of the last century concerning gender and age-group roles and statuses. Not even "Mom" and "Dad" are what they used to be. All our direct, intuitive impressions are mediated through patterns and schemata that are not of our own individual making. We do not fabricate them in solitude but they come to us as socio-culturally powerful stimuli with unavoidable impact and found everywhere from fashion trends to legislative measures.

## The "Intervention" of Language

In these socio-cultural pressures just mentioned, we can see what I call the "intervention" of language. Language makes its appearance in social interaction between individuals, and it comes between us and the things of and in the world. The furniture of the universe is available to us only as symbolically mediated—and primarily so in language. When we are born, we enter a social world that readily embraces us, as Cozolino emphasized: there are family, norms, language, schools, social institutions and so much else that we take for granted but which is truly quite remarkable. One can call it "tradition," or as it was called in the French sociological school of Émile Durkheim and Marcel Mauss: "social facts." Tradition must be learned and reproduced, but traditions also constantly change. As the philosopher John McDowell noted, "Even a thought that transforms a tradition must be rooted in the tradition that it transforms" (1994, 187). According to the social context, food, fashion and slang may change quickly (in the late modern world at least), whereas

10. On the development of the idea of "normative cognition," see Jensen, forthcoming.

some things change very slowly, e.g. our ideas about individuality and responsibility. But irrespective of the speed of such changes, we are all embedded in traditions, and we do not remain untouched emotional and cognitive machines for very long after birth.

This is only possible, however, because we are symbolizing animals who are able to learn to use images, metaphors and symbols in order to get by in social life, for instance by knowing that some things are signs of something else. Any individual restricted to thinking only in concrete terms and who understands everything literally would be severely socially handicapped. Again, it is narrativity and story-telling that are important bearers of opportunities and perspectives. One may think of how teenage girls fabricate stories about themselves and others, in an age where "self" is a rather new discovery, where embarrassment looms large over almost everything and when almost everything is emotionally processed and typically so beyond meta-cognitive emotion control. On the other hand, we do depend on our emotions to navigate both cognitively and socially, for if all things were equal, it would be very difficult to choose direction. These functions also seem to be reciprocally related, for navigating cognitively is social, and social navigation is also cognitive. I suggest we see these discrete conceptual domains as connected and overlapping in reality.

An additional important human skill is the ability to use matter from one domain or context in another, often termed "de-coupling." This competency is demonstrated, for instance, when hunting behavior is used in religious ritual. It is a basic skill in social life and very frequently employed in religious communication and practice. The same ability surfaces in children's play where they blend possible and impossible things, fact and fiction. In such cases, it is absolutely important that we have the ability to refer to non-existing entities. What an empiricist philosopher would see as a problem of lacking reference is, however, a valuable asset in a social and cultural world, that is to say, as a result of our creative human faculties. Matters are not only important if they can be found in the material world; the most important is whether they can become entities in our imagined universes. Unicorns have no place in the zoologist's taxonomy of existing (and extinct) animals, but they are highly salient in narrated worlds where they have their own symbolism and "meaning-history." For in all discussions about religion, culture and narrativity, it is obvious that the question of meaning is important and that the relations between sense and reference are closely related to the issue of what meaning is altogether. With philosopher Karl R. Popper, we may speak of there being three "worlds" (1982). The first is the physical and material world, which in spite of relativism and advanced physics we know comparatively well through scientific insight and practical knowledge. The second world is that of the mental, of which we do not yet know as much as we would like to (we are working on it as this volume demonstrates). It does seem reasonable to presume that we have "inner" life, because that is at least how we perceive it as thinking animals. The third world is that of meaning, semantics and signification, that which Popper called "objective knowledge." It consists of those products of meaning that humans have created and presented to the world as "objects," and which therefore acquire a measure of independent, objective existence.[11] This applies to matters as diverse as symphonies, money and gospels—as parts of worlds of significance and signifying objects. But does the same not hold for the perception of having an inner mental life? That perception most likely demands precisely that we share "objective" conceptions of our having an inner mental life. The experience of "me" is in all likelihood not at all as given as it might appear at first sight. Thus conceived, there are important

11. On the issue of the objectivity of social facts, see also Searle (1995).

unsolved issues on the question of our own subjectivity (as also noted above). A further issue is the matter of *what* it is that confers meaning, signification and sense onto these three worlds. It cannot be solved here, but I will sketch a possible interpretive frame.

It is the hypothesis here that we ought to distinguish between three kinds of knowledge concerning each of these three worlds. Concerning the first, the physical and material world, we must presuppose some measure and kind of realism in order to at all obtain knowledge about the conditions of that world. This is done in different theoretical, scientific (and often formal) languages that may be corrected as our knowledge expands and is corrected. We are not infallible; scientific, epistemic knowledge is a piecemeal affair and we know that our comprehension is provisional. Concerning the second world, the mental, it appears that the mental is only accessible to our own comprehension and the awareness of others in so far as it can be expressed and interpreted. Analyses of the meaning, sense and significance of our linguistic expressions demonstrate that meaning is not "in the head" as our intuitive introspective impression might lead us to think. Meaning, sense and significance are at our disposal because they are in the world into which we are born, and it is so because we need to share them with others in order to "make sense." Concerning the third world of meaning, it seems to be the case that linguistic meaning is objective in a sense, that is, in as far as it is publicly available. In the material sense, letters of the alphabet are "nothing but" inkblots on paper, but for us humans, the meaning (if it is a language we have learned to read) is just as real as the inkblots. So, a certain measure of symbolic or linguistic realism is inevitable: the signs are *there* and we are there with them.[12]

Part of the aforementioned problems in the cognitive sciences of religion appear because of a widespread disinclination to understand what we might call the "meta-cognitive," that is, that which is suspended "over" individual brains as "webs of signification," as Clifford Geertz once called it (1966).[13] Many cognitive materialist hard-liners want to put it all back into the brains of individuals. But by doing so, we miss a point: for it is at the level of the meta- or supra-cognitive that we find the world of meaning. Many consider this to be an airy and rather strange idea and quite metaphysical too, for how can this third world of meaning be real? To be quite honest, we do not really know. It is so far one of the real mysteries in human life; but if it were not real, then this contribution—as well as the rest of the volume in which it appears—would not make much sense. You would not even be able to disagree with the point just made. We should remember, here, that there is no genuine scientific or philosophical consensus on the issue of how thought and language, our thinking and speaking, are related.[14] Recent cognitive semantics present interesting proposals as to how this might be conceived, as this line of research manages to combine body and action as necessary preconditions and correlates for thinking.[15] But that is not enough, since the production of meaning must transcend the individual and become externalized in order to become part of the collective production of consciousness. In a loose analogy, we may speak of meaning-production as the making of a kind of human "intentionality-

12. I have subjected this last point about the objectivity of meaning to more detailed exposition and debate in Jensen 2003 and 2004. My views are based on Popper's arguments (Popper 1972, 1982).

13. Recent cognitive science findings support Geertz' classic views, see Deeley (2004).

14. There have been many attempts, however, such as Whorf (1956).

15. See, e.g. Coulson and Oakley 2005. Now classic is Varela, Thompson and Rosch 1992. Recent brain-imaging experiments demonstrate that the same areas in the brain 'light up' when we use a hammer, see a hammer, imagine or hear about a hammer (Shinkareva *et al.* 2008). Thus, there are connections between language, mind and body.

traces," as when dogs both leave smells and are led by smells "deposited" by other dogs (they probably do not do this very intentionally). Humans can produce signs for others to interpret, we can synchronize our intentionalities, so to speak, and this is highly relevant for the "making of ourselves," both phylogenetically and ontogenetically.[16] Being human, we not only feel and perceive the world, that is "note" it, but we can also "denote" and mark it in particular ways so that it becomes our world—which is ever so different from the world of dogs in many ways. For example when we indicate distances in the metric system, a rather convenient (for most) invention which also consists of things that "do not exist" in raw materiality. The metric system was devised by humans in order to measure, divide, and conceive of the properties of the physical world. On the basis of such examples, there is all the more reason to insist that competent use of the brain demands the assistance of external symbolic systems (Clark 1997).

"Culture" is a convenient designation for all that which is a precondition for us to acquire in order to become "normal" persons in our social environment. Just think of how children need to learn when and when not to laugh in more normative and complex social situations and contexts—you do not (where I come from) laugh at traffic accidents or the handicapped. It is precisely this which is the miracle of the human brain: to be able to transcend the limitations of the individual brain (some higher animals can do the same to a more limited extent). Humans have a special competence in cognitive regulation, and this is especially effective because it enables us to perceive and understand other members of our species as intentional agents, even at very complex levels and, even more remarkable, in a range of modalities from wishes, promises and warnings, to prohibitions, admonitions and commands.[17]

Thus, it is through the intervention of language and other symbolic systems that we may take the step from individual cognition to mediated, distributed and collective cognition and, not least, to normative cognition.[18] Culture is, consequently, not just a product of cognition but just as much the precondition for cognition in an ontogenetic perspective. Whichever culture we come across, it is always the result of the work of many generations: the production of knowledge and meaning as well as other inventions and artifacts by which we cope with the world. Today most children in the late modern world interact with their physical and social environments almost exclusively through already existing cultural artifacts and activities, from toys to TV. The neurologist and psychologist Merlin Donald noted on his idea of the "collectivity of mind" that our cultures invade us and posit our agendas for action, and "the creative spark of cognition still depends on the individual conscious mind, but even this statement has to be qualified because creativity cannot be exploited, or even defined, without a cultural context" (2001, 299). So, Donald's conclusion is that "the triumph of consciousness will be complete when it can finally reflect on the collective process itself and see only itself, in the mirror of its own reflection" (236). In a puzzling way, this resembles the "Triumph of the human spirit" in the philosophy of Hegel—but now on a neurobiological basis.

---

16. On this, see especially Michael Tomasello (1999).
17. As far as we know, no other species is capable of performing "speech-acts." Modalities in this connection means the modes that we know from the grammatical and semantic ranges of modal verbs. Psychologists, for instance, think of the senses when they use the word "modality."
18. On the theory of distributed cognition, see Hutchins (1995).

## On Narrative and "Other Worlds": Individually and Collectively

In our individual cognition, we are able to imagine things that are really strange and improbable, we can twist and turn our representation in many ways, but these imaginations do not become shared, collective imaginary worlds without a cultural and social background or the intervention and use of language and narrative. There is so much that humans can say in and about fiction and fictive universes; several academic disciplines devote their time and energy to its study.[19] Most of what they find applies equally well to studies of religious worlds and the narratives in and about them. Broadly speaking, narrative and narrativity are ways of expanding consciousness. Narrative "spaces" provide options for thinking about possible events and for potential orientation in possible scenarios. We can use narrativity to adjust to situations which have not yet arisen, things that have not happened and through this kind of high-level "predictive coding," we reduce contingency and the range of possible interpretations. This certainly applies to our mutual, interpersonal understanding of where and when we expect to find meaning in the behavior of others. Narratives are as such a form of social practice that is eminently important to our cognitive development and for the manners in which we create coherence and integration of ourselves and our experiences. Narratives have direct influence on how our brains become organized and function (e.g. Siegel 1999). This also indicates how it is that religion can hold such extreme power over human minds.

## Narrative in Individual Integration

Already from the age of three, children begin to develop an "author-self," so that they can tell about themselves and others. By doing so, they can embark on reflexivity and start to create their understanding of themselves in social contexts. Narratives of all kinds, and especially stories favored by children, play formative and crucial roles in the relations between children and adults. Combined with experiences of (secure) attachment, narratives and stories are basic to the development of the child's sense of self and the growth of autobiographical awareness. Such awareness helps produce a sense of cohesion in life so that existence may become meaningful. The narrative self is an abstract construction that functions as a system of general intentionality where the underlying cognitive mechanisms advance intuitive abilities to look for purposes and causes of own experiences in life on the basis of relations to others. It is an open question how much "self" we would have at all without our ability to share with others. The capacities of humans to collaborate as well as to self-monitor and self-control depend on the ability to share. The impression that there is a "me" is fundamental for social interaction. Such ideas about social interaction we also find in religious traditions where they can be expressed in many different ways, from the most indirect and complicated myths about primeval beings and their deeds to the most elaborate philosophical tracts. For instance, one could mention Buddhist notions about the nonexistence of the self (which may very well be true), but none the less "it" might be a useful fiction concerning the elements necessary for Buddhists to come into being as agents. It is important that we are able to narrate about ourselves and our experiences so that we can think about how we should act. We create consistency in our lives on the basis of schemata based on our experiences, and in the re-telling, coherence appears. As the most radical hypothesis, we can claim that narrative practice and neural integration mutually mold each

19. Instructive is the volume edited by Herman *et al.* (2005).

other in the individual as well as it has presumably happened in our species' evolution. As psychologist Daniel Siegel emphasized:

> Natural selection has thus enabled our minds to evolve mindsight capacities. Being able to navigate our way through such an intense social world requires the ability to make sense of other minds. Communicating the more elaborate and intricate learned aspects of this mentalizing knowledge to others allows the benefits of one individual's wisdom and experience to be shared with others in the group. This knowledge is transmitted from one individual to others in the group by means of storytelling. *Making sense of other minds is the essential stuff of narratives.* (1999, 329, italics in original)

It is a vexed question whether the understanding of self or the understanding of other comes first, but there can be no doubt that they are closely related. "Autonoesis" (self-understanding) and personal identity come into being on the basis of inherited neuropsychology, but it takes practice in social and emotional attachment for the individual in order to become a fully functioning person with a responsible emotional structure.[20] Recurrent patterns of emotions and memories about them shape the roles and scenes in our personal dramas, and these are based (to a great extent) on experience in the earliest and most brain-plastic periods of our lives. Our affect systems are biological and inherited, but our feelings comprise not only affects but also greater and lesser awareness of them and in combination *with other* affects and emotions. Pride, guilt or shame, for example, are feelings about other feelings. The neural affect-system shapes experience into emotional patterns through the socio-cultural "scripts" that we go through in our dependence on and relations to others. All this is organized in moral spaces which we create through narrative—also in narratives that are private. There are things we do not tell others, but even as private as they may be, they are told in the same formats and in the same words as public narratives. Not even privately do we speak private languages.

Narratives provide us with the possibility of simulating and having vicarious experience, and that is a most fantastic ability to have. As noted above, the experiences of others, even if they are imagined, can become our experiences, and then in a strange way, when we come to think about it: first they are stored in our semantic memories, and then they appear in our episodic memories. This is the reverse of all other experience in the animal kingdom. But then again, our human brains are remarkable biological devices. What we know, that is, our epistemic, cultural and personal competence, we mostly retrieve from our semantic memory, and that includes what we have heard from and about others. This we use in our interpretations of the world and all that is in it. The ideas of others (reflective beliefs) may become very convincing so that they govern our perception (intuitive beliefs) and cognition entirely (Deeley 2004). Religion is a prime example of such mechanisms. We use interpretive schemata from narrative in almost everything we do. We do not even need to think about it, as long as "all is well," but when things go wrong and our souls hurt, then we certainly become aware of this. This is where cognitive therapies demonstrate in all clarity what traumas, depressions, dysfunctional interpretations and narratives may do to us. But the therapies also reveal how we may use language as "medicine" for the purpose of changing negative, cognitive and emotional patterns to the better. When language and narrative may be used as instruments of healing, it also shows us just *how* fundamental language and narrative are in our social and individual lives.

---

20. Although I know of no comparative anthropological analyses of the universality of this issue, it is nevertheless a reasonable (and to a certain extent testable) hypothesis that all cultures have normative notions not only about moral but also emotional responsibility.

## Narrative in Collective Internalization and Integration

Humans are social animals, and with other primates we share a great deal of our genetic pools. We resemble them and they resemble us in interesting ways, and the study of primates has contributed profound insights into human nature, individually and collectively. With the intervention of language, however, we have become hyper-social because we could break free of our biological solipsism and begin to share our mental states with others much more than other animals can do. We have been able to create something like collective brains with the aid of external communication technologies and symbol systems by which we can objectify and store meaning and intention. We have been able to produce objective, shared memories and identities with the aid of language. Humans have provided and stored meanings, narratives and patterns in consciousness by referring them to transcendent authorities. We then imagined that the same transcendent authorities had established those meanings, etc. as normative, axiomatic truths which we all have to conform to. Religion and religious narrative are precisely such matters, and they have contributed to the communal projects of organizing and maintaining societies. For better or worse, humankind has created itself in and through religion. We may hypothesize whether religion has provided human groups with advantages in evolutionary adaptation. It seems likely that the functions of religion in evolutionary adaptation and socio-cultural development is better viewed at group level, rather than in terms of individual coping (although this certainly also must play a role). A pragmatic theory of religion would hold that shared norms and values, shared classifications such as local, cultural cosmologies and anthropologies, stable social and cultural institutions—all other things being equal—would have been beneficial for the social cohesion of groups and thus for their survival.[21] Not least so in conflict with other groups.

The cultural and social evolutionary developments gained both speed and momentum after the "symbolic revolution"—whenever it happened, it was probably a very slow "revolution." With that "revolution" came language, symbolization and, ever since humans have been able to collectively imagine the most counter-factual and counter-intuitive matters, also matters with which to imagine and mirror oneself. Even the most bizarre facts from the worlds of religions are evidence of that. We may therefore expand the interesting philosophical problem concerning "what there is" with the probably even more interesting opposite question (at least for the study of religion) of "what there is *not*." This seems to be a very fascinating question, for all things point to the fact that our species simply cannot live without deep concerns for all that which "is not," and that in many cases it is more important to relate to "what there is not" in order to find out about ourselves and the world.

It is precisely here that the study of religion and anthropology have an almost unbounded reservoir of knowledge about how humans have created their realities through fiction, in all imaginable and idealized notions about the world and which they have projected onto that same world. But this insight was already discovered a long time ago by Ludwig Feuerbach, Karl Marx and Émile Durkheim—to name some of the most well known ancestors in the theory of culture and religion. As a consideration concerning method, I must confess that I think that we should not forget the old masters. Their answers may not have retained their validity, but they sensed that the questions were important. We still think so too, and we should realize that we live under the basic condition of history: all science is provisional but that does not render it a futile affair. We can say more now than the old masters could.

21. This view is expressed in more detail in Wilson (2002).

We have new methods, theories and vocabularies so that we now are much better equipped with the recent involvement of the cognitive sciences to investigate the relations between religious narrative, cognition and culture. On a biological basis, too—when it comes to that.

## References

Barthes, Roland. 1994. *The Semiotic Challenge*. Berkeley: University of California Press. Translation of *Aventure sémiologique*. Paris: Éditions du Seuil, 1985.

Boyer, Pascal. 2003. "Religious Thought and Behaviour as By-products of Brain Function." *Trends in Cognitive Sciences* 7(3): 119–124.

Bremond, Claude. 1966. "La logique des possibles narratifs." *Communications* 8.

Byrne, Ruth M.J. 2007. *The Rational Imagination: How People Create Alternatives to Reality*. Cambridge, MA: MIT Press.

Cassirer, Ernst. 1955. *The Philosophy of Symbolic Forms. Vol II: Mythical Thought*. New Haven, CT: Yale University Press.

Clark, Andy. 1997. *Being There: Putting Brain, Body, and World Together Again*. Cambridge, MA: MIT Press.

Coulson, Seana and Todd Oakley. 2005. "Blended and Coded Meaning: Literal and Figurative Meaning in Cognitive Semantics." *Journal of Pragmatics* 37: 1510–1536.

Cozolino, Louis J. 2002. *The Neuroscience of Psychotherapy: Building and Rebuilding the Human Brain*. New York: W.W. Norton.

Davidson, Donald. 1989. "The Myth of the Subjective." In *Relativism. Interpretation and Confrontation*, edited by Michael Krausz, 169–173. Notre Dame, IN: University of Notre Dame Press.

Day, Matthew. 2004. "The Ins and Outs of Religious Cognition." *Method and Theory in the Study of Religion* 16: 241–255.

Deeley, Peter Q. 2004. "The Religious Brain. Turning Ideas into Convictions." *Anthropology and Medicine* 11(3): 245–267.

Donald, Merlin. 2001. *A Mind So Rare: The Evolution of Human Consciousness*. New York: W.W. Norton.

Frankenberry, Nancy and Hans. H. Penner, eds. 1999. *Language, Truth, and Religious Belief*. Atlanta, GA: Scholars Press.

Frith, Chris. 2007. *Making up the Mind: How the Brain Creates Our Mental World*. Oxford: Blackwell.

Geertz, Armin W. 2008. "How *Not* to Do the Cognitive Science of Religion Today." *Method and Theory in the Study of Religion* 20: 7–21.

———. 2009. "New Atheistic Approaches in the Cognitive Science of Religion: On Daniel Dennett, *Breaking the Spell* (2006) and Richard Dawkins, *The God Delusion* (2006)." In *Contemporary Theories of Religion: A Critical Companion*, edited by Michael Stausberg, 242–263. New York: Routledge.

Geertz, Clifford. 1966. "Religion as a Cultural System." In *Anthropological Approaches to the Study of Religion*, edited by M. Banton, 1–46. London: Tavistock.

———. 1973. "The Growth of Culture and the Evolution of Mind," in *The Interpretation of Cultures*, idem. New York: Basic Books, 55-85.

Godlove, Terry F., Jr. 1997. *Religion, Interpretation and Diversity of Belief: The Framework Model from Kant to Durheim to Davidson*. Macon, GA: Mercer University Press.

Goodman, Nelson. 1978. *Ways of Worldmaking*. Hassocks, Sussex: The Harvester Press.

Greimas, Algirdas J. and Joseph Courtés. 1979. *Sémiotique. Dictionnaire raissonné de la théorie du langage*. Paris: Hachette.

Habermas, Jürgen. 1987. *The Theory of Communicative Action*. Vol. 1, *Lifeworld and System: A Critique of Functionalist Reason*. Boston: Beacon Press and Cambridge: Polity Press in association with Basil Blackwell, Oxford.

Hale, Bob and Crispin Wright, eds. 1997. *A Companion to the Philosophy of Language*. Oxford: Blackwell.

Harré, Rom and Grant Gillett. 1994. *The Discursive Mind*. London: Sage Publications.

Herman, David, Manfred Jahn and Marie-Laure Ryan, eds. 2005. *The Routledge Encyclopedia of Narrative Theory*. London: Routledge.

Hutchins, Edwin. 1995. *Cognition in the Wild*. Cambridge, MA: MIT Press.

Jensen, Jeppe Sinding. 2002. "The Complex Worlds of Religion: Connecting Cultural and Cognitive Analysis." In *Current Approaches in the Cognitive Science of Religion*, edited by Ilkka Pyysiäinen and Veikko Anttonen, 203–228. London: Continuum.

———. 2003. *The Study of Religion in a New Key. Theoretical and Philosophical Soundings in the Comparative and General Study of Religion*. Aarhus: Aarhus University Press.

———. 2004. "Semantics in the Study of Religion." In *New Approaches to the Study of Religion*, vol. 1, edited by Peter Antes, A.W. Geertz and R. Warne, 219–252. Berlin and New York: Walter de Gruyter.

———. 2005. "Structuralism—Further Considerations." In *The Encyclopedia of Religion*, Rev. ed., vol. 13, 8757–8760. Detroit: Macmillan Reference.

———. forthcoming. Normative Cognition. In preparation.

Jensen, Jeppe Sinding and Luther H. Martin, eds. 2003. *Rationality and the Study of Religion*. London: Routledge.

Keane, Webb. 1997. "Religious Language." *Annual Review of Anthropology* 26: 47–71.

Laidlaw, James. 2007. "A Well-Disposed Social Anthropologist's Problems with the 'Cognitive science of Religion'." In *Religion, Anthropology and Cognitive Science*, edited by Harvey Whitehouse and James Laidlaw, 211–246. Durham, NC: Carolina Academic Press.

Langer, Susanne K. 1951 [1942]. "Life-symbols: The Roots of Myth." In *Philosophy in a New Key: A Study in the Symbolism of Reason, Rite and Art*, 148–174. New York: A Mentor Book.

LeDoux, Joseph. 2002. *The Synaptic Self: How Our Brains Become Who We Are*. New York: Viking.

Lévy-Bruhl, Lucien. 1987. *How Natives Think*. With a new introduction by C. Scott Littleton. Princeton, NJ: Princeton University Press.

Lincoln, Bruce. 1999. *Theorizing Myth: Narrative, Ideology, and Scholarship*. Chicago, IL: University of Chicago Press.

Lyotard, Jean-Francois. 1984 [1979]. *The Postmodern Condition*. Manchester: Manchester University Press.

McDowell, John. 1994. *Mind and World*. Cambridge, MA: Harvard University Press.

Nair, Rukmini Bhaya. 2002. *Narrative Gravity: Conversation, Cognition, Culture*. Oxford: Oxford University Press.

Polkinghorne, Donald E. 1988. *Narrative Knowing and the Human Sciences*. Albany: State University of New York Press.

Popper, Karl R. 1972. *Objective Knowledge. An Evolutionary Approach*. Oxford: Oxford University Press.

Quine, W.v.O. 1953. *From a Logical Point of View*. New York: Harvard University Press.

Quinn, Philip L. and Charles Taliaferro, eds. 1996. *A Companion to the Philosophy of Religion*.

Oxford: Blackwell.

Ricoeur, Paul, 1991. "The Function of Fiction in Shaping Reality." In *A Ricoeur Reader: Reflection and Imagination*, edited by Mario. J. Valdés, 117–136. Hemel Hempstead: Harvester Wheatsheaf. Originally published in *Man and His World* 12(2), (1979): 123–141.

Searle, John R. 1995. "Language and Social Reality." In *The Construction of Social Reality*, 59–78. New York: The Free Press.

Shinkareva, Svetlana V. *et al.* 2008. "Using fMRI Brain Activation to Identify Cognitive States Associated with Perception of Tools and Dwellings." *PlosOne* 1, e1394: 1–9.

Siegel, Daniel J. 1999. *The Developing Mind: Toward a Neurobiology of Interpersonal Experience.* New York: Guilford Press.

Sinha, Chris. 2005. "Blending Out of the Background: Play, Props and Staging in the Material World." *Journal of Pragmatics* 37: 1537–1554.

Slone, Jason, ed. 2006. *Religion and Cognition: A Reader*. London: Equinox.

Stiver, Dan. 1996. *The Philosophy of Religious Language: Sign, Symbol, and Story*. Oxford: Blackwell.

Tomasello, Michael. 1999. *The Cultural Origins of Human Cognition.* Cambridge, MA: Harvard University Press.

Turner, Mark. 1996. *The Literary Mind.* Oxford: Oxford University Press.

Varela, Francisco J., E. Thompson and E. Rosch. 1992. *The Embodied Mind: Cognitive Science and Human Experience.* Cambridge, MA: MIT Press.

Whorf, Benjamin Lee. 1956. *Language, Thought, and Reality*. Cambridge, MA: MIT Press.

Wilson, David S. 2002. *Darwin's Cathedral.* Chicago, IL: Chicago University Press.

Wilson, Robert A. and Frank C. Keil, eds. 1999. *The MIT Encyclopedia of the Cognitive Sciences.* Cambridge, MA: MIT Press.

Wittgenstein, Ludwig. 1973. *Philosophical Investigations,* third ed. New York: Prentice-Hall.

# —3—

# Language as an Emergent Function: Some Radical Neurological and Evolutionary Implications[1]

Terrence W. Deacon

## Introduction: The Fallacy of Reverse Engineering

There can be little doubt that the development of formal generative linguistic analysis has been the most important advance in the study of language in the last half century. These formal models predict the structure of linguistic output both accurately and compactly, including its open-ended generativity. The elegance and compactness of these formal theories are a function of good reverse-engineering. It would be odd if these theories exhibited poor predictive power, considering that they were consciously designed to yield results consistent with vast amounts of known (and self-generated) linguistic data. Thus it seemed reasonable to extrapolate from this model to the evolutionary and neurological origins of language. But after nearly a half-century of unsuccessful efforts to find the neurological and genetic counterparts to universal grammar or the algorithms comprising the principles and parameter-setting mechanisms of language, it may be time to reflect on why neurobiology and formal linguistics have not converged.

Paradoxically, the successes of formal generative linguistic theories may have impeded progress toward understanding language neurology and language evolution, even while they have provided such remarkably sophisticated tools for the description of language structures. This is because the apparently remarkable adequacy of formal models to account for the complexities of language structure have contributed to an unwarranted assumption that language can be studied as though its structure was designed by a kind of instruction logic, as are other formal systems. Yet despite compelling evidence that language has a formal structure consistent with top-down rule governed systems, its status as an evolved —not designed—biological phenomenon raises serious questions about the plausibility of extrapolating from this descriptive analysis to a theory of language processing.

This problem shares some ironic parallels with mid-20th century debates about the immune response. The generativity and flexibility of the immune system was then thought

1. This chapter was originally published in *Theoria* 54, 2005, 269–286. Grateful acknowledgement is extended to the publisher.

to require some mechanism for instructing immune cell molecules to conform to the specific structure of disease antigens in order to be able to bind to them selectively. Immune potential could be encoded in the genome, but the complexity of this code would need to be enormous. Because specific immunity can be acquired, antibody structure might also be "learned" by matching to antigen structure on exposure. The solution required abandonment of both instruction theories. Generation of adapted antibodies results from blind variation, serendipitous matching, and differential proliferation of the best fit immune molecules in response to antigen exposure. With respect to language, this is a cautionary tale about the (in this case erroneous) assumption that only an instructional mechanism can accomplish this degree of functional precision and generativity.

Probably the most compelling reasons to avoid assuming a simple correspondence between linguistic models and neural processing logic, however, comes from comparative neuroanatomy. Despite the obvious human uniqueness of language, no unprecedented new brain structures distinguish human from nonhuman primate brains. Even the so-called language areas of the cerebral cortex (e.g. Broca's and Wernicke's areas) have been shown to derive from primate homologues sharing, positional, cytoarchitectonic, and connectional patterns with their human counterparts, despite not subserving language or even vocal control (e.g. see Deacon 1997; 2004). Although brains also have functional hierarchies that divide into interdependent functional systems, this reflects the constraints of embryology and phylogeny and not any linguistic functional role. So language processing is ultimately carried out with the same brain structures and functional logic that other primates use for nonlinguistic functions.

## Mechanical Versus Organic Analogies

The difficulty may, however, be more serious. There is reason to believe that these formal models predict generative architectures that substantially *reverse* the order of the critical generative processes.

Linguistic logic is based on rules for manipulating basic units (e.g. phonemes, morphemes, grammatical operators, etc.) to create larger complexes (e.g. phrases and sentences). Linguistics is thus largely an enterprise of top-down analysis. For linguists, full-fledged language phenomena are the starting points of analysis. The basic units that comprise patterns of use and with respect to which linguistic judgments are made are morphemes, grammatical operators, syntactic relationships, and so forth. Most linguistic theories assume the primacy of these linguistically defined language units, and differ primarily in their analysis of operations applied to them. This is equally true for algorithmic and constraint-based approaches to grammar, and to a lesser extent for cognitive grammars. An exemplar of a classic phrase structure analysis based on rewrite rules is depicted in Figure 3.1. This approach is characterized by instructions for generating phrases and sentences by combining lexical elements. It assumes a separation of syntactic from semantic operations, and these from phonological operations.

As a result, there must also be a system of bridging rules for linking these three derivational systems. Some of the computational complexity of this approach is avoided by constraint-based approaches to grammar. A system of combinatorial instructions is unnecessary if one assumes that the basic lexical building blocks come with implicit structure—analogous to chemical valence—that limits their combinatorial possibilities (see Figure 3.2). Thus phrases and sentences can self-assemble without extrinsic instruction,

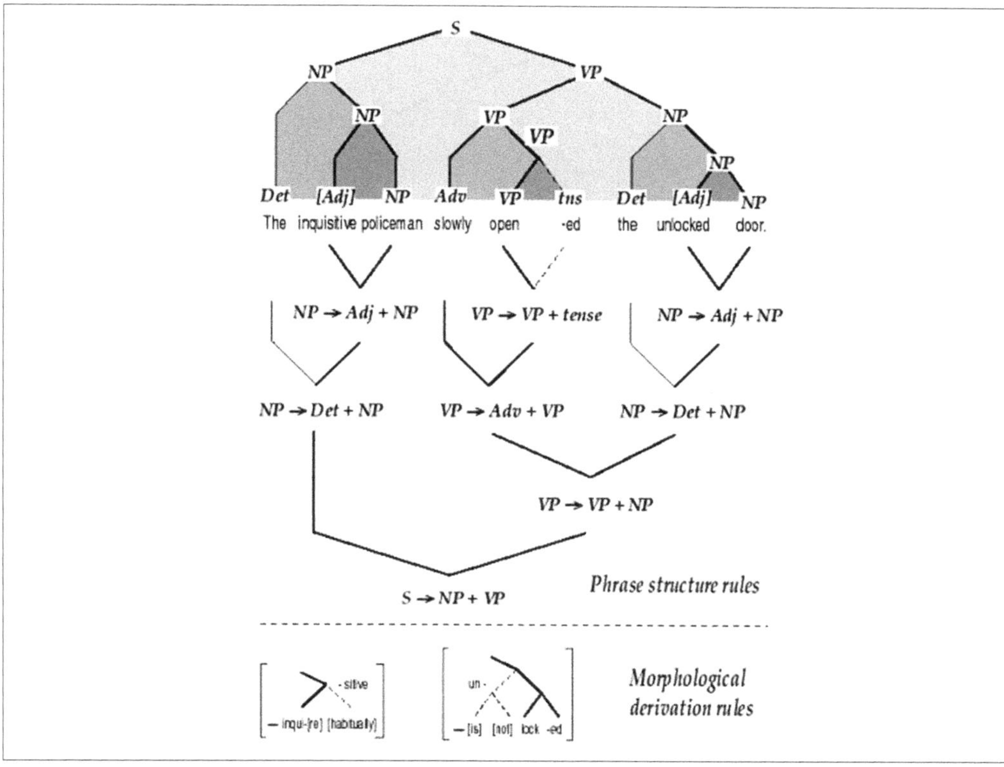


Figure 3.1. Algorithmic hierarchic "re-write" conception of sentence generation. Rules of composition and analysis are imposed top-down, allowing complete arbitrarity of the characteristics of lexical units. Lexical units still need to be identified with respect to grammatical categories, and then separate derivation of reference and phonology. Such an architecture is analogous to a Turing machine, and if completely described, can be modeled as a determinate machine as well as a formal system.

driven only by elicitation of the lexical units and a generic "merging" operation. This shifts the burden of structural information to the lexical level, and is in accord with the psychological experience of unmonitored spontaneous sentence generation in which most of the conscious attention is directed toward content word choice. But it even more emphatically requires an explanation of the basis of these constraints (e.g. innate grammatical categories) and the mechanism by which possible merges are allowed or disallowed. All approaches that assume that sentences are neurologically built-up from more basic non-composite grammatical units, must provide a separate account of the source of these prior rules or constraints and must incorporate a mechanism for sentence construction or combinatorial constraint, respectively. Ultimately, we must map all to neural processing units.

The assumption that minimal lexical units are primary, and linguistically composite structures are derived from combining these more basic units seems too obvious to question. Most sentences are novel productions, while all words and grammatical markers have ancient origins. How could sentences ever be prior to the words that compose them, either historically or mentally? This compositional analysis is consistent with a formal or engineering analogy for sentence generation. In such systems, a set of design instructions (or assembly constraints, or both) and a set of component parts precede all operations. Complex structures are accurately modeled by combining components according to these rules.

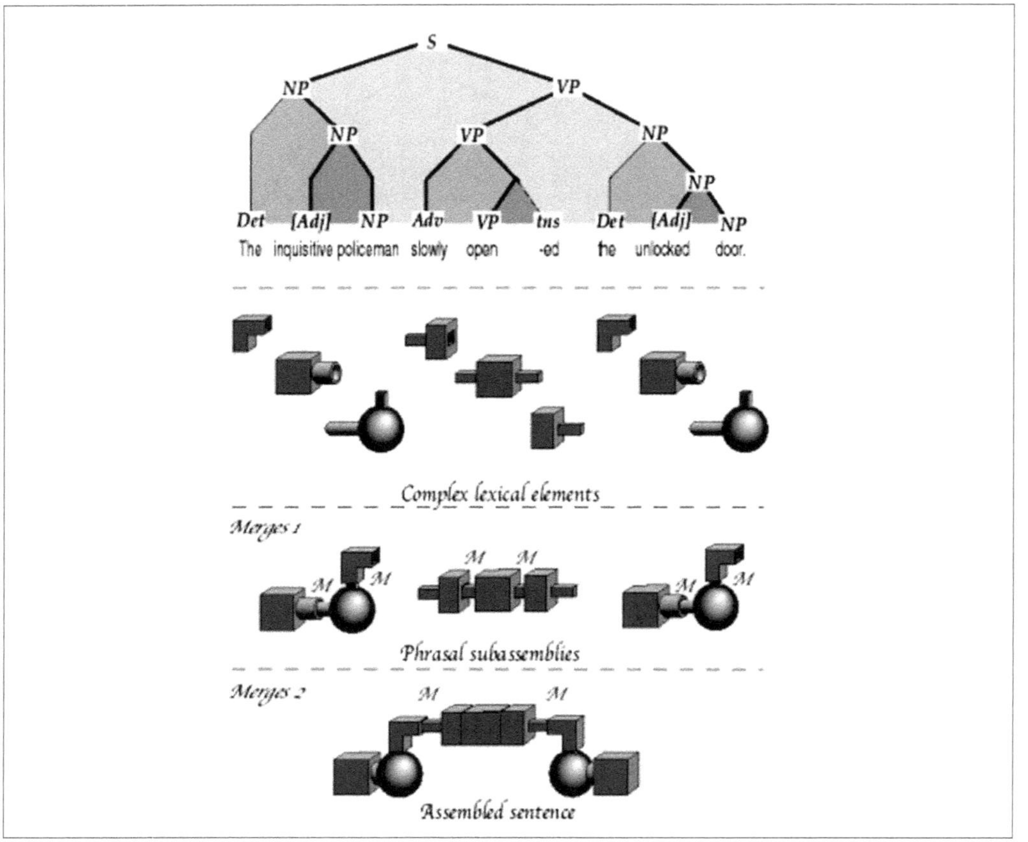


Figure 3.2. Constraint-based conception of sentence generation. By assuming that lexical units have constrained binding possibilities (on the analogy of specific types of atomic valence) phrasal assembly and sentential architecture restrictions can be derived spontaneously without top-down instructions (though order of operations may still require extrinsic constraint). However, this requires explaining how lexical items inherit this valence structure and how this is implemented other than by rules.

So it seems natural to assume that the neural production of sentences should also proceed this way.

Now contrast this logic with the logic of evolution, embryonic development, or the development of brains (discussed below). Biological structure generally arises from bottom-up differentiation processes. This is the inverse of top-down instruction and the imposition of combinatorial order on component parts. In biological processes, functional parts (e.g. cells, organs, limbs) are not collected together and then assembled to make larger functional wholes. Undifferentiated wholes (e.g. a fertilized ovum, a gastrula, and embryo) precede their differentiation into composite structures, even though these also involve multiplication and subsequent differentiation of similar components from this beginning. The very few special exceptions (e.g. the eukaryotic cell) involve symbiotic relationships (e.g. mitochondria evolving from endosymbiosis) of already functionally complete wholes. In biology, functional integration is primitive, not imposed. Even if genes are construed as instructions, they are only instructions for molecular structure providing biases and constraints on potential molecular interactions. To get from this to higher-order functional organization, many levels of emergent cellular-molecular interactions must intercede. Thus in developed organisms earlier less differentiated stages (whether in phylogeny or embryology) are visible only as a sort of dim palimpsest, buried in and modified by many stages of subsequent differentiation. So although we often describe the structures and functions of bodies as though they are marvels of engineering, they achieve this appearance by means almost diametrically opposed to engineering design.

Noam Chomsky's (1957) early demonstration that syntactic manipulations of language could be modeled with considerable adequacy as rewrite operations (as could be implemented as a Turing machine) was a compelling indication that language is special, and it suggested an almost irresistible comparison with other formal systems of computation. Moreover, language is a social phenomenon that is potentially responsive to pressures for rule-governed standardization (of course this typically imposes one naturally evolved language regularity on the behavior of speakers with slightly different variants). But as compelling as this is, the fact remains that language is a spontaneously evolved phenomenon produced by a biological organ, and biology is not engineering.

## Brain Development as Self-Organization and Micro-Evolution

It has become increasingly clear that the vast majority of the "design information" expressed as brain structure is not encoded in genes. Mouse brains, which are roughly the size of the last digit of your little finger, develop using *roughly the same number of corresponding genes* as does the human brain, which is roughly the size of a pineapple. With each neuron interconnected with roughly a thousand others in both kinds of brains, the information embodied in these two structures differs by many orders of magnitude. So where does the extra information come from to build human brains?

The answer does not come from reapportioning influences of nature versus nurture, but rather from neither. The extra structural information arises spontaneously, not by magic or divine intervention, but by a process that, like evolution, can spontaneously generate adaptive correspondences and novel complex structure and function without external guidance. Indeed, brain development resembles a kind of microevolution in many important respects. And both evolution and embryology take advantage of a variety of spontaneous ordering tendencies as well.

The way that open dynamical systems fall into orderly patterns without extrinsic imposition of these regularities has come to be called *self-organization*, and the dynamical structures that arise from such processes (as well as from evolutionary processes) are generally described as *emergent* (for an overview see Deacon 2003b). Examples of autonomous pattern-generation can be seen in the formation of hexagonal convection cells in a heated liquid (Benard cells), the individually idiosyncratic yet hexagonally symmetrical growth of snow crystals, and the interwoven spirals of seeds, leaves, and petals that spontaneously organize to exemplify the Fibonacci number series (which is not explicitly encoded in the plant's genes). These regularities of structure and function are not prefigured in any antecedent instructional process but come into existence dynamically as the repeated interactions of components gradually reinforce some structural biases and dampen others. Self-organized regularization can lend itself to selection dynamics as different patterns of interaction are pitted against one another in a larger "ecology" that is more concordant with some compared to others. This interplay of self-organization and selection processes is responsible for much of the emergent structure and adaptive complexity of brains.

For example, in the process of brain development, patterns of connection are initially generated by the self-organizing interactions of axonal growth processes between regionally organized distinct cell populations (these in turn arose from self-organizing and competing cell proliferation and interaction processes). Converging axons then compete with one another in an 'ecology' of signal-processing demands. The resulting selection process culls many and preserves other cells and connections that are more synergistic in activity patterns. In this way, much like phylogenetic evolution, the developing nervous system can augment the biasing influences of the genes by using this as a base from which to 'explore' adaptive correspondences between different neuronal populations, between regionally different signaling patterns, and between organism and environment. This contextually sensitive sculpting of cell populations and connections results in the spontaneous emergence of complex functional synergies as the developing brain adapts to the body it finds itself in.

This should not be misunderstood to mean that brains of humans and brains of chimpanzees, for example, mostly differ with respect to these plastic connections. Human brains resemble other human brains, not chimp brains, because of the influence of genetic differences. Self-organization can in fact augment the subtle biases contributed by gene differences affecting cell-cell interactions to produce large-scale systemic change. Physical systems, like snow crystals for example, share remarkable family resemblances despite diverse conditions of origin, because of (not in spite of) the regularizing affect of self-organization. Though stochastic factors may make the fine details unpredictable, general patterns are reliably produced even though this emergent structure is not predetermined. Similarly, selection processes tend to produce convergence toward common forms (e.g. streamlining in diverse aquatic species) despite diverse origins and substrates. This is an important point, which is often misrepresented by overzealous critics of genetic determinism. Emergent structures are often highly predictable and can converge on universal features, even from quite different initial conditions. This is of course relevant to language regularities, as well.

When brain development is viewed through the prism of self-organization and selection processes, little remains of the engineering logic that is so familiar, and yet the result can still be precisely organized. Brain development demonstrates that just because a biological structure is highly predictable, complex, and systemic in organization we need not appeal to algorithmic or instruction logic to explain this fact. Moreover, the extensive role played by self-organizing dynamics in brain development should make us highly suspicious of engineering analogies used to explain brain functions like language production.

## The Emergent Architecture of Neurolinguistic Processes

Languages and human language abilities have evolved spontaneously. Like other naturally evolved complex systems we should expect to find that they exhibit the hallmarks of this undirected process reflected in an emergent architecture. Language has an emergent architecture to the extent that its structure is a product of spontaneous bottom-up self-organizing interactions, not top-down imposition of structure or constraint by any pre-existing template. This requires conceiving of basic linguistic units as differentiated end-products of a cognitive process rather than as fundamental atoms of analysis.

To do this we need to analytically *de*-differentiate the many levels of a speech act, but this introduces a conceptual difficulty. We cannot start with the familiar components of expressed language, but must derive these from something far less concrete that has troubled psychology since its inception: the form of a thought before it is put into words.

We find it difficult if not impossible to gain introspective insight into the nature of a word before it is formed, or the idea that a sentence conveys before it is encoded into words. Remarkably, for all the difficulty we have describing this, it is probably fair to say that a good deal, if not most, of our mental life is lived in this not quite articulated not quite formulated state.

This stage of cognition that serves as the nearly unconscious and automatic ground of language use is quickly and effortlessly resolved into words and sentences. We are usually entirely focused on aiming for and achieving expressive goals, not on selecting function words or following grammatical rules, and so long as these results are achieved without any serious hitch (e.g. because of content word-finding difficulties) the antecedent generative processes go unnoticed.

But if sentence structure is produced analogous to the way embryos develop, not as machines are built, then words and sentences must also begin as some less differentiated cognitive structures. Following this analogy we should consider language processing as though words and phrases differentiate from more inclusive generic precursors, which are the cognitive ancestors to structures with multiple component parts. These precursors are not, then, linguistic units, but rather more general cognitive, semiotic, and pragmatic structures. The apparently most elementary phonetic and morphological features of language are, in comparison, the most elaborated fractional distinctions to emerge from the progressive differentiation of these larger semiotic-neurological speech-act "embryos" (see Figures 3.3–3.5).

So we must begin with the most difficult step first: identifying these most abstract levels of what can be called the "infra-linguistic" hierarchy. The questions "What is a concept before it is expressed in words and phrases?" and "What is a proposition or request before it is phrased as a sentence?" must eventually be given serious attention as linguistic issues. The linguistic paradigm calling itself Systemic Functional Linguistics (e.g. Halliday 1994) probably comes closest to this with respect to the analysis of communication processes in general, but parallels need to also be devised for the neurology of language processing (but see Brown, 1979 for a similar analysis).

Let's begin by considering a simple declarative sentence produced with the intention of describing some state of affairs. It may have a social context, which directs and constrains

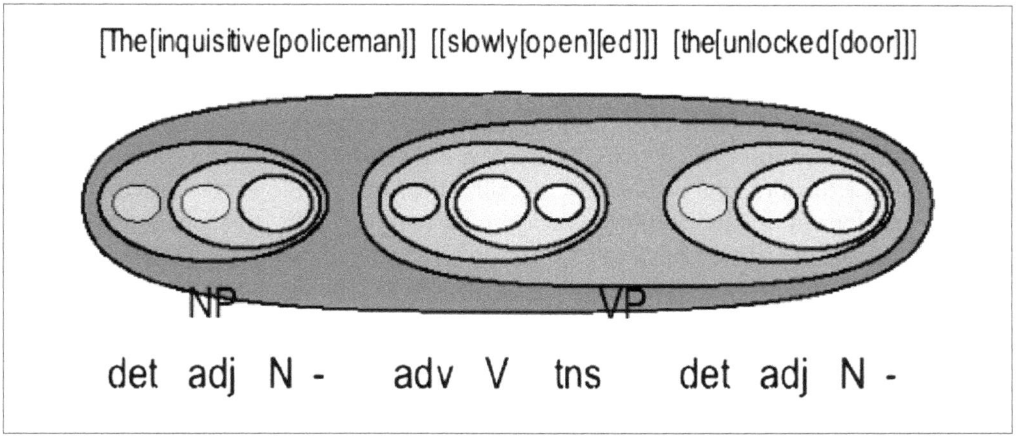


Figure 3.3 Hierarchic compositional analysis of sentence structure modeled by a Venn diagram (isomorphic with tree diagrams).

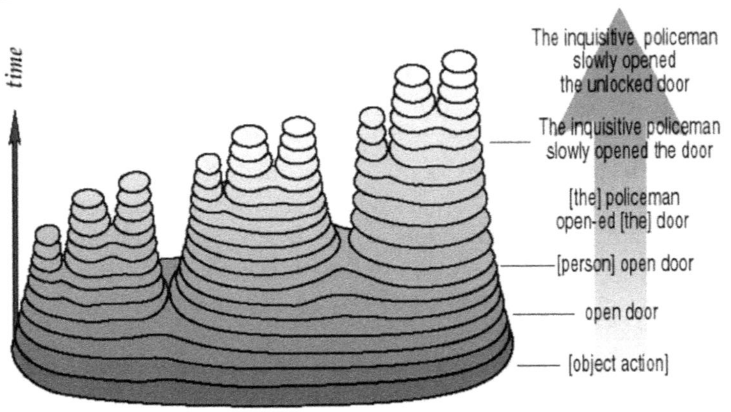


Figure 3.4 Expanding the embedding logic of the Venn diagram in a third dimension to represent a possible differentiation hierarchy helps in visualizing infra-linguistic units of processing that may have their own derivation logic at that level of processing. These levels of differentiation can be glossed (right) as progressively differentiated sentences or holophrases.

the kind of information that will best fit, such as being produced in response to a request or expectation. This will play a role in promoting one's change in arousal to speak and in eliciting certain memories or shifts of attention to relevant events. The social-pragmatic context includes a shift to a specific mode of communicating as well—e.g. providing or seeking information—and may consequently involve activating habits associated with this social role. This first stage is thus a social and pragmatic orientation stage that creates both a specific communicative frame and an arousal to act within it. Even merely interpreting someone else's speech requires establishing such a communicative framing of activity along with attention focused on certain expected content. It is what amounts to the act before it is initiated, the content before it is developed, the perception before it is sensed. It is a focused readiness and expectation with respect to social interaction.

Within this framing of social-communicative arousal what might be described as the "mood" of the speech-act or interpretive-act is differentiated. This forms the minimally differentiated space of options from which further differentiation of content and expression can proceed. But the neural trace that constitutes this extrinsic framing does not get "passed on" to some subsequent brain structure for processing (as might be imagined for a component assembly process). Rather this "mood" needs to be maintained in parallel throughout the sentence generation processes and is updated only with a shift in communicative role. This requires a distinct neural substrate specialized for maintaining social orientation and arousal, and for monitoring conditions that would require a shift from this state.

Neurologically we should not expect to find that the areas of the brain associated with this phase of language behavior is specifically associated with speech production or comprehension *per se*. The arousal process almost certainly involves limbic structures and adjacent perilimbic cortical regions, as well as deeper brain structures associated with social arousal. Cortically this includes the anterior cingulate cortex, which mediates the arousal and monitoring process. These earliest phases are also comparatively slow-changing, so that many sentences may be differentiated within a single generalized communicative mood. Later phases will correspond with progressively smaller units of a communicative act.

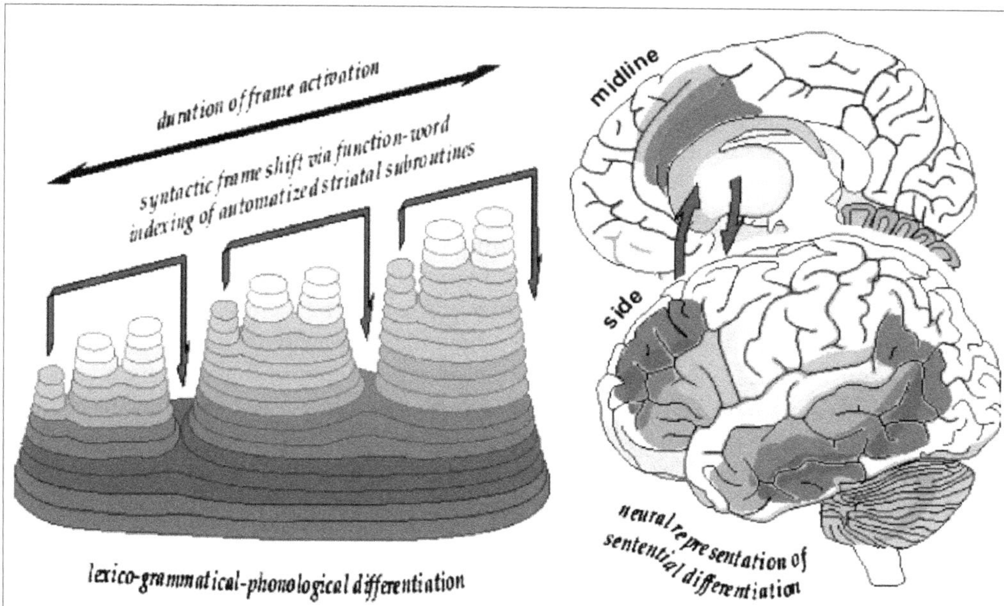


Figure 3.5 Mapping a differentiation hierarchy onto cortical areas (and some subcortical structures) that likely correspond to each given level of sentential differentiation. Note that differentiation converges towards classic language areas and ultimately terminates in auditory and motor areas. Each level of processing is complete in itself and does not produce or pass on a partially constructed sentence to the next. Rather each nested level is supported by level-specific habits acquired by language experience and these tendencies are differentially catalyzed or constrained by the state of the preceding level of differentiation and also by the biases introduced by the just preceding lower level process. Higher-order (less differentiated) processes also operate on a time scale that is slower and encompasses many lower level operations. These time domain differences may also express in language as short-term memory limitations that differ for different classes of language operations.

Damage affecting the brain regions associated with this arousal and monitoring of communication produces deficits that are seldom described in terms of aphasia, since word-choice, grammar, and phonetic decisions are unaffected. But more global disturbances of language are typical. Most notably, damage to midline frontal regions, including anterior cingulate cortex and supplementary motor area (especially bilaterally) is known to produce akinetic mutism. This has often been described as an inability to generate sufficient arousal-to-act because it can suddenly and transiently abate under high arousal conditions. *In vivo* imagery has shown that differentially elevated anterior cingulate activity is also associated with many language generation tasks, even when motor speech functions are controlled for (e.g. see Deacon 1997). It is notable that midline frontal regions are also the only cerebral cortical regions from which primate vocalizations have been elicited by stimulation. Although speech cannot be elicited by cortical stimulation in humans (only blocked or modified), there have been reports of expletives being elicited by stimulation of limbic structures in human patients and such automatic arousal-correlated word production is often all that is spared in cases of severe global aphasia with massive damage to left perisylvian areas.

Within this mood frame, the first specific orientation and expectations of the semiotic process are next generated. These include the intended goal of the communication and the selection of major categorical orientations relevant to this context. This is a phase of neural processing in which specific sensori-motor orientations relevant to the communication become more specified and need to be sustained in the face of competing and interfering alternatives.

These processes appear to involve sub-regions of prefrontal, parietal, and middle temporal cortex. These areas of cortex are mostly polymodal, judging from their primate homologues, and in classic (misleading) terminology were described as "association cortex." Consequently, damage to parietal and temporal areas tends to result in difficulties of maintaining attentional focus and distinguishing distinct objects of interest, respectively, especially when multiple modalities of discrimination are involved. Damage to lateral prefronal areas disturbs the ability to manage multiple competing sensori-motor attentional options, irrespective of salience, recency, or immediate reward contingencies. Prefrontal functions are sometimes referred to as 'working memory' because they involve the ability to hold many simultaneous competing predispositions in mind at once so that they can be assessed with respect to one another. *In vivo* imagery studies consistently show differential activation of the left ventral prefrontal region (anterior to what is more commonly defined as Broca's area) during word-association and memory tasks that involve rapid symbolically mediated decisions (see examples in Deacon 1997).

With respect to language differentiation processes, I think this can be understood as the phase in which predication differentiates out of a more general mood of communicative arousal. In many ways, the structure of predication has much in common with specific sensori-motor orientation. The logical structure of predication, formally symbolized by a function $F$ operating on one or more variables, in the form $F(x)$, or $F(x,y)$, etc., can be interpreted in semiotic terms as a symbolic relationship $F$ indexed to (pointing to) some locus or loci $(x,y)$ in the world, which may include loci within the physical communication process. In computational terms these comprise an operation and one or more "slots" for pointing to the "addresses" for the data to be operated on. In terms of semantic theory, $F$ is a predicate (e.g. verb) and $x$ and $y$ are arguments to which it applies (e.g. subject and object). In neurology we may see an analogy to the trace of a sensory or behavioral association and an attentional orientation toward some locus or loci to which it currently applies (either external or internal to the organism).

In an essay concerning some semiotic constraints on language structure (Deacon 2003c), I have argued that, because symbolic reference is "reference mediated by a system of symbol-symbol relationships," each symbolic function requires an associated indexical operation (described as a bound index) to point outside this system to objects of reference. Thus the utterance of a lone noun or verb almost never constitutes a completed act of communication. In other words it *predicates* nothing. It does not refer outside its implicit web-locus of symbol-symbol relationships. It is a fragment, lacking a necessary functional component that if present could constitute a sentence, proposition, request, order, and so on. But notice that when coupled with a pointing finger or uttered in a context where a specific object or state of affairs is obviously salient to the message recipient, this missing role is filled. The missing piece was thus playing an indexical role. This suggests that a complete symbolic act, so to speak, consists of a least one grammatical unit playing a symbolic role and another playing an indexical role. In terms of the $F(x,y)$ formalism, $F$ is the symbolic operation and $x$ and $y$ are serving as its bound indices. Even though these variables can be filled with other symbolic operations, these too must be individually indexed or reference fails.

To the extent that prefrontal, parietal, and middle temporal systems play critical roles in maintaining and deciding among specific orientation and action options, they are also relevant to the parallel operations on the virtual objects of symbolic reference as well. So, for example, the traces of object-attribute associations—and thus also the correlates for symbolic operations—are likely generated in posterior (temporal-parietal-occipital and polymodal cortex), while the traces of orientational dispositions with respect to them—and thus also the neural correlates of indexical operations—are likely generated and maintained in prefrontal cortical areas.

A classic descriptor of a global deficit common to patients with damage to one or more of these cortical systems is the "loss of abstract attitude." Though difficult to characterize, it generally refers to the overly literal way these patients approach language and behavior in general. This can be understood as an impairment of the ability to inhibit prepotent orientation to concrete sensori-motor associations compared to those that are more indirect and symbolically mediated. In classic aphasia terminology, the so-called transcortical aphasias offer relevant correlates of damage to these still early stages of language differentiation. These leave perception, repetition, and production of speech intact but diminish control of these higher order content orientations. Posteriorly, this can produce confusions of associative analysis (transcortical sensory aphasia, and semantic aphasia) in which spurious interpretive substitutions may arise despite minimal nonsense paraphasia (non-word substitutions). For example, there may be word substitutions that reflect wildly divergent and incompatible categories, though they represent real words of the language and may even be appropriately inflected. Anteriorly, this can involve a weakened control of speech by predication with respect to merely reactionary or echolalic speech (transcortical motor aphasia) (for a classic review of aphasiology, discussed here and below, see Lecours *et al.* 1983).

Historically, these aphasic syndromes have come under critical scrutiny and have been considered by some to be of questionable validity as discrete syndromes or specific language disorders. This ambiguity reflects the semiotically more general character of this early phase of processing before word choice, syntax, and phonological realization are relevant. In this sense, these kinds of deficits are often described in terms of ideational difficulties rather than linguistic impairments. Persistent transcortical motor aphasia also appears to require damage to underlying white matter and basal ganglia structures, but as we will see, this is a feature that appears common to all frontal language deficits. This is often considered an argument against identifying frontal cortical structures with language processing per se, but this is an overreaction. The relatively greater importance of deep brain structures to frontal systems almost certainly reflects the far greater elaboration of cortical-basal-ganglia-thalamic-cortical circuits in frontal as compared to posterior cortex, and not a reduced importance of anterior cortical areas to language processing.

This predication-orientation phase of differentiation establishes the frame in which the first distinctively linguistic differentiation functions emerge, and for which there is general agreement that correlated impairments constitute true aphasias. These involve superior temporal regions of Wernicke's territory and the ventral frontal and prefrontal regions of Broca's territory (I prefer the designation "territory" to the more common "area" to avoid the implication of anatomical and functional homogeneity). In this phase of language differentiation, the distinctions between frontal and posterior functions become more divergent and their functional interdependence decreases. This is in part a function of the decreased time domain for these functions to be performed, which limits the possibilities for complex interactions. Functional integration and coordination have however already been established by earlier phases that are more globally coupled.

The temporal regions adjacent to Heschl's gyrus (the site of the primary cortical auditory map) that comprise Wernicke's territory are extensively interconnected with middle temporal and inferior parietal polymodal areas. The predication-orientation established in these polymodal systems (which also activates parallel differentiation of more specific sensory imagery in modality specific areas to which they are also connected) superimposes corresponding constraints and biases on these more auditory specialized regions. This facilitates the activation of relevant classes of phonological traces for words (in anticipation of producing them or of the high probability of hearing them). Many cycles of word elicitation may occur within the frame of a single predication-orientation. This frame must also impose agreement constraints on the subsequent elicitation of words. Damage to cortical areas at this level of processing effectively interrupts this constraining and facilitation of word-sound expectation by predication relationships. Thus Wernicke's aphasics typically are not confused about their intentions to communicate and may understand much of the intention of others trying to communicate with them, but they tend to make both production and comprehension errors at or just below the word-formation level. While content words (carrying much of the load of predication) are inaccurately differentiated in these patients, resulting in frequent nonsense words and word substitutions that are in category but wrong in detail (e.g. "chair" for "bed"), function words ("this" "is" "it") tend to be retained and used fluently in appropriately structured phrase and sentence frames despite the paraphasias. We can describe this as a failure to differentiate the primary symbolic tokens despite maintenance of both the predicate frame and the indexical supports.

Before analyzing these temporal functions further, consider the complementary role of frontal cortical systems associated with Broca's territory. The functions of this layer of processing take place within the predication-orientation frame established by prefrontal and polymodal posterior systems. The prefrontal contribution (as distinct from the parietal-temporal contribution) to establishing this predication frame is primarily with regard to the orientation or indexical component of this frame; i.e. the orienting with respect to different conditions and objects of attention. In general this aspect of the predication frame can be described as a schema for conditionally shifting orientation and redirecting attention. This is a fundamental feature of most complex learned behaviors in general. The subsequent phase of processing, characteristic of the cortical regions associated with Broca's territory, involve the differentiation of the elicitation and sequencing schemas regulating word production. This is accomplished by using the orienting constraints of the predicate frame to regulate predispositions activated by phonological cues from posterior processes and high frequency word-association habits keyed by prior word production. But this is one stage less differentiated than the motor production of the content words themselves, and can probably best be envisioned as generating the ordered slots into which these words will be inserted. But the indexical orientations of the previous level differentiate into at least one form of overt motor output at this level: the production of function words, pronouns, and articles that serve as the markers for these syntactic slots and phrase transitions. These syntactic markers are the overt trace of an orienting and pointing schema that has differentiated in parallel with the posterior differentiation of word phonology.

These classic language areas thus represent the first level in which there is differentiation of linguistic units *per se*, but this interface should not be construed to be discontinuous from earlier stages. The various "component" linguistic units derive their referential power and combinatorial constraints from these many levels of infra-linguistic processes. These lexical units may appear to self-organize as though they possessed intrinsic structure, but this apparent structure is rather a reflection of the fact that they are merely surface markers for the end stage of a cognitive differentiation process. Well before these specific lexical units are crystallized into aural-vocal traces, their combinatorial options have been prefigured by prior differentiation processes involving larger less differentiated semiotic units.

Broca's aphasia is typically identified with labored production of words and non-fluent speech, despite relatively spared vocabulary and comprehension. The non-fluency is also in part due to the absence of function words and grammatical markers. Although difficult production may indicate adjacent motor area damage, it may also simply be the result of the absence of sufficient cuing by syntactic markers whose function is to mark and point to the space where symbolic operations are required.

The subsequent, most differentiated levels of language processing—phonological analysis and vocal articulation of sentences—constitute a surface map on which linguistic tokens mark transition points of these deeply rooted continuous processes. The features that are articulated provide just enough cues of this hidden differentiation process as to allow others to independently recapitulate the process in parallel. Their efficacy is entirely dependent on a rich matrix of assumed interpretive commonality and anticipatory processing that will spontaneously differentiate with only minimal priming.

In summary, language is a neurologically emergent function because its structure is a product of the complex synergy that develops between these multiple systems and through levels of progressive neural differentiation. Each stage of differentiation involves correlated processing in corresponding levels of anterior (intention-action) and posterior (attention-sensory) cortical systems. Each establishes constraints and biases that are the ground from which the succeeding stage of differentiation begins. Subsequent phases involve both more specific and more rapidly differentiating and shifting processes, so that often a number of higher-order developments will occur under the relatively more persistent state of differentiation of the prior phase. This produces a multilevel nesting of more specific phases of differentiation within less differentiated frames, embedding shorter within longer time domains of operation. The processing of the exceedingly rapid phonological transitions of speech perception and the production of the rapid and subtle motor transitions of speech articulation are thus highly constrained by this nesting in a way that minimizes both the diversity of alternatives that must be anticipated with the demands of maintaining large scale functional integration.

## Implications

So how does this translate into linguistic terms? Most current linguistic paradigms analyze grammar and syntax at one level—the level of differentiated linguistic structures—and derive rules and principles for handling all possible relations among these units. The consequence is that complex instructional architecture and massive processing capacity must be postulated to handle all the presumptive operations. This has further fueled the expectation that a language "processor" must be present in the brain and that only special language mutations of the genome could have made this possible. From a neural differentiation perspective, however, language processing develops through nested levels of operations on linguistic units of different orders of differentiation. The early stage levels are not explicitly represented as distinct linguistic units and largely involve neural systems that are strongly homologous in function to their nonhuman primate counterparts. Only the very last levels of functional differentiation correspond to linguistic compositional features, but this composite structure is a *post hoc* re-presentation of the entire differentiation hierarchy, not a recapitulation of it. The kind of question we must ask of our linguistic theory, then, is something like "What kind of semiotic unit—not linguistic unit—is a sentence or phrase?" This is a question of referential function and communicative pragmatics more generally. It suggests that a more sophisticated embedding of linguistic theory in semiotic theory is necessary in order for progress to be made in bridging the gap between linguistics and neurology.

The inversion of logic distinguishing generative formal analysis from biological differentiation logic also infects the study of language origins. The two paradigms of language generation correlate with two very divergent approaches to the evolution of its neural substrate. Did evolution produce an innate language faculty, in the form of a grammatical template or a set of grammatical categories with intrinsic combinatorial constraints (e.g. Pinker 1994), or did it produce a synergistic combination of tweaks to the learning, mnemonic, automatization, and attentional processes of the human brain that together merely constrain and facilitate the emergence of language regularities (e.g. Deacon 1997; 2003a)?

A synergistic evolutionary account is at odds with many currently popular scenarios that suggest that language appeared rather recently and suddenly (50,000 to 100,000 years ago) as a result of one or a few critical mutations. Such "big bang" scenarios are only conceivable to the extent that the language faculty is treated as a monolithic instructional mechanism. But language is thoroughly integrated with other cognitive functions; it is developmentally canalized and robust in the face of early brain damage and genetic diversity; it is supported by many parallel supportive modifications in multiple cognitive systems; it takes advantages of unprecedented changes in the vocal system; and it exhibits a functional complexity that is unmatched by any other adaptation in ourselves or in other species. These well-known facts offer overwhelming evidence that the so-called "language faculty" is a multifaceted system of adaptations that could only have been collectively shaped by selection with respect to language and each other over an extensive period of hominine evolution (Deacon 1997; 2003a).

Unfortunately, the study of language is not made easier by shifting to an emergence paradigm—indeed, it begs new questions, demands new methodologies, and requires a far more interdisciplinary approach than before. Moreover, this is not yet a theory, but merely a first speculative suggestion of another way of tackling this mystery. It is lacking in the linguistic details and in the descriptive power that is offered by even the simplest formal generative approaches. But the comparison is not fair. Formal theories have a *post hoc* tautological character—reverse-engineered by millions of person-hours from linguistic data. Their derivational logic is thus inevitably more elegant, their predictions are more accurate, and their domain of application is more comprehensive, so long as neural processing and evolvability considerations are not at issue. But ultimately these 'organic' considerations must trump predictive adequacy and formal elegance.

The formal design metaphor has diverted scientific attention and research resources from the implications of these biological considerations for a half a century, but refocusing attention on these factors does not mean turning our backs on formal linguistics, only abandoning the search for its literal counterparts in genes and the neural processes that generate language communication. I imagine that future linguists will look back upon this period in the history of the science and wonder why we didn't see the obvious utility in reserving engineering logic for the study of machines, formal logic for the study of computation, and organic logic for the study of brains and language.

## References

Brown, Jason W. 1979. "Language Representation in the Brain." In *Neurobiology of Social Communication among Primates*, edited by H. Steklis and M. Raliegh, 133–195. New York: Academic Press.

Chomsky, Noam. 1957. *Syntactic Structures*. Cambridge, MA: MIT Press.

Deacon, Terrence W. 1997. *The Symbolic Species: The Co-evolution of Language and the Brain*. New York: W.W. Norton.

———. 2003a. "Multilevel Selection in a Complex Adaptive System: The Problem of Language Origins." In *Evolution and Learning: The Baldwin Effect Reconsidered*, edited by B. Weber and D. Depew, 81–106. Cambridge, MA: MIT Press.

———. 2003b. "The Hierarchic Logic of Emergence: Untangling the Interdependence of Evolution and Self-Organization." In *Evolution and Learning: The Baldwin Effect Reconsidered*, edited by B. Weber and D. Depew, 273–308. Cambridge, MA: MIT Press.

———. 2003c. "Universal Grammar and Semiotic Constraints." In *Language Evolution*, edited by M. Christiansen and S. Kirby, 111–139. Oxford: Oxford University Press.

———. 2004. "Monkey Homologues of Language Areas: Computing the Ambiguities." *Trends in Cognitive Science* 8: 288–289.

Halliday, Michael A.K. 1994. *An Introduction to Functional Grammar*. 2nd ed. London: Arnold.

Lecours, Andrè-Roch, F. Lhermitte and B. Bryans. 1983. *Aphasiology*. London: Bailliere Tindall.

Pinker, Stephen. 1994. *The Language Instinct: How the Mind Creates Language*. New York: William Morrow.

# — 4 —

## The First Hybrid Minds on Earth[1]

Merlin Donald

We acquire our symbolic skills from the outside in. Therefore, we had to evolve them in the same way. Symbolic thought and language are inherently network phenomena. Thus their existence cannot be explained in a solipsistic manner. The problem calls for a paradigm shift, away from our mainstream theories of human evolution, which tend to assume that language evolved inside the brain box; that is, from the inside out.

### Abandoning Solipsism

The theory outlined in this paper might be called biocultural. It places the origin of language in cognitive communities, in the interconnected and distributed activity of many brains. Such communities have their own independent dynamics. A brain that evolved specifically for this type of environment would not necessarily resemble one that evolved to produce language on its own. This is partly because human culture includes a great deal more than language and partly because the minute you embed a brain into a cognitive community, you change what it must do in order to remember, think, and represent reality.

The evolution of human cognitive communities can be taken as an excellent example of emergent evolution, in which a series of small incremental changes at one level can result in a radical change on a different level. The formation of cognitive communities was undoubtedly one of the most extraordinary events in the history of the biosphere, yet it seems to have been caused by a relatively simple expansion of the executive brain, with a corresponding change in developmental plasticity. The specific form of human consciousness was fixed by the demands of this adaptation. We are culturemongers, driven by the very nature of our awareness to seek refuge and solace in community. We connect with and learn from others to a unique degree. Symbolic thought is a by-product of this fact, and so is language. Both result from the collision of conscious minds in culture.

The evolutionary origins of language are tied to the early emergence of knowledge networks, feeling networks, and memory networks, all of which form the cognitive heart

1. This paper is an abridged reprint of chapters seven and eight in *A Mind So Rare: The Evolution of Human Consciousness* (New York: W.W. Norton, 2001), 252–326. Permission by the publisher is gratefully acknowledged.

of culture. Language was undoubtedly produced by Darwinian selection, but it evolved indirectly, under conditions that favored those hominids who could make their shared cognitive networks more and more precise. To judge from its impact on the development of infants and apes, the emergence of language could not have initially been an end in itself and would not have initially changed the nature of cognition. It would merely have made the operations of the same system more exact and less ambiguous. But since language obviously had great survival value, natural selection would have increasingly nudged hominids in the direction that humans finally took, with the emergence of a vast, high-speed speech capacity. The strangeness of this final direction (primates are poor vocalizers) should not distract us from the need for smooth continuity in the early transitions from primate to human. The first priority was not to speak, use words, or develop grammars. It was to bond as a group, to learn to share attention and set up the social patterns that would sustain such sharing and bonding in the species.[2]

It may seem odd, but the almost inconceivable complexity and beauty of language are better explained this way than by the reverse scenario, which demands the evolution of a complex language module first, and then the use of language to create cultures. The machinery of language is far too exquisite to have been encoded entirely inside something as unpredictable as the genome of a developing brain. It was more probably shaped by the demands of a communicative universe that was much larger than one contained inside a single brain and was instead provided by a community of brains. Under the right circumstances, such a community could generate a virtual infinity of possibilities. Just as the physical environment drove the evolution of perceptual capacities, so cultural energy drove the evolution of sophisticated communicative capacities. It also generated forward pressure on the developing mind. Recent developmental research has provided some important clues to the origin of language. The mental abilities most central to enculturation, those involved in joining cognitive communities, develop very early and are logically and empirically prior to language, both in development and in evolution.

This makes the mechanism of language highly dependent on culture. The same may be said of all forms of symbolic processing. This is the ultimate irony: We could not even begin to think our most private thoughts without satisfying our brain's cultural habits first.

## Consciousness and Community of Mind

The relationship between consciousness and culture is a reciprocal one. While culture emerges from the attempts of an expanded awareness to connect with others, it is immersion in culture, rather than any feature of the brain, that defines our truly human modes of consciousness. Enculturation dominates human cognitive development. Moreover, this principle must have been in play from the start of our cultural adventure. Enculturation necessitates the early differentiation of working memory into multiple fields of awareness, as well as a linking of its powers of recall to the appropriate cultural subroutines. These are crucial in preparing a brain to find the attentional flags and markers that will eventu-

2. There are too many social theories of language origins to cite here in detail. A representative sample can be found in recent books edited by Hurford, Studdert-Kennedy and Knight (1998); Jablonski and Aiello (1998); and Corballis and Lea (1999). I also strongly recommend recent works by Dunbar (1996, 1998); and Tomasello (1999), which make a strong case, from two very different research directions, for the cultural origins of human higher cognition.

ally lead it to the heart of culture. Although we still do not know enough about how these sequences unfold in detail, we know, from observing the limitations of wild-reared apes and feral children, that if these are not programmed into the brain early enough, the system will later become impossible to train.

A key step in this process is the interlinking of the infant's attentional system with those of other people. Reciprocal eye contact, or gaze, is one of the channels through which this develops. Voice and touch are also common channels. These are all involved in familiar circular routines, such as greeting, hugging, and playing. Moreover, they must become reciprocal, so that the infant is an active participant in these routines. These interactions are learned by the infant primarily through intimate personal encounters, starting with early mutual imitations with the mother and broadening to include more elaborate exchanges of facial expression, voice, and gesture, which lead to playacting and various social games. Such encounters serve to build a complex repertoire that will regulate shared attention for the rest of the child's life. The cultural repertoire constitutes a scaffolded system, with each new level adding to a vast cultural edifice of control. The result is to interlock the infant's growing mind with those of its caretakers and ultimately the broader society. A hierarchy of habits for shared noticing, caring, feeling, and remembering are cultivated in this way and prepare the child's growing mind for more ambitious forays into culture at a later age, when the learning of new skills will depend on more intricate interlinkages of attention, such as those that allow us to notice, and sometimes to share, group intentions. They are also vital for verifying that one's interpretations are indeed aligned with those of others.[3]

These patterns of behavior are transmitted across cultural generations by repeating the same basic sequences in the context of pedagogy. The distributed texture of this shared pedagogical process constitutes a kind of cultural memory, which records the huge unwritten fabric of shared feeling, group bonding, and common behavior that underwrites the deep enculturation of every infant. Despite its apparent position at the bottom of the cognitive hierarchy, this system of cultural bonding operates at a very high level of abstraction and is well beyond the capacities of other primates. It demands an extraordinary degree of conscious control, especially during the learning and teaching process, but it does not need to be executed in a self-aware, or deliberately guided, manner. We need to keep in mind the distinction between controlled processing and explicit self-awareness, which is entrenched in very old cultural habits that are implicit in their influence.

These basic tools can lock a child into a largely invisible framework of attentional control that shapes every subsequent encounter with culture. Every nuance of gaze, gesture, tone of voice, or facial expression acquires a cultural meaning and, if understood, might serve as a cue to what is important. If the child can learn how to read these signals, evaluate them, and respond to them, it can learn to navigate the cultural labyrinth. Once acquired, the same skills will also enable a person to become a pedagogue later in life and assume the role of cultural guide for someone else. This will happen without reflection since it will simply involve the reapplication of basic habits learned in infancy.

Under normal circumstances this process constitutes a distributed social learning device that goes far beyond the old-fashioned concept of a general-purpose learning ability. It is

3. Paul Bloom (2000) has made the telling point that children, and even infants, usually discover the meanings of words by watching others talk; that is, by tracking their intentions, rather than by listening to child-directed speech. This argues for the existence of very abstract social sensitivities, and an extraordinary capacity for remembering exactly where to direct attention, in infants less than two years old. See also Nelson and Shaw (2000) for a discussion of how socially shared symbolic systems emerge.

very different from the learning capacities of other mammals, in that it is design-built for living in culture and incorporates group cognition into its routines. This early connection with culture enables human infants to track and understand many of the complexities of cultural survival before they learn to speak. These basic routines are the basis on which more difficult self-supervisory habits can be acquired later in childhood. During their first eighteen months, infants are already beginning to learn how to review their own actions over the longer run, in the light of what others think and what the consequences of these actions will be. This is a generic metacognitive skill whose importance cannot be exaggerated.

Infants must also develop an early familiarity with other cognitive domains that matter in a social context. For instance, to prepare for language, they must first learn about such things as the significance of different tones of voice, of taking turns, and of pauses in conversation. They must learn the overall sound contours of their native tongue and the links between these prosodic aspects of speech and the events they perceive in the social world. In the process they must also learn something about society's attitudes toward emotional expression and various kinds of role playing. Infants learn to accept these things as integral parts of the theater of human life. In turn, this will enable them to find the significances that lie behind the bewildering behavior of adults. In the familiar scenarios of childhood, children repeat and reenact the basic rules and expectations of social life until they have been written deeply into memory.

Note that this is all domain-general training. Virtually any cognitive module might prove relevant to cultural life, and thus the habits of social interaction tend to engage the generalist aspects of brain function. Without this training, children will not acquire language normally. In cases of extreme deprivation, they will remain deficient in symbolic thinking and autobiographical memory, for life. Helen Keller and Laura Bridgman were able to learn language long after the critical period precisely because they enjoyed a normal stream of experience during their first eighteen months and were able to acquire these basic executive skills. This gave them the broad foundation they needed for their later reconnection with culture. Without it they would have failed. There are no recorded cases of successful late language development in anyone who has been deprived of this vital early stage. The early programming of shared attentional habits and working memory skills is an absolute precondition, not only for the development of language but for achieving fully human consciousness later in life.

## The Cultural Relevance of a Multifocal, Multilayered Consciousness

In order to acquire our deepest cultural linkages, humanity needed to evolve a major change in the nature of working memory. Human cultural evolution could not have progressed very far if we had not been able to subdivide working memory into several subdomains. It would have been impossible to develop a collective mentality of any kind without some means of subdividing working memory into at least two fields: self and other. Even in a simple two-person interaction, it is important to control and monitor the attention of the other person. This applies even to nonverbal communications, including gesture, pantomime, imitation, and skilled rehearsal. To make even a simple gesture, such as turning the hand while reaching toward a doorknob to indicate a desire to open the door, infants must be able to keep the gesture itself distinct from the reactions of others to it in their working memories, while switching back and forth between the two perspectives. There is considerable attentional dexterity at the heart of the self-other distinction. Moreover, pedagogy, so vital in human culture, relies on our ability to subdivide our working memory system into even more compartments, juxtaposing in awareness not only self and other but also past and present, all within the same time frame.

Thus it is clear that fairly early in our cognitive evolution, hominid working memory must have differentiated into a system that was more flexible than that of apes, so that it could split into several open-ended subregions, each with an extended time span. This adaptation would have helped sustain the pedagogical interactions needed to teach the difficult skills practiced by early hominids, such as spear throwing and the manufacture of stone implements. Early hominids must also have evolved a capacity to switch between several of these subregions at once. This would have allowed them to parallel process and thus compartmentalize their conscious mental activities, so that they could run several trains of activity concurrently, while reviewing and comparing experiences. This would have been necessary even for relatively simple achievements, such as group coordination, organized big-game hunting, and long migrations.

The existence of this adaptation is confirmed in the nature of modern human consciousness. Although our experience is subjectively unified, our awareness stream is rarely one-dimensional in a structural sense. Under the right circumstances, we can maintain several parallel lines of thought, each in a different mode. You and I can read the newspaper, eat breakfast, respect the social niceties, make conversation, take a few phone calls, plan our day, and ruminate about a possible trip to Argentina all more or less simultaneously and within the same intermediate time frame. Running frames within frames concurrently is routine for our species. Moreover, it is a flexible process. Our conscious activities seem to be stacked, or arranged in cascade, whereby our working memory system can be subdivided into several simultaneously active zones, or narrowed to one, as the situation demands.

Our human cognitive style is linked to this multifocal consciousness, and language, in particular, is highly dependent on this feature. Many other adaptive changes were important for the evolution of language, including a modified vocal apparatus and an expansion of memory capacity. But the executive system was the key. It was always at the center of the cognitive storm because with the growth of culture, our conscious universe started to move at a new speed, generating many more facts and encompassing many more players and a more complex social system. Above all, the speed at which knowledge could accumulate depended on better working memory management.

Modern symbolic technology capitalized on this multifocal capacity and radically transformed our knowledge networks in the process. It also changed the rules of the cognitive game by plugging us into more powerful distributed systems. Thanks to those changes, we can now play faster and more complex games in awareness. But the increasing number of potential foci, the higher turnover rates of information, and the speed with which we can shift time perspectives or change our locus in the memory stream have driven our conscious capacity to the wall.

## The Stages of Human Cultural and Cognitive Evolution

In previous work, I proposed that the present form of the human mind evolved over the past two and a half million years, in three major cultural stages, or transitions (see Donald 1991, 1993a, 1993b). My central hypothesis was that during that time, hominid cognitive evolution was increasingly tethered to culture. Our remarkable evolutionary drive was presumably sustained by the many advantages of having a collective mentality, and our brains went through a series of modifications that gave them this strong cultural orientation. That orientation was something new in evolution and has made us very successful as a spe-

| Stage | Species/Period | Novel Forms | Manifest Change | Governance |
|---|---|---|---|---|
| EPISODIC | Primate | Episodic event perceptions | Self-awareness and event sensitivity | Episodic and reactive |
| MIMETIC (first transition) | Early hominids, peaking in *H. erectus* 2M - 0.4 Mya | Action metaphor | Skill, gesture, mime, and imitation | Mimetic styles and archtypes |
| MYTHIC (second transition) | Sapient humans, peaking in *H. sapiens sapiens* 0.5 Mya - present | Language, symbolic representation | Oral traditions, mimetic ritual, narrative thought | Mythic framework of governance |
| THEORETIC (third transition) | Modern culture | External symbolic universe | Formalisms, large-scale theoretic artefacts, massive external storage | Institutionalized paradigmatic thought and invention |

Table 4.1 Successive layers in the evolution of human cognition and culture. Each stage continues to occupy its cultural niche today, so that fully modern societies have all four stages simultaneously present.

cies. It has profoundly affected everything we do for a living, whether hunting, gathering, toolmaking, fighting enemies, organizing migrations, adapting to climate change, or protecting ourselves from perils. The close linkage between brain and culture has accelerated the rate of human evolution. As communities became better able to store and disseminate knowledge, they evolved a Baldwinian strategy that hijacked the normally slow-moving process of natural selection and caused it to speed up. Cultures are more efficient than individuals at exploiting the fitness value of genetic variations, which might otherwise have a negligible impact. This is the basis of Baldwin's effect. In this manner the speed of evolution of the human brain was doubly increased not merely by its extraordinary ability to learn but by its capacity to share knowledge in distributed networks that can accumulate knowledge at blinding speeds.

There are three major stages, or transitions, in my version of our cognitive emergence. These are shown in Table 4.1. Each of these transitions changed the nature of human consciousness in a major way. The scenario of human evolution seems to be one of tension between culture and conscious capacity, with culture steadily pushing that capacity to the edge, so that it continuously expanded. Culture was a radically new presence, and the mind kept adjusting itself to the new reality of distributed cognition. The result of that tension, in the long run, was the emergence of a symbolizing mentality.

The first transition started a little more than two million years ago, when the species *Homo* first appeared on Earth. It is not absolutely certain which of several hominid species made the first great intellectual breakthroughs, but there was a general increase in hominid brain size during that transition period, and all these early hominids were more humanlike than the australopithecines. The first stone tools appeared simultaneously with the species *Homo*,[4] along with evidence of a drastically changed diet. *Homo* was an omnivorous species from the start. There is evidence that this species ate much more meat than any of its predecessors. Some of this meat was obtained by hunting big game, a remarkable achievement for a nearly naked, relatively small creature. Early hominids could not have achieved this feat, or obtained so much meat, without improved tools. To this end they invented stone tools. The archaeological record also shows some fossilized wooden spears from this early period, but it is likely that they also had tools made from other perishable materials, which vanished in

4. This view has now changed; stone tools preceded Homo by at least half a million years.

the intervening two million years. They had a communal style of living and used home bases as a fixed point from which to venture forth, to hunt and gather foodstuffs. They left behind toolmaking sites, seasonal hunting camps, and continuously occupied fire sites, all of which indicate a group-oriented way of life, in which both material and cognitive resources were shared. Thus, very early in our evolution, hominids had become highly social and evidently used a cultural strategy for remembering and problem solving. In other words, although we have no evidence of language at this time, there is very good evidence of culture.

The achievements of early hominids revolved around a new kind of cognitive capacity, mimetic skill, which was an extension of conscious control into the domain of action. It enabled playacting, body language, precise imitation, and gesture. It also acted as a mode of cultural expression and solidified a group mentality, creating a cultural style that we can still recognize as typically human. Mimesis enabled early hominids to refine many skills, including cutting, throwing, manufacturing tools, and making intentional vocal sounds. Although not yet language, these sounds were nevertheless expressive. We call such vocal modulations prosody. They include deliberately raising and lowering the voice, and producing imitations of emotional sounds.

The second transition started with the arrival of archaic *Homo sapiens*, about half a million years ago. It culminated in the evolution of our particular subspecies, *Homo sapiens sapiens*, about 125,000 years ago. During this time the brain and vocal tract underwent a great change. Sapient humans started with the rather primitive material culture they inherited from their predecessors but then began to innovate at a much higher rate. They invented a wider range of sophisticated tools and produced beautifully crafted objects, improved shelters and hearths, and elaborate graves. Within another 10,000 years, they had started to use several forms of self-adornment and were manufacturing a very large variety of multipart objects, including weapons, hafted tools, boats, complex dwellings, ritual quasi-symbolic artifacts, and simple musical instruments. They had also migrated over much of the world, using various technologies to adapt to a variety of climates and ecologies. They came to dominate the Earth, and spoken language was undoubtedly the special power that favored them over their rivals and predecessors. Spoken language produced oral culture, which was the universal form of human culture until very recently.

The third transition started about forty thousand years ago, and revolved around a revolution in the technology of symbols. Cognition continued to evolve, but this time it was mostly driven by technology and culture itself. The main cognitive driving force underlying this transition was the externalization of memory. Whereas earlier humans had to depend entirely on their biology—that is, on their brains—to remember, modern humans can employ a huge number of powerful external symbolic devices to store and retrieve cultural knowledge. This revolutionized the way humans think and the kinds of distributed cognitive systems we could construct. Thus modern culture contains within it a trace of each of our previous stages of cognitive evolution. It still rests on the same old primate brain capacity for episodic or event knowledge. But it has three additional, uniquely human layers: a mimetic layer, an oral-linguistic layer, and an external-symbolic layer. The minds of individuals reflect these three ways of representing reality.

In effect, these three transitions are major checkpoints on a long evolutionary road, which may have had many stops and starts. They caused three shifts in the nature of consciousness during our evolution: (1) more precise and self-conscious control of action, in mimesis; (2) richer and faster accumulation of cultural knowledge, in speech; and (3) much more powerful and abstract reflective cultures, driven by symbolic technology.

## The First Transition: Establishing the Mimetic Framework of Human Culture

Unfortunately, we have no way to time-travel and actually observe the first members of our species, but we know a lot about their nervous systems, lifestyles, and major cultural achievements, and when we place them on a curve, extending from primate to human, we can narrow down the number of feasible theories about their mental capacities and social organization. By drawing on our knowledge of cognitive science and neuroscience, we can infer what their cognitive options were. On the basis of these considerations, the first cognitive transition seems to have revolved around one central issue, the invention of culture as a collective means of accumulating experience and custom.

The first humanlike culture was associated with a new species, *Homo*. This was a culture of public action, without languages or symbols but equipped with mimetic expressive skills. This was the birth of the actor, the tribe, and the gesture. Given its evolutionary proximity to primate social styles, it must have been dominated by direct expressions of emotion, and indeed, the control of emotion by public action and gesture would have been one of its first priorities. Thus contrived public displays, public competition for control, and a great deal of deception would have been the rule. It would also have been possible to share attention and knowledge, albeit indirectly, by means of gesture, body language, and mime, any of which can communicate an intention quite effectively, without words or grammars.

Mimesis must have come early in hominid prehistory because it was a necessary preadaptation for the later evolution of language. It provided the underpinnings of social connectivity and conventionality. It took the primate mind one step farther in the direction of improved social coordination and collective cognition. The group was primary, and thus having an accurate sensitivity to group feelings was a survival-related skill. Mimesis is still the elemental expressive force that binds us together into closely knit tribal groups. Of all our human domains, mimesis is closest to our cultural zero point. It is also closest to emotion. Mimetic capacity has huge emotional ramifications because it involves both the conscious elaboration and the suppression of emotion.

Mimesis is the result of evolving better conscious control over action. In its purest form, it is epitomized by four uniquely human abilities: mime, imitation, skill, and gesture. These are direct offshoots of the expansion of the human executive brain system and the Executive Suite. The most basic form of mimetic action is mime, the imaginative reenactment of an event. It is exemplified by the pretend play of children, in which they might, for example, feed a doll in exactly the same way they were fed themselves, a few minutes earlier, by their parents. Pretend play is not a fully symbolic mental representation, but it has implicit reference in that it refers to something other than itself (see Harris 1998).[5]

We can also reenact emotional events, and here we can see how emotional mimesis differs from the innate emotional communications of other species. Innate expressions, such as crying and laughing, are stereotyped, universal, and part of the actual emotional experience itself. This gives such expressions a signal, or cue, value, but it does not make them into intentional acts. Mimesis is different. A child miming the act of crying is deliberately acting out a role; that is, pretending to be sad. This is a case of cognitive, not emotional, control over action. It is also a quasi-intentional act and a very abstract representation, independent of the emotional state to which it refers (even though it may restart the emotion).

5. For more background, see Greenfield and Smith (1976); Trevarthen (1979, 1980); Bruner (1983); and Suddendorf (1999).

The second product of mimetic capacity, precise imitation, is more challenging than simple mime, because it is usually seen as an attempt to replicate an event that has some instrumental purpose.[6] To copy a purposeful action, we need to understand the other person's objective. For example, to imitate the sharpening of a spear accurately, we have to understand what the spear is used for and why the point needs to reach a certain degree of sharpness. For certain kinds of cultural transmission, it is not enough simply to mime the movements themselves. A child can mime spear sharpening long before it can actually copy the act of sharpening a spear, and this is not always explainable by the child's lack of strength and coordination. It takes time to develop an understanding of purpose, and some purposes are easier to grasp than others. These distinctions are a matter of quantity, not quality.

The third product, skill, is closely related to mime and imitation. Skill results from rehearsal, systematic improvement, and the chaining of mimetic acts into hierarchies. Thus, to learn to play tennis, we must master a number of elementary action chains and then piece them together into a very complex contingent arrangement, so that the right action will result on very short notice in every possible future situation. The same rule applies to any skilled craft. Whether we are learning to weave, manufacture tools, or cook food, we must learn a set of basic action sequences, generalize them, and rehearse them until they become second nature. To do this, we must mime our previous actions and modify those productions until we approximate an ideal. We must usually go well beyond rote mime or imitation to do this well. We may actually have to modify the action sequence for ourselves and generate new ways of executing it. We take for granted our ability to do such things because we are so good at it. Children commonly practice actions for no apparent reason other than the pleasure of refining them. They might practice hopping on one foot, skipping stones, or kicking a ball for an entire afternoon just to become better at doing these things. By its very nature, our capacity for skilled rehearsal is potentially creative because it can generate variation; that is, contribute to the overall range of actions in our social group's repertoire.

The fourth product, gesture, is a natural derivative of the first three. Gesture and mime are closely related, except that we usually define gesture as an explicitly communicative, or intentional, act. The easiest gestures for infants to learn are simple mimes that refer to specific things or people by miming events. Developmental psychologist Linda Acredolo has shown that a child as young as nine months can easily learn, for example, to represent a fish by miming its puckered mouth. Gesturing becomes much more subtle and complex as children grow older, and they eventually become able to express complicated things in sequences of iconic gestures, which might trace the path of objects in an imagined space or mime the opening of a door or a box. Susan Goldin-Meadow, who has studied gesturing in depth, has shown that it can also contain implicit grammars, which resemble the iconic grammars of sign languages (see Acredolo and Goodwyn 1988; Goldin-Meadow *et al.*, 1996, 1998, 1999). That particular kind of gesturing is very abstract and was probably a late development in mimetic culture, possibly the main lead-in to the evolution of language. It is an advanced form of mimesis and must have been preceded by simpler reenactive mime, imitation, and skilled practice.

Mimetic capacity produces a layer of cultural interaction that is based entirely on a collective web of conventional, expressive nonverbal actions. Mimetic culture is the murky realm of eye contact, facial expressions, poses, attitude, body language, self-decoration,

6. See Tomasello 1999, for imitation's relationship to social cognition. See Meltzoff and Moore (1997), for a theory of infant imitation that has key similarities to the model of mimesis I have presented here. See also Donald (1991, 1993a, 1999).

gesticulation, and tones of voice. It is still the primary public dimension that defines our personal identity, and in it, style and tradition matter, to the degree that these things establish who we are, who our friends are, and where we stand in society. Mimetic styles vary tremendously from one cultural group to another. The mimetic elements of culture are so distinctive that they are often the most difficult features for foreigners to master.

Mimesis is also the cultural layer that children encounter first. The distinguished developmental psychologist Katherine Nelson has assembled an extensive body of evidence that shows how mimetic event representation is the universal means of interacting with very young children. In fact, she has called the world of the two-year-old an almost purely mimetic culture (see Nelson 1996). The exaggerated facial expressions, tones of voice, and role-playing games that children enjoy are mimetic in character. A child's life is really a series of mimetic set pieces, in which various cultural scenarios are being learned, acted out, rehearsed, and varied at a very fast pace. As a child goes through a typical day, it is expected to act out various roles, such as "eating at table," "washing," "dressing," and "going out" (see Harris 1998). Other societies might have very different scenarios, but family life provides the first stage on which our actor minds become locally famous, interlocking with other minds, engaging in roles, joining attention with others in common causes, sharing emotions, customs, and action memories. By means of mimesis, we can share tacit knowledge to a remarkable degree. Mimesis is the level of cultural interaction on which we first assume a basic tribal identity and become conscious of ourselves with reference to our primary social group.

Mimetic capacity was primarily the result of merging the executive brain with the action brain, when the hominid executive brain system extended its anatomical territory into the frontal and subcortical regions that control voluntary action. This anatomical change had major social ramifications. Active social networking, even of the exclusively mimetic variety, makes heavy demands on attention and memory, and we can corroborate this with evidence drawn from studies of child development.

Even in adults, mimetic role playing and group identification remain in place as powerful forces for regulating social coordination. As long as the members of a culture understand the roles they are playing in a situation, the basic frame of group activity will work well. Without that frame, groups become unstable and quirky, to say the least.

Mimesis stretches far back into prehistory, and we can confirm this by the nature of the cultural skills our archaic forebears evidently had. For over two million years hominids shared food, disseminated skills by imitation, created mini-industries for activities such as tool manufacture, and held group behavior to very stable long-term patterns, such as the continuous habitation of a site and the maintenance of fire. These things are not easy to achieve, and there is no good evolutionary rationale for attributing them to language. On the contrary, language is thought to be the main reason why our particular species, *Homo sapiens*, broke away from this early pattern. But to achieve what they did, archaic hominids needed an expressive skill that enabled them to bond, coordinate group activity, transfer and refine skills, and establish a network of custom and convention. Such things take considerable management, and even though they did not have the modern cognitive tools we enjoy, they had mimesis, which is a surprisingly effective management device. The emergence of mimesis was our first step toward evolving an effective distributed knowledge network, which could coordinate the actions of groups of people.

Because of the complexity of the motor brain and its many semi-independent subsystems for actions such as walking, running, gripping, looking, and vocalizing, the physiological challenge of evolving such an integrative motor capacity in the brain is not trivial. But that is precisely what hominids did. They evolved a mimetic controller, a whole-body mapping capacity that is truly domain-general in its reach and able to bring a variety of action systems under unified command. Unified voluntary action schemes of this sort would not be possible without a centralized brain map, a virtual space where the actor can deliberately review and modify every action in imagination. Think of this map as a "model of models," an image of the self-in-world that is connected to the motor systems. Since mimetic performances are often very fast-moving and complex, involving the cross-modal coordination of many different muscle groups, this map must have the capability to override our lower-level reflexes.

Before their motor productions could become more skilled, archaic hominids had to evolve direct executive governance over action. *Attention had to be redirected inward, away from the external world, and toward their own actions.* This was a significant break with the primate past. Whereas primate consciousness had occupied a largely perceptual domain and was directed mostly at the outside world, hominid conscious capacity invaded the domain of action. The executive brain system improved its ability to monitor the state of the physical self and gained access to much more detail, so that precise attention could be paid to the body's own movement patterns. Thus the first rung in our distinctively human ladder of awareness is the physical self, a supraordinate form of body awareness, born of a need to refine action.

The hominid mind achieved this stage of primal culture by turning further in on itself and evolving better self-representations (see especially Varela *et al.* 1995, Damasio 1999 and Lakoff and Johnson 1999). The result, our capacity to imagine the self-in-its-environment, was fundamental to human cultural change at that stage. This improved the hominid's ability to imitate, gesticulate, and acquire complex skills and generated waves of novel activity, which reflected back and forth, creating an increasingly complex and unpredictable public theater of convention. This relatively simple event triggered the long give-and-take between brain and culture and established a universal mimetic framework for human life.

Moreover, it achieved this cultural change without languages or symbols, which were still far beyond the zone of proximal hominid evolution. Mimesis fell very comfortably within that zone, so much so that it may have started to evolve long before the appearance of *Homo*, perhaps in australopithecines, without leaving a trail at that early stage. The revolutionary impact of mimesis should not be underestimated. All previous brains had been designed to fend for themselves, not to form cognitive networks with other brains. Mimesis reversed that tendency, started a distributed cognitive process, and triggered a series of cultural revolutions.

## The Second Transition: The Spiraling Co-Evolution of Thought and Symbol

Language is different from mimesis, but it has mimetic roots. It is a collective product and must have evolved as a group adaptation, in the context of mimetic expressive culture. Given the conventional, collective nature of language, it could not have emerged in any other way. Language is above all a cultural product. Just as our bodies conform to the mimetic conventions of culture, so our minds conform to its linguistic ones. By internalizing these conventions, we gain new powers and can trade, amplify, and crystallize thoughts, remember with greater clarity, share memory, and enjoy membership in an enduring cognitive entity that transcends the individual. Like mimesis, language proceeds from outside to inside, and children must first master its public forms before internalizing its use. Like mimesis, it imposes massive conformity and welds a group of people into a cohesive entity. But here the resemblance stops. Language changed the rules of the cognitive game. It introduced an infinitude of new possibilities. It borrowed much from mimesis but started a whole new level of cognitive-cultural interaction.

The surface form of language is subsidiary to a deeper process, which cannot be, in its origins, linguistic or even symbolic. We could never detect our common failures in capturing thoughts in words if these were nothing but strings of symbols in the first place. If thoughts were simply words or their functional equivalent, our detection of expressive failures would itself have to depend upon properly formulated propositions, expressed in symbols, and so on ad infinitum. Thought would be locked into a self-enclosed circle of symbols. However, this cannot be. How could a reader anticipate a hypothetical state of understanding never before experienced, or validate that experience when it arrived, if it were only words that resulted from the reading? In their creative origins, symbols are a product of thought, not vice versa, and in their interpretation, symbols get their meaning from thought, not vice versa.

Language cannot "explain" thought at any level. Primates can obviously think, but they have no natural language. They can solve some fairly difficult problems without symbols. Humans can also think without language. This can be seen in very young children. The same is true of some aphasic patients who lose language altogether yet continue to be able to think in an artistic, musical, mechanical, social, or spatial sense. More important, language cannot even explain the existence of verbal thought. As I have just pointed out, it cannot explain our own innovative uses of words and other symbols or our ability to judge their success. In every case thought is the arbiter, and language is the child of thought, invented in the service of thought, employed forever as the amplifier and mediator of thought. The human mind is infinitely wider and more supple than all its languages and symbols.

The quality of thoughts can be improved with language. But thoughts do not start there or end there, nor are they judged there. Words and sentences define and clarify knowledge that resides elsewhere, in foundational semantic processes that we share with other primates and where the motive force for the evolution of language must have originated. It is easy to swim in our symbolic-cultural sea without noticing this, but our ability to perceive the inadequacies of spoken and written language reveals its deep cognitive roots. Our conscious mind strives after clarity to placate its own inner semantic processes. Somehow, it can judge when clarity has been achieved.

We evaluate all symbolic expressions from outside the symbol system, from a region of mind that, in its principles of operation, is different from, and much more powerful than, the reach of any consensual expressive system. Consensual symbol systems exist for the purpose of satisfying our deeper semantic (metaphoric?) intuitions. We can even judge the failure of an expression without having words or symbols for the as-yet-unexpressed ideas that we know we have failed to express! We can recognize the value of ideas whose mere possibility we suspect, and whose reality still eludes us completely, to the point of pursuing them passionately for a lifetime. Language is harnessed to a thought skill. It amplifies the primary drivers of thought and serves as the mediator, but never as the inspiration, interpreter, or final judge. Words and languages are mysterious things if we insist on closing the linguistic circle and keeping it closed. But once we open the circle, we are left with a system that is at least solvable in principle.

The creative tension between a symbol and its cognitive domain is the source of the cognitive work the symbol does. This tension works in both directions. Boston linguist Ray Jackendoff has explored the complex standing relationships between languages and the various domains they represent.[7] He and his colleagues have uncovered the implicit grammars that apply to symbolized musical and pictorial vision, and these are presumably representative of the kinds of rules and conventions that inevitably follow, once a group starts to construct a symbolic consensus in some domain. Symbols are produced by the tension between the two sides of a symbolized cognitive system, the form of the symbol itself, and the meaning systems to which it is attached and on which it acts.

Our neural models of the process whereby forms are mapped onto meanings are at best superficial, and we have almost no detailed knowledge of what is going on inside the vast neural networks that produce these complex states of tension. Nevertheless, it is evident that states of tension between large neural networks seem to parallel the state of our conscious semantic awareness. Awareness watches over and initiates all major symbolic performances (there are also small, repetitive, automatic performances, but these do not usually draw on conscious capacity). Consciousness is always on watch.

## Piggybacking Language on Culture

There are several good reasons for developing a culture-first approach to language evolution. First, it allows us to remain consistent with what we know about the evolving hominid brain. Second, it provides both continuity and a radically new evolutionary factor, enculturation, that might account for the qualitative shift in human cognition. Third, it accounts for more than language and explains the robust mimetic framework within which language exists. Fourth, it explains why language does not self-install in the minds of isolated human beings. Fifth, it gives us a semantic base, a means of referring language outside itself. This is important because hominids could not have evolved a language capacity as an isolated adaptation. They had to maintain a self-sufficient lifestyle at all times. Hominid symbolic capacity cannot be disconnected from the cultures of which it has always been part. Language is only a part of our larger cultural adaptation.

Finally, a culture-first approach provides us with an efficient mechanism, call it a piggybacking strategy, for the self-replication of the entire human cognitive system. Early computer programmers used the term "piggybacking" to refer to a strategy whereby they stored as much information as possible on digital tape and imported it into memory only when necessary. This allowed them to overcome the limitations of the very small memory systems available thirty years ago. Over the infinitely longer stretches of evolutionary time, hominids have off-loaded as much storage as possible to the cultural attic and brought it onshore (into the brain) only when necessary. With that strategy in place, the hominid brain had to evolve the equipment needed to read the current state of the cultural storage system. It could thereby replicate the current version of the hominid mind, in both its individual and distributed aspects. It is crucial to remember that language is only a secondary level of storage in this system. We cannot even replicate the language mechanism itself without using some of our other culturally stored information.

Customs, languages, and semiotic codes are distributed across many individuals and

7. Jackendoff's views on the relationship between language and other domains, such as visual space, music, and thought are briefly summarized in his book *Patterns in the Mind* (1994).

across a multitude of external devices that constitute much of our material culture. The individual brain has only to track the emerging complexity of this entire matrix, not to copy it. The most efficient way to solve a distributed web is undoubtedly to become part of it and evolve with it, rather than try to simulate or duplicate the whole mechanism in each mind. Cultures and their brains have advanced by interlocking with each other. Language can be evolved and replicated only by means of this piggybacking strategy. This framework determined the way in which languages first emerged. Otherwise, we would have very different brains from the ones we have.

Our cultural storehouses far exceed what any brain could possibly copy or remember, and no single brain can totally comprehend every aspect of a symbolic cultural environment, let alone generate one. Cultural richness accumulates over time, from the interactions of many brains over many generations. In this, cultures are more like ecologies than they are like the millions of individual organisms that make up ecologies.

## Our Cerebral Boxing Match with the Cultural Matrix

This places the creative source of language in a unit that might be called the individual-conscious-mind-in-culture. To shift back to the vantage point of the individual, symbols of all kinds are the playthings of a fantastically clever, irrational, manipulative, largely inarticulate beast that lives deep inside each of us, far below the polished cultural surface we have constructed. That passionate and devious intelligence drives all our collective enterprises, lobbing its semiotic hand grenades at various imaginary targets, while it lurks in the background, just out of sight, self-conscious, self-deprecating, self-doubting about its own value, and largely unaware of the detailed complexity of the cultural matrix. This intelligence is isomorphic with our conscious experience of the world, just as it is in other mammals. This is why the human brain cannot symbolize if it is isolated from a culture. This is also why the brain has always been embroiled in an ongoing boxing match with culture. This match has been filled with tension from the start, and there have been unexpected shifts in the balance of power between the brain and its cultures or rather between the culture and its brains. Now the mind leads with gestures and words to push one way, and now the culture pushes back in another direction, perhaps one that no one would have predicted. The sly semanticist in the head now pushes culture to provide the clarification it seeks, and now it is dazzled into acquiescence by the collectively generated result, sliding into temporary equilibrium.

The tension between cultural symbolic systems and the underlying intelligences that use them determines the quality of our uniquely human modes of consciousness. Awareness never resides entirely in what is said or in the ephemeral sensations that it captures and holds for ransom. It resides in the intermediate-term governor of mind that we endow with intellect and soul and call "me." But it is more than this since it also runs an elaborate cultural machine for the purpose of its own satisfaction. Culture shapes the vast undifferentiated semantic spaces of the individual brain. The brain takes on its self-identity in culture and is deeply affected in its actions by culturally formulated notions of selfhood. This is what theories of cognitive evolution must explain. Explain the knowing beast within, and we shall have a chance to build a scientific theory of symbolic communication. Ignore the beast, and we shall always fall short because without it, the rest of our theoretical edifice is left without a framework to support it.

The notion that consciousness is locked in an eternal struggle with huge cultural forces is acknowledged implicitly in Freud's concept of the unconscious, which is still important to those who study culture. Much of what Freud attributed to the unconscious is truly unconscious only in the cultural sense of the word—that is to say, things that were unexpressed or actively repressed at the level of culture, which is not the same thing as what we now mean by the term. In the terminology of modern cognitive science, the unconscious is a strictly solipsistic notion, not a cultural one. It refers to a region of the individual mind that lies permanently and by definition outside the reach of consciousness. The lines between consciousness and the mind's inaccessible unconscious modules are drawn very deep in the sand, and these lines are inviolable. By definition, a module, such as object vision, is permanently inaccessible to awareness. It serves up all the richness of the three-dimensional visual world to awareness, gratis and fully formed. But we can never gain access to the mysterious region of mind that delivers such images. It lies on the other side of cognition, permanently outside the purview of consciousness.

The cultural unconscious of psychoanalysis is another thing altogether. Its intuitions, drives, and emotional complexes are represented in art, writing, and theater. They are lived out on the cultural stage and are never, in principle, inaccessible to awareness. They can always surface in individual consciousness, even when poorly expressed at a cultural level. Repression is, more often than not, a failure of apperceptive or representational capture in the public arena.

The most common way to bring something into cultural awareness is to find the right symbol to express it publicly. Language is often the principal source of such expressions. From early childhood it is the handmaiden of cultural awareness and mediates our thoughts. Linguistic idea laundering is essentially like any other hierarchy of interdependent cognitive operations. A child loops a thought into primitive symbolic form, produces an external utterance or gesture, evaluates the feedback obtained, modifies the expression, and continues until "clarity" ensues in the networks that generated the idea; that is, until they stop striving to improve their final states. This is another piggybacking operation, but an internal one; we learn words by hanging them on a mimetic fabric of action, based on experience.

The process of self-clarification and verification is not unique to humans. It is what most large waking brains do for a living. But the loop through a collective cognitive process is new to our species. Culture adds great power to our typically mammalian brain systems by filtering our cognitive activities through an elaborate public process that serves as a source of constant clarification and definition. The public loop is, metaphorically (and simplistically), a bit like the cleanup circuits used in neural net computing, which are employed to sharpen the state of another network. Like culture and language, they are parasitic on the networks that do the real semantic work. But they are also invaluable. A consciousness immersed in culture resides ultimately in the temporary state of a semantic network honed through a public process. The network itself stands in judgment of the success or failure of this whole operation. In effect, the semantic beast can stand back and evaluate its own expressive performances, just as it judges everything else about culture. The difference between us and other primates is only that our conscious-beast-within is much better equipped than theirs to understand and navigate the complex environment needed for public idea laundering. Our brains have been modified by millions of years of natural selection to do this well, and as a result, we can invent and employ languages to run our ambitious cognitive schemes.

Symbolic idea laundering takes a lot of management. A creative, interactive cultural process increases the sheer quantity of knowledge available to a person, and that knowledge has to be filtered, sorted, stored, and made retrievable. Moreover, as cultural knowledge networks become larger and more complex, they demand greater conscious capacity, especially with regard to voluntary recall from memory and selective attention. Our realtime, on-line mental lives become much busier because in addition to having to deal with the cognitive challenges that confront other mammals, we have to deal with a parallel cultural context every moment of our waking lives. We have to solve concrete problems, such as how to move about in space without injuring ourselves, find things, and fight battles, at the same time as we play out mental scenarios in imagination, talk to people, think about the social consequences of our actions, and plan ahead. This takes executive management. It also takes plasticity because cultures constantly reorganize the inner universes of their members and multiply their mental models. This loads us down with thousands of ideas and facts, and above all, it demands our close attention to the rules and codes that allow symbolic cultures to operate.

## Symbolic Invention and the Growth of the Lexicon

Where do words stand in all this? Every word is a cultural invention, and individuals must learn the consensual maps that every culture uses to graft word forms onto meanings. The first vocalizations need not have been specifically intended as parts of speech—that is, nouns, verbs, adjectives, or pronouns—as we know them. Rather they might have been intended to specify entire chunks of an event representation, like the first utterances of infants, which often encompass full sentences in themselves, even though they may contain only a single word.

On a cultural level, language is not about inventing words. It is about telling stories in groups. *Languages are invented on the level of narrative, by collectivities of conscious intellects.* They went through many stages of evolution before they arrived at their present level of refinement, and categories such as nouns, verbs, and adjectives emerged only after episodic representations had been specified, first in mime and gesture and then in a less ambiguous form, in consensual symbolic systems. These took very long to evolve and emerged coextensively with the hominid brain. One might say that the ultimate complex structure of language was implicit even in pantomime and gesture but took time to evolve. Hominids needed considerably better attentional and working memory capacity and metacognitive skills to construct and maintain public systems of communication, with complex rules and thousands of words.

Language came late. It coincided with the last great expansion of the hominid brain, which further increased the importance of the executive brain system, the same prefrontal-cerebellar system, with its many linkages, that expanded during the first transition. In addition, there was an expansion of other areas associated with language. These areas include the entire cortical region surrounding the base of the hippocampus and the peri-Sylvian cortex. We do not yet know exactly how languages are distributed within these enormous brain networks, but it is certain that the executive requirements of language were tremendous at every stage of its evolution. This demanded, at the very least: (1) a greatly extended and differentiated working memory; (2) a capacity for multifocal attention; (3) lifelong plasticity; (4) a huge expansion of long-term memory capacity, able to store thousands of neural word systems, in instantly retrievable form; and (5) a great increase in the amount of brain space devoted to semantic representation, since semantic space would have become more and more finely differentiated under the growing pressure of an emerging language capacity. These are all domain-general requirements. The only modular element in language evolution appears to have been the vocal apparatus, but even this was not a truly modular change at the cognitive level since phonology, in its strictest definition, is not restricted to the vocal system (sign language has phonemes).

But it was the nature of conscious experience itself that changed most dramatically. Language differentiates experience for us. It refines the thought process and embeds it in increasingly precise culturally imposed algorithms. We know intuitively what words and sentences do for us. They define reality. They focus our attention. They elevate our awareness of whatever they specify. Moreover, when stories and ideas are juxtaposed, so that their meanings collide, they can shift our focus to new semantic spaces. Once the circular process of linguistic definition has been triggered, these spaces are constantly opening up. Once we have a sentence or a word for something, we seem to own it. We can roll it around in our minds, reflect on it, and conduct experiments with it. In a word, languages clarify the experienced world. They placate the internal beast of consciousness. The drive that energizes our use of language today, and invented it in the first place, is that same intermediate-term governor, constantly seeking greater clarity, conjointly with other conscious beings, in culture.

## The Virtual Realities of Oral-Mythic Culture

Once we have leaped into a narrative mind-set, our worlds become virtual ones. Language gives us a cognitive zooming facility, with which we can voluntarily alter the scale of an experience, plunging into microscopic detail or pulling back for the big picture. This is an awesome power, unequaled in the history of cognition. It allows us to bracket experience and survey our cognitive realm, with remarkable detachment.

If mimesis is our cultural glue, stories are the main by-product, as well as the principal organizing force, behind the classic form of human culture, the oral tradition. We are dominated by our stories, whether of magic, witches, devils, demons, great heroes, or genesis. They are the imaginative fodder of self-identity, morality, class, and authority. Stories can become so influential and so deeply rooted in the daily operation of the culture that they assume a special cognitive status, that of myth. Myths are standard versions of very old and widely shared stories. A mythic narrative surrounds great cultural heroes, such as Ulysses, Moses, and Christ. And mythic collections such as the Vedic and Homeric myths can influence every phase of life in their societies. People model their lives on mythic figures. Western religion, through its domination of the mythic resources of Europe, was able to influence every date on the calendar, every major architectural and engineering project, every library, and every philosophical and literary work. It even financed the rise of science, often against its own better judgment. Advances in astronomy were pursued aggressively in seventeenth-century Europe mainly because the papacy needed more precise ways of determining the exact date of Easter. Similarly, the Crusades were fought, and new continents conquered, largely because of an idea fixed in a narrative framework of what the world was like or should be like. Stories of imaginary places, like the kingdom of Prester John, or of the coming of the Apocalypse have inspired people to give up their roots to live in danger and often to die.

Myths have driven the rise of every empire, from Islam, India, and China to Venice and London. Narrative traditions ruled the minds not only of poets and artists but of emperors and foot soldiers. Local narratives ruled every tribe and village, through their elders, chiefs, shamans, and bards. They have been a governing force in human thought and often move people to undertake the most incredible projects. They not only can spell out the project of a life but can provide a sense of group identification that drives people to attempt almost anything.

Human conscious capacity has been enormously expanded by this collective leap into the narrative domain. The process of mapping the form of any symbol onto meaning gives us much more than clusters of associations, or fast-moving maps of forms with discrete meanings, or ways of parsing and segmenting events. Words and grammars are merely the entry-level skills without which narrative traditions could not exist, but once they are acquired, they are secondary to the stories themselves. The cognitive impact of language can be measured primarily through evaluating the cognitive value of narrative skill, rather than vocabulary size or grammatical elegance. Narratives, especially shared life narratives, are the basis of autobiographical memory itself. Stories and myths can completely reshape our semantic spaces, leading to a consensual definition of a shared virtual reality that is the core of oral culture.

Language is secondary to mimesis in its early acquisition and does not emerge as an autonomous level of representation until children, somewhere around age four, become competent at telling their own stories and start developing an active autobiographical account of their own lives. This ability allows them to time-track their lives and to place various events in an all-encompassing model of their own experiences. This ability has introduced a new layer of culture into the traditions we inherited and has given us a new means of invading one another's minds. But language did not devalue or displace mimesis. It serves a different function in society. The mimetic body language, prosody, and ritual of life continue, but under the control of a new mythic mantle. Thus the storied king, with his divine right and his power protected by an armamentarium of validating myths, nevertheless had to rely on coronations, elaborate weddings, customs, such as bowing and curtsying, and all the formalities of a courtly hierarchy to consolidate and implement his power in daily life. We have retained everything that works: a narrative tradition, mimesis, and the episodic cognitive foundations we inherited. Each serves its own function in society. We still have craft, custom, skill, and ritual, all mimetic in their roots. But myth has come to dominate ritual, and ultimately narrative has come to dominate and surround mimesis. There is no doubt where the power lies in traditional societies. It lies with language and the common cultural myths it generates.

## Collectivity of Mind

Collectivity has thus become the essence of human reality. Although we may have the feeling that we do our cognitive work in isolation, we do our most important intellectual work as connected members of cultural networks. This gives our minds a corporate dimension that has been largely ignored until recently. The word "corporate" usually refers to institutional entities, such as banks and governments. This is not so anomalous a label as it might seem because corporations are unified in a cognitive sense, just like the bodies of living organisms. But unlike organisms, which are locked in on themselves, corporations can distribute their intellectual work over many minds and employ various external symbolic devices, such as writing systems and computers, to facilitate this distribution process. They can develop corporate perceptions, ideas, agendas, and even personalities. Individual minds are thus integrated into a corporate cognitive process, in which single individuals rarely play an indispensable role.

This gives us a good reference point for how we think as cultural beings. Our cultures invade us and set our agendas. Once we have internalized the symbolic conventions of a culture, we can never again be truly alone in semantic space, even if we were to withdraw to a hermitage or spend the rest of our lives in solitary confinement. Big Brother culture owns us because it gets to us early. As a result, we internalize its norms and habits at a very basic level. We have no choice in this. Culture influences what moves us, what we look for, and how we think for as long as we live. We work out the vectors of our lives in a space that is defined culturally.

Conscious minds have always lived at the cognitive cutting edge of culture. This is true of everyone, not only of geniuses. The larger cognitive-cultural system embeds our minds into a distributed network, and it is this collective framework that directs the conscious process in certain creative individuals. After many centuries, this process has generated the complex web of habits, customs, and beliefs that define human culture. These are now unconscious, of course; in cultures, as in individuals, automatization is the other side of advanced consciousness. And the conscious mind is thus part of a larger fabric, much of which exists outside its grasp. Nevertheless, the essence of our human mentality is to harness outselves to the tremendous collective organizing energy of culture. This is our obsession and our inevitable fate. For many, in an ironic twist on the notion of self-identity, culture has become a form of self-justification. We have created a collective organism that appears ominous at times. Our interlinked nervous systems, newly powerful in their electronic extensions, are now challenging the supremacy of the natural world.

But this masks the larger role of humanity in evolution. Through us, nervous systems in general have broken free of their autochthonous solipsism. We are the means whereby the physical universe has finally reflected back on itself and reaped the effects of our painfully slow evolutionary gestation. But ironically, our elaborate cultural games are subjecting our brains to forces that are far beyond our control, and the future direction of our civilization is not clear. We tend to ignore what is going on inside us and pay attention mostly to our cultural universe, rather than the natural world, which we now experience primarily through one of the primary filters of our new religion, science. We are a culturally bound species and live in a symbiosis with our collective creation. We seek culture, as birds seek the air. In return, culture shapes our minds, as a sculptor shapes clay.

## The Triumph of Consciousness

The evolution of languages and oral cultures left our species with a formidable strategy for representing reality and elevated consciousness to a new plane. But the emergence of the human mind did not slow down following the arrival of our species. Rather it accelerated. We have travelled a great additional distance. Something about our mentality changed in the past few millennia, something that made us able to construct such exotic things as symphonies, philosophies, oil refineries, nuclear weapons, and robots. Do such achievements have implications for theories of consciousness? Many would deny that they do. They would claim that the parameters of mind were surely fixed long ago, when we emerged as a species, and that culture can add nothing to an equation written deeply into the human genome.

But that common belief does not stand up to scrutiny. The human mind has been drastically changed by culture. In modern culture, enculturation has become an even more formative influence on mental development than it was in the past. This may be a direct reflection of brain plasticity, rather than genetic change, but that does not in any way diminish the importance of the change from a purely cognitive standpoint. The human mind is so plastic in the way it carries out its cognitive business, individually and in groups, that the core configuration of skills that defines a mind actually varies significantly as a function of different kinds of culture. This is especially true of the most conscious domains of mind, such as those involved in formal thinking and representation.

Let me be very clear about what I mean here. I am not speaking of trivial cultural changes, such as variations in custom or language usage. These are by far the most common and have no proven cognitive impact. But there are other innovations that have a significant impact. The most important of these is literacy. Literacy skills change the functional organization of the brain and deeply influence how individuals and communities of literate individuals perform their cognitive work. Mass literacy has triggered two kinds of major cognitive reorganizations, one in individuals and the other in groups.

To become fully literate, the individual must acquire a host of neural demons that are completely absent from anyone who lacks literacy training. This involves massive restructuring. There is no equivalent in a preliterate mind to the circuits that hold the complex neural components of a reading vocabulary or the elaborate procedural habits of formal thinking. These are unnatural. They have to be hammered in by decades of intensive schooling, which change the functional uses of certain brain circuits and rewire the functional architecture of thought. This process can be very extensive. For example, multilingual people might have hundreds of thousands of lexical items, linked to a variety of grammars. Consider the impact of twenty or more years of schooling on the brain of someone who has acquired full symbolic literacy in several different technical, mathematical, scientific, and musical fields. These skills encumber neural resources on a vast scale and change how the person's mind carries out its work.

In such individuals, the demons of literacy reside in large neural networks that, complex as they are, form a conceptual entity that might be called the literacy brain. The literacy brain is a cultural add-on to the normal preliterate state of the brain. It determines a great deal about how the operations of the conscious mind are carried out. It affects the relative growth and synaptic richness of certain regions and structures of the brain. Even though the neural circuits of literacy are far from fully understood, we know that they engage both hemispheres, albeit asymmetrically, with more circuitry on the left than on the right in most people. Sometimes these networks occupy at least as much space as the networks of spoken language. They also break down in a parallel manner to disordered language, or aphasia, but independently of it. Damage to the literacy brain causes dyslexias (reading problems) and dysgraphias (writing and spelling problems) independently of other cognitive loss. This shows that there are several new physical systems involved in setting up a literacy brain, depending on the number and complexity of the symbolic skills the brain has acquired. These systems qualify as functional components in exactly the same way as the subsystems of spoken language, and they are unquestionably a product of enculturation, rather than genetic predisposition.

Literacy skills are a response to the invention of external symbols. As the technology of symbols evolved, literacy skills became more complex. But the effects of external symbols did not stop with the reorganization of the individual brain. They transformed the collective architecture of cognition and changed how the larger human community thinks and remembers. They also enabled many new forms of mental representation. There are now entire classes of cognitive work that cannot take place without external symbols. If we define symbolic technology very broadly, including everything from musical and mathematical notations to art, circuit diagrams, and maps, it is clear that most of our major cultural institutions and a high percentage of our cutting-edge work are completely dependent on symbolic technology.

Symbolic technology has changed the way we think, remember, and experience reality, individually and as a collectivity. In fact, it has triggered a third cognitive transition, equal in importance to the first two.

## The Third Transition: The Invention of Symbolic Technologies

Symbols can be internal or external to the brain. For example, the words of spoken language are internal symbols stored in the brains of speakers and listeners. The same words typed on a sheet of paper become external symbols, stored on the printed page, rather than in the brain. Symbolic technology is the enterprise of manufacturing and crafting external symbolic artefacts and devices. These have enabled us to build a vast cultural storehouse and an external symbolic storage system, which serves as a permanent group memory and includes such things as books, museums, measuring instruments, calendars, and computers. These are extensions of what archaeologists call material culture. But unlike most aspects of material culture, they are designed specifically to help us think, remember, and represent reality. They have gone through thousands of years of development, and the end result, in our time, is a plethora of very sophisticated technological devices that are interwoven with our minds. Books, paintings, maps, musical notations, and a huge number of cognitive machines, such as clocks and observatories revolutionize what we can do with our minds.

Symbolic technologies serve many different purposes, but they all share a common underlying principle. They liberate consciousness from the limitations of the brain's biological memory systems. The new physical media of symbolic technology have enormous advantages over brain-based memory media. One of their chief advantages, as we shall see, is that they are fully accessible by awareness. From the standpoint of evolution, this was a revolutionary development. It carried the great hominid escape from the nervous system to its logical conclusion. Because of the limitations of biological memory, conscious thought was enormously difficult when contained entirely inside the brain box. External storage changed this and gave thinkers new strategic options.

The power of symbolic technology can be attributed largely to its impact on the conscious process. External symbols are revolutionary because they transform the architecture of conscious mental activity. This can be seen most clearly in the way they are displayed to awareness. The region of symbolic display may be called the external memory field. This is where the brain-symbol interface begins and ends.

The external memory field includes everything in use at the moment and possibly some items not in use that might later become relevant. Although this arrangement constitutes a very ordinary work environment in our highly literate society, it is an extraordinary historical development because *it changes the long-standing relationship of consciousness to its representations*. We can arrange ideas in the external memory field, where they can be examined and subjected to classification, comparison, and experimentation, just as physical objects can in a laboratory. In this way, externally displayed thoughts can be assembled into complex arguments much more easily than they can in biological memory. Images displayed in this field are vivid and enduring, unlike the fleeting ghosts of imagination. This enables us to see them clearly, play with them, and craft them into finished products, to a level of refinement that is impossible for an unaided brain. Thus the display characteristics of the external memory field expand the range of mental operations available to a conscious mind.

External symbols introduce more powerful memory media into human cognition. The biological memory records of the brain, known as engrams, differ from external symbols, or exograms, in most of their computational properties. Engrams are impermanent, small, hard to refine, impossible to display in awareness for any length of time, and difficult to locate and recall. In fact, to find a natural memory, we tend to rely on ancient associative principles, such as similarity and contiguity. In contrast, external symbols give us stable, permanent, virtually unlimited memory records that are infinitely reformattable and more easily displayed to awareness. Moreover, exograms are much easier to search, and we can recall them with a variety of retrieval methods. The availability of powerful external media increases the number of ways we can represent reality. Just as better memory media can transform what a computer can achieve, so they can revolutionize what a conscious mind can achieve. A cognitive change this revolutionary could never have occurred simply by evolving a larger brain. But it has been achieved, relatively quickly, with improved symbolic technology.

The external memory field creates a mirror world for consciousness. It reflects the architecture of biological memory back into the symbolic environment, and this mirror image is then reflected back into the brain. The conscious mind is thus sandwiched between two systems of representation, one stored inside the head and the other outside. An image might help clarify what this means and how the external memory field fundamentally changes the architecture of consciousness. Imagine the preliterate state of the conscious mind as a space (not literally, but in a metaphorical sense) with a simple structure, as shown in Figure 4.1.

In this arrangement, sensory inputs fall into the vivid core of awareness, the focus of level-l and level-2 processing. This core is surrounded by a less vivid conscious space, which is the broader realm of level-3 processing. This processing can endure over a much longer time span and incorporate many layers of ongoing activity. It also draws from the brain's long-term memory systems and recalls ideas or images into awareness as needed.

This is the canonical structure of the conscious mind, our "normal" state of mind. The high-turnover vivid core is supplemented by the many low-turnover, slower-moving networks running in the intermediate-term background of awareness. The "thinker" resides in the latter, although the high-turnover focus might dominate awareness in the short run. For example, while walking through the forest with my son, the vivid core of my awareness would probably be filled with sights, smells, and sounds. His voice and mine would trigger a second level of awareness, a barrage of ideas and images related to the conversation, which would inhabit my level-3 awareness and play out on more than one level. Consciousness would thus be influenced by both the vivid core and the background conversation.

This arrangement is drastically altered when I plug myself into a powerful symbolic device, such as a book or computer. When I confront such a device, my operational mental architecture is temporarily altered. Awareness now finds itself juxtaposed between two simultaneously present storage systems, one internal and biological, the other external and technological, each with long-term and short-term aspects. Two memory systems are thus available, whereas before there was only one. A mirror image of natural working memory is now available to awareness in the external memory field, as shown in Figure 4.2.

In this case, the conscious mind receives simultaneous displays from both working memory and the external memory field. Both displays remain distinct in the nervous system. The external memory field is not just another sector of working memory. It taps directly into the neural networks of literacy, located in brain regions that are distinct from those of working memory. Working memory and the external memory field thus complement each other, and this allows the brain to exploit their distinct storage and retrieval properties.

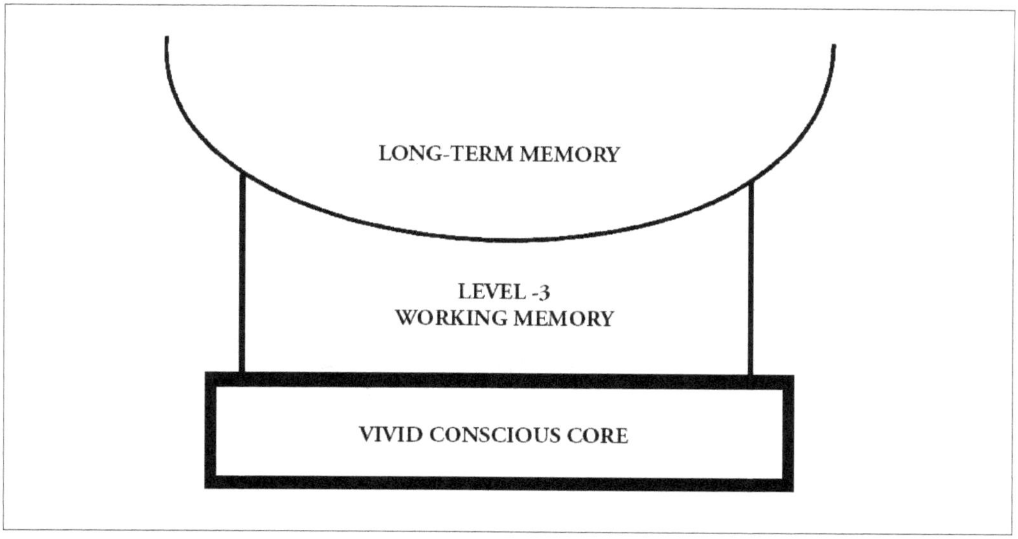


Figure 4.1 The preliterate state of the conscious mind.

This gives awareness a much richer structure. These two displays, the one externally and the other internally generated, are processed by the conscious mind at the same time and serve different functions in the same cognitive sequence, as when a literate thinker reads about something and at the same time reflects on what he is reading.

The vivid core of awareness, where it all started, still sits at the geographical center of the phenomenal universe, but not in a position of control. It is now between two much more powerful cognitive fields, both capable of condensing and focusing a host of culturally generated algorithms. Each has its own role to play in energizing and directing awareness. To maintain control, a fully literate consciousness must navigate a way through this complex of ideas and images, optimizing its use of resources.

If this sounds familiar, it should. It is surely obvious that the creative bob-and-weave boxing match between consciousness and spoken language is also played out with external symbols. It is the same conscious mind, after all, simply transposed to a new medium, that generates the domain of external symbolic storage. The mechanism the human mind uses to generate languages also equips it to generate external symbols. It is not the underlying creative mind that has changed. We are no more creative, in this sense, than our Stone Age forebears. It is the evolving change in the memory media themselves that have made such a huge difference in what the mind can finally achieve.

Improved memory and reflective power have complicated the predicament of the conscious mind. Internal and external memory can now be played off against each other. Indeed, this has become one of the main activities of modern thinkers, who have to switch between internal and external representational systems and resolve tensions between them. This is the conscious mind in its new role as arbitrator. Multifocal awareness, forged in the furnace of oral-mythic culture, is stretched to the limit by the infinitely changeable, infinitely dense vagaries of symbolically literate culture. Its dance with symbols keeps pushing back the frontiers of thought, and the symbol-work itself, along with the cultural products that come from it, has assumed a tremendous presence in our lives.

Culture is now in the business of manufacturing virtual realities. The kinds of realities

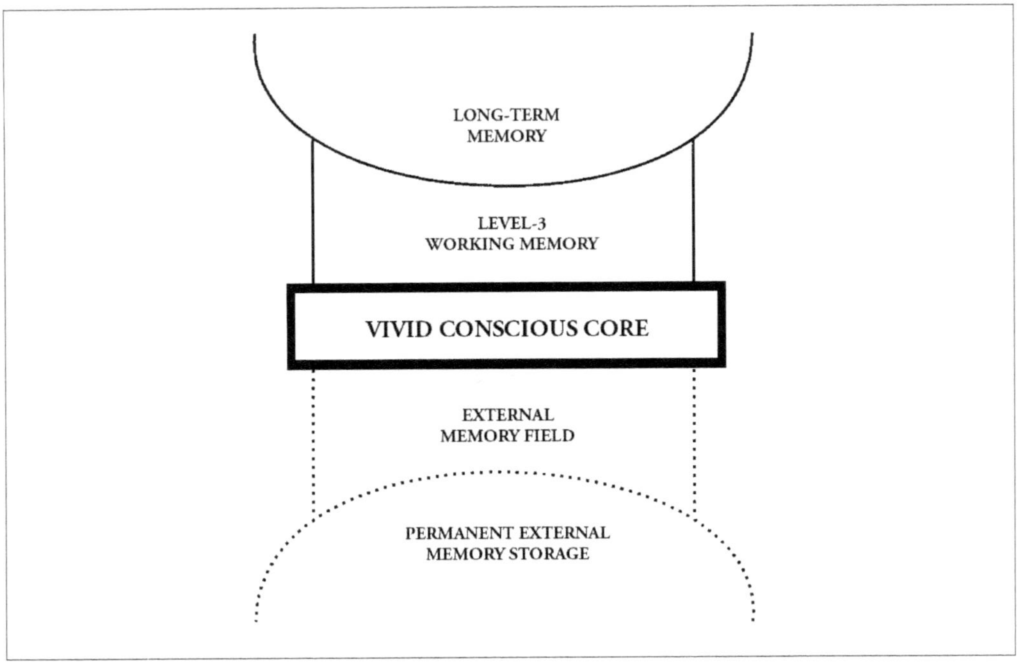


Figure 4.2 The literate state of the conscious mind.

imposed artificially by culture can, in theory, outstrip natural experience in both intensity and variety. In modern society, the sequencing of experience by means of symbolic technology has developed into an exact science. It has gone far beyond the static image into the scripting of many images in series. This process is the basis of complex media such as film and the novel. Scriptwriters of various stripes have become the chief cognitive engineers of mass culture. They design many of our experiential trips into virtual reality.

The availability of external symbols thus makes culture infinitely more self-reflective and feeds back onto the texture of individual consciousness itself, making experience the object of its own deliberate capacities to plan. This might point to the ultimate triumph of consciousness, the control of consciousness itself, the scripting of individual awareness by culture to enhance our understanding of the universe. Unfortunately, the same shared technology can also be used to deceive, confuse, regiment, and even to prevent thought, as it was during the Second World War and behind the Iron Curtain. However it is used, it confers a terrifying power on the conscious process.

## A Cerebral Trojan Horse

External symbols can always be translated into a series of internal physiological events. The external memory field gains its foothold on the mind through its ability to program brain activity. To enter the cognitive system, an external memory display must project into the brain of one or more conscious recipients. Otherwise, it is a mere lifeless object. It acquires psychological reality only by gaining control over a set of automatic demons in someone's brain. This engagement seems to occur initially in the parietal and temporal lobes of the cortex, where high-level perceptual analysis is carried out. If the literate viewer has the appropriate demons, there will also be a higher-level cognitive response distributed throughout many areas of the brain, which opens the doors of semantic memory, calls ideas and scenarios into awareness, and juxtaposes experience. We are not yet sure how anything this complex translates into neural activity on a microscopic level, but it does. It harnesses the basic executive brain systems. In addition, it engages huge amounts of memory. Thus the external memory field provides a fast channel from the cultural universe to the control centers of the brain. It is a shortcut from the engineered symbolic universe into the deepest, most complex networks of the mind. It externalizes control and allows the outside world to pull our mental strings directly. There are gains in this, of course. But there are also losses.

To an extraordinary degree, this makes the human mind externally programmable.[8] In symbolic technology, consciousness has gained a means to reflect its own activities back on itself and program its own operations. It can redirect the narratives and images of culture back into the flow of individual awareness, where it can take advantage of the tremendous resolving power of our perceptual apparatus. Through the external memory field, external symbolic devices create stable, vivid displays of many things that were previously undisplayable. Ideas, theories, strategic plans, even imaginary events are left suspended in the mind, to be processed as we please. Or to be processed according to the dictates of platoons of cognitive engineers. The essence of the modern condition is the proliferation of deliberately engineered sources of experience. Modern conscious experience has drifted very far from the natural streams of events that we imagine to be the normal source of experience. We now have entire industries that deliberately design and manufacture experience.

One might argue that it has always been thus. Perhaps. Certainly, tribal society is permeated by religion, which provides a formidable means of controlling how life is experienced and what kinds of thoughts and images are acceptable. In traditional urban societies, religion sums up the whole hierarchy of expression and ideas offered by that society and provides a framework within which life is experienced. However, it is doubtful whether traditional religion as a whole could be characterized as a deliberately designed means of controlling experience. It is more often a kind of cultural attic, containing the major ideas and images that influence a people's history. Everything, from the architecture of such societies to their costumes, music, habits of prayer, and formal rituals, has an impact on experience, as it sets the main emotional markers in the cultural landscape. But these are usually not so much designed as they are spontaneous outgrowths of tradition and the popular will.

Modern society is different. It has hundreds of very powerful external symbolic technologies that provide a medium for collaborative cognitive work, and these are used routinely to design the cultural environment in a more deliberate manner.[9] Entire industries do this for individual consumers, following the directives of advertisers and programmers. They offer more choice than ever before, and many more alternatives than those offered by tra-

8. Recent research on child development shows that the cognitive enculturation of modern children is highly complex. Their introduction to formal cognitive operations is through the educational system, as they are led through a tangled web of representational modes and complex institutionalized algorithms, all of which are tied to symbolic literacy. See, for example, Nelson (1996); and Karmiloff-Smith (1992). In effect, children's minds are being fashioned into multilayered cognitive systems, and thus they carry within themselves, as individuals, the evolutionary heritage of the past few million years.

9. Cognitive studies of the modern workplace—for instance, Hutchins (1995); Kitajima and Polson (1997); Suchman (1987); and Olson and Olson (1991)—examine the way that electronically distributed knowledge representation and computer-coordinated planning and problem solving are affecting the relative roles of individual minds and external memory devices in this collective enterprise. There is no doubt that these roles are changing as the technology supporting distributed networks evolves and that the nature of cognitive work itself is being transformed in the process.

ditional society, but the entire menu is highly contrived. Symbolic devices are absolutely essential to conceive, set up, and execute such schemes. A proscenium stage, or a large-screen monitor, or a television network can coordinate experience on a scale that was previously impossible to contemplate. These technologies are designed to induce a common end result, a conscious state of great subtlety, with textures and complexities of raw feeling that are almost impossible to describe.

## Multilayered Cultures, Multilayered Domains of Awareness

The fate of human consciousness is thus tied to all the stages of human cultural evolution. Each major transition has increased the scope of basic awareness by looping it through a new kind of public process. We now have three distinct layers of cultural representations, each of which has a magnifier effect on the previous layers. This has increased the number and variety of conscious modes. However, the core brain structures involved are exactly the same. On the basis of what we know of our brain's evolution, we must conclude that the raw feel of being human is probably not qualitatively different from the raw feel of being any kind of primate. The same fundamental brain functions negotiate all primate experience: binding, short-term memory, and intermediate-term governance.

But if the raw feel is the same, consciousness has gained far more control over mind and action in humans than in any other species. It has also enriched the content of individual awareness, through its immersion in communities of mind. The differences between us and other primates can be attributed largely to deep enculturation and the resultant reprogramming of conscious experience. Our unique relationship with culture gives us a hybrid awareness and several channels for transmitting cultural influence. We are married to culture and fated to play out its algorithms in our conscious acts. The course of our evolution has been marked by an almost incredible series of intellectual revolutions, whereby conscious capacity has been modified to cope with the overwhelming presence of culture. The collective matrices of culture have exploded, impressing their memory records into our rapidly evolving brains. As a species we have exploited the conscious resources of the individual mind to build a new collective process. This has resulted in a multilayered cultural framework for modern consciousness, which might be conceptualized as a set of concentric rings, as shown in Figure 4.3.

The center ring is the episodic core we inherited as members of the primate family, the bedrock experience base of a precultural mind. At this level consciousness consists of a stream of discrete events, served up as the natural units of experience. Surrounding that central ring are three culturally defined levels of consciousness, each connected to a corresponding cultural matrix, or pattern of convention. The mimetic matrix is the theatrical domain of human life, the actor's realm, with its attitudes, gestures, postures, stances, and unspoken nuances. This matrix defines both actor and audience and sets out the basic rules for all communication and expression. The linguistic matrix is a highly precise and efficient system of knowledge representation, encoding countless stories, myths, and traditions. The external matrix is an even more precise collective system, which provides more powerful media for formulating and displaying knowledge as well as for operating institutions, governments, and formal symbolic systems of thought.

All three of these domains are irrevocably tied to changes in the nature of awareness, especially self-awareness. Mimetic skill is all about self-representation and self-definition, but mimetic style has meaning only in relation to the actions of others. Language is a powerful means of constructing autobiographical memories, but our sense of self takes on meaning only

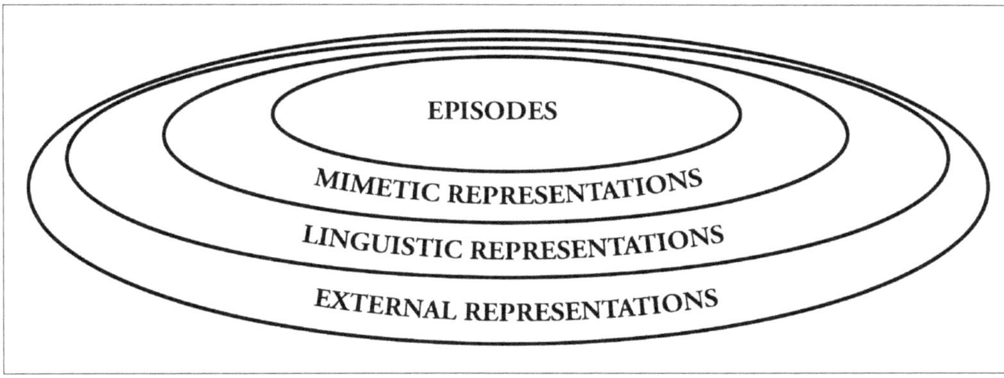


Figure 4.3 Multilayered consciousness.

within a shared oral tradition. External symbols define us as individuals in a thousand ways, but they too acquire meaning only in the context of a collective sense of social structure.

The nature and range of human conscious experience are no longer a biological given. Rather they depend on a somewhat unpredictable chemistry of brain and culture, whereby the processes of mind can be endlessly rewritten and rearranged by cultural forces. A mind isolated from culture is deprived of the outer layers of our representational map, despite having the brain mechanisms required to support them. Culture must provide the necessary demons before consciousness can exploit them. We can achieve full awareness only if culture is able to write its key algorithms onto our minds. Evolution has nudged us in this direction, interlocking brain and culture, toward the absorption of individuals by communities of mind. The human conscious process is a specialized adaptation for navigating the turbulent waters of culture as well as the primary channel through which cultural influence can be transmitted back to us.

## The Essential Unity of the Conscious Hierarchy

Consciousness has served as the engine of our evolutionary adventure, and still does. Each stage in human evolution was constructed by expanding the same set of foundation skills and reprogramming the mind as we evolved more and more elaborate cultural networks. This enriched the human intellectual universe without sacrificing the basic integrity of cognition and awareness.

Human conscious capacity has thus been transformed in its function into an integrative device. It is the final common path of the labyrinthine cognitive-cultural system we evolved. Ultimately the entire distributed network relies on a set of simple operations, performed one at a time, in a specified order. Out of simplicity, complexity. The capacity of the executive brain system thus sets a limit on the kinds of operations that a mind can assimilate from culture and on the complexity that a cultural system can tolerate. Human conscious capacity must be regarded as a distributed resource that limits the rate at which culture can accumulate knowledge and determines what kinds of representational systems a culture can successfully construct and maintain.

The unity of this sprawling cognitive-cultural system, which has been a central consideration in constructing this theory, can be seen in the flow of control through the three evolved layers of human mind and culture. The core episodic layer in Figure 4.3 is inherently other-directed. It observes the world from an unselfconscious perspective. In contrast, the mimetic layer is acutely self-conscious; its own actions are subjected to relentless comparison and review. The mimetic imagination is a dramatic actor. It gives culture shape, by using essentially the same set of conscious operations as an episodic mind, in a wider theater of action.

The narrative layer adds greatly to this cognitive-cultural system. It gives ideas a certain autonomy from personal experience and creates the possibility of abstract beliefs and public discourse. But narrative constructions can govern the larger cognitive-cultural system only if they are successful in dominating the mimetic layer of the system. Narrative imagination wins control by altering the mimetic templates of culture and bringing them in line with myth. The chain of command is clear: words and stories will dominate only if they successfully influence role playing, skill, and style. Myth gains power only when it directly controls the mimetic imagination. Mythmaking has become an important industry in our cultural hierarchy, and it usually emerges victorious. But its victories never destroy the unity of conscious experience because as we have seen, language floats on the surface of a very deep cognitive ocean. It has no autonomy. It cannot shatter the basic unity of experience and imagination because like episodic and mimetic cognition, it is rooted there.

External symbolic storage significantly alters the powers of our larger cognitive-cultural system by creating conditions within which formal theoretic systems can emerge. But symbolic technologies do not shatter the basic unity of conscious experience either. Symbols and theoretic systems are subject to the same centrifugal forces as every other medium of symbolic thought, within the larger system. Theoretic representations can win governance over culture only if they successfully operate the levers of the hierarchy and assume direct control over narrative, mimetic, and episodic imagination. Theoretic skills are important for survival, but in isolation they never endure. They will govern human culture only if they can dominate the entire cognitive-cultural hierarchy. The governing mode must attain the peak of the collective system, the capstone of the cognitive pyramid. There is no way to short-circuit this elaborate social-cognitive mechanism, and no reason to do so. It works wonderfully well.

The unity of this system is closely related to the question of teleology or purpose. Purpose implies conscious deliberation. There is no such thing as unconscious purpose or unconscious will. There may be unconscious motives and drives, but purpose, by definition, implies conscious design. However, its proven reach is very limited. Human beings are capable of micromanaging very small events, such as wars, fish stocks, and small economies. If we were able to control the Earth's climate, we might make a credible claim that purpose was now a larger player in the universe, but this would still amount to only a small victory in the context of a universe of chance.

Although purpose is anchored in consciousness, it can never be truly attributed to a single conscious mind, in isolation. It is the conscious mind *in culture* that constitutes the source of teleology in the affairs of the human world. There might be a race of beings somewhere out there that is capable of conceiving purpose in a truly solipsistic manner, but human purpose always has a cultural dimension and is inherently distributed in its origins. Ours is a collective teleology, and its creative engine is a conscious mind that has assimilated the algorithms of a culture and is thus a vehicle through which the collective cognitive-cultural hierarchy can act.

## Conclusion

Fate has given us this hybrid nature, by which we are joined to communities of our own invention. We remain distinct from them but are never fully autonomous. The human brain is a poor thing on its own, an inarticulate, undifferentiated, metaphorizing beast like any other. But joined to a community of its fellows, it has this remarkable capacity to create a community of mind, acquire symbolizing powers, and vastly expand the range of its own awareness, in proportion to the depth of its enculturation. If this appears strange to us, this is surely only a reflection of our conventional notions of, among other things, strangeness. We have lived comfortably with the myth of the isolated mind throughout most of our history. We like to think of ourselves as self-complete little monads dwelling inside our sealed biological containers, peering out at the world from the safe haven of consciousness.

But we are edging closer to the truth. We are collective creatures, even to the texture of our awareness. When we face that fact, the irrational beast that drives our cognitive enterprise will finally be able to see itself as it really is, as part of a collective process that has freed us from our solipsistic prison, if not from our materiality. The triumph of consciousness will be complete when it can finally reflect on the collective process itself and see only itself, in the mirror of its own reflection.

## References


Acredolo, L.P. and S.W. Goodwyn. 1988. "Symbolic Gesturing in Normal Infants." *Child Development* 59: 450–466.

Bloom, Paul. 2000. *How Children Understand the Meanings of Words*. Cambridge, MA: MIT Press.

Bruner, Jerome S. 1983. *Child's Talk: Learning to Use Language*. New York: W. W. Norton.

Corballis, M.C. and S.E.G. Lea, eds. 1999. *The Descent of Mind: Psychological Perspectives on Hominid Evolution*. Oxford: Oxford University Press.

Damasio, Antonio. 1999. *The Feeling of What Happens: Body and Emotion in the Making of Consciousness*. New York: Harcourt Brace and Company.

Donald, Merlin. 1991. *Origins of the Modern Mind: Three Stages in the Evolution of Culture and Cognition*. Cambridge, MA: Harvard University Press.

———. 1993a. "Human Cognitive Evolution: What We Were, What We Are Becoming." *Social Research* 60: 143–170.

———. 1993b. Précis of *Origins of the Modern Mind* with multiple reviews and author's response. *Behavioral and Brain Sciences* 16: 737–791.

———. 1999. "Preconditions for the Evolution of Protolanguages." In *The Descent of Mind: Psychological Perspectives on Hominid Evolution,* edited by M.C. Corballis and S.E.G. Lea, 138–154. Oxford: Oxford University Press.

———. 2001. *A Mind So Rare: The Evolution of Human Consciousness*. New York: W.W. Norton.

Dunbar, Robin. 1996. *Grooming, Gossip and the Evolution of Language*. Boston: Faber and Faber.

———. 1998. "Theory of Mind and the Evolution of Language." In *Approaches to the Evolution of Language: Social and Cognitive Bases,* edited by J.R. Hurford, M. Studdert-Kennedy and C. Knight, 92–110. Cambridge: Cambridge University Press.

Goldin-Meadow, S., D. McNeill and J. Singleton. 1996. "Silence Is Liberating: Removing the Handcuffs on Grammatical Expression in the Manual Modality." *Psychological Review* 103: 34–55.

Goldin-Meadow, S. and C. Mylander. 1998. "Spontaneous Sign Systems Created by Deaf Children in Two Cultures." *Nature* 391: 279–281.

Goldin-Meadow, S. and D. McNeill, 1999. "The Role of Gesture and Mimetic Representations in Making Language the Province of Speech." In *The Descent of Mind: Psychological Perspectives on Hominid Evolution,* edited by M.C. Corballis and S.E.G. Lea, 155–172. Oxford:

Oxford University Press.

Greenfield, P. M. and J. Smith, 1976. *The Structure of Communication in Early Language Development*. New York: Academic Press.

Harris, P.L. 1998. "Mimesis, Imagination and Role-play." In *Cognition and Material Culture*, edited by C. Renfrew and C. Scarre. Cambridge: McDonald Institute for Archaeological Research.

Hurford, J. R., M. Studdert-Kennedy and C. Knight, eds. 1998. *Approaches to the Evolution of Language: Social and Cognitive Bases*. Cambridge: Cambridge University Press.

Hutchins, E., 1995. *Cognition in the Wild*. Cambridge, MA: MIT Press.

Jablonski, N.G. and L.C. Aiello, eds. 1998. *The Origin and Diversification of Language*. San Francisco: Memoirs of the California Academy of Sciences.

Jackendoff, R. 1994. *Patterns in the Mind*. New York: Basic Books.

Karmiloff-Smith, A. 1992. *Beyond Modularity: A Developmental Perspective on Cognitive Science*. Cambridge, MA: MIT Press.

Kitajima, M. and P. Polson. 1997. "A Comprehension-Based Model of Exploration." *Human-Computer Interaction* 12: 345–389.

Lakoff, George and Mark Johnson. 1999. *Philosophy in the Flesh: The Embodied Mind and Its Challenge to Western Thought*. New York: Basic Books.

Meltzoff, A.N. and M.K. Moore. 1997. "Explaining Facial Imitation: A Theoretical Model." *Early Development and Parenting* 6: 179–192.

Nelson, Katherine. 1996. *Language in Cognitive Development: Emergence of the Mediated Mind*. Cambridge: Cambridge University Press.

Nelson Katherine and L.K. Shaw, 2000. "Developing Socially Shared Symbolic Systems." In *Language, Literacy, Cognitive Development*, edited by J. Byrnes and E. Amsel, 1–52. New York: Erlbaum.

Olson, J.R. and G.M. Olson. 1991. "User-Centered Design of Collaboration Technology." *Journal of Organizational Computing* 1: 61–83.

Suchman, L. 1987. *Plans and Situated Actions: The Problem of Human-Machine Communication*. Cambridge: Cambridge University Press.

Suddendorf, T. 1999. "The Rise of the Metamind." In *The Descent of Mind: Psychological Perspectives on Hominid Evolution,* edited by M.C. Corballis and S.E.G. Lea, 218–260. Oxford: Oxford University Press.

Tomasello, Michael. 1999. *The Cultural Origins of Human Cognition*, Cambridge, MA: Harvard University Press.

Trevarthen, C. 1979. "Instincts for Human Understanding and for Cultural Cooperation: Their Development in Infancy." In *Human Ethology: Claims and Limits of a New Discipline*, edited by M. Von Cranach, K. Foppa, W. Lepenies and D. Ploof, 530–571. Cambridge: Cambridge University Press.

———. 1980. "The Foundations of Intersubjectivity: Development of Interpersonal and Cooperative Understanding in Infants." In *The Social Foundations of Language and Thought*, edited by D. Olson, 382–403. New York: Norton.

Varela, F.J., E. Thompson and E. Rosch. 1995. *The Embodied Mind: Cognitive Science and Human Experience*. Cambridge, MA: MIT Press.

# — 5 —

# Iconology and Imagination in Human Development: Explorations in Sociogenetic Economies[1]

Chris Sinha

## Introduction

"Iconology" is a term introduced by the art critic Erwin Panofsky (1939), who distinguished the general study of the meaning of imagery from the more traditional "iconography," or classification of religious images in art. W.J.T. Mitchell (1986) extended the notion to encompass the reflective understanding of the concept of the image:

> If Panofsky separated iconology from iconography by differentiating the interpretation of the total symbolic horizon of an image from the cataloguing of particular symbolic motifs, my aim here is to further generalize the interpretive ambitions of iconology by asking it to consider the idea of the image as such. (Mitchell 1986, 2)

Iconology, then, in Mitchell's interpretation, is the study, not just of images, but also of the representation in thought and word of the image: of its relations to its cognates—imagery, imagination, figure—and to its Other—text, discourse, word. As Mitchell makes clear, the "idea of image" (remembering that for generations of philosophers, following Aristotle, an idea was itself a mental image) is far from being "clear and distinct." Rather, it is "essentially contested;" it has been, throughout recorded history, a focus and a site of political, social, religious, and intellectual conflict, of a struggle between *iconophilia* (idolatry) and *iconophobia* (iconoclasm). Iconology, then, is about the relationship between image and text, and about the different discourses through which this relationship has been thought and contested.

This chapter focuses on the iconology of human cognition and rationality, in terms of the historical development of specific modes of rationality, the history of the representation in theory of "iconologies of rationality," and the ontogenesis of the personal and social imaginary. The chapter therefore explores the sociogenesis of iconological concepts, taking as its object a process which is both reflexive (it involves the *relationship* between thought, and

1. This chapter is a revised and edited version of a paper originally published in *Cultural Dynamics* 5 (1993): 57–82.

thought about thought), and productive (it involves processes which make possible new kinds of thought). I further propose that the specific site, or space, which this sociogenetic process inhabits and populates is what I shall refer to, following Castoriadis (1987), as the *social imaginary*.

## Exchange, Equivalence and the Social Imaginary

We may begin our exploration of the social imaginary by focusing on the notion of *exchange*, which found wide currency in late twentieth century French structuralist and poststructuralist approaches to anthropology, linguistics, psychoanalysis and social theory. The most explicit formulation of an exchange-based sociogenetic theory is to be found in the work of Jean-Joseph Goux (1990). The intuition underlying his work was already expressed, however, in the early part of the twentieth century by the anthropologist Marcel Mauss, who proposed that "exchange is the common denominator of a large number of apparently heterogeneous social activities" (Goux 1990, 2). By the 1960s it was becoming almost a commonplace to suggest that a unification (in terms both of delineating the conceptual field, and of understanding the historical conditions of emergence) of the human sciences might be provided by a generalized notion of exchange, since "the semiotic, economic and psychoanalytic horizons all emphasized the question of substitution and its correlative, value" (Goux 1990, 2).

Yet, as many critics were to point out, the analogy between linguistic communication, monetary exchange, and the symbolic structures proposed by psychoanalysis (especially in its Lacanian variant) may be nothing more than an analogy, and a misleading one at that. That Saussure (1966) employed economic and monetary metaphors in formulating the principles of structural linguistics, is perhaps nothing more than an instance of a "new" science borrowing the terminology of an already established one, for heuristic and didactic purposes. That Lacan (1966) chose a linguistically based exposition of the "symbolic economy" of the unconscious, may be accounted for by reference to the exhaustion and anachronism of the standard "hydraulic" metaphors. And "exchange," considered as an economic-anthropological category, is so far from homogeneous in its membership that the inevitable question arises as to whether the very concept is not a universalistic projection (or fetish) of the generalized commodity form of the advanced capitalist mode of production. In short, there are good reasons to doubt that the concept of "exchange" is capable of revealing any more than superficial commonalities between the different human sciences.

Here, though, Goux (1990) takes another tack. Instead of "exchange," he posits a notion which is both more "abstract" and more "historicist" (that is to say, a socially and historically emergent concept): the concept of the *general equivalent*. This, following Aristotle, he glosses as a standard "which, by making things commensurable, renders it possible to make them equal" (Goux 1990, 3). The "discovery" or "invention" of the general equivalent (or generalized concept of exchange and exchangeability) first occurred, probably, in Ancient Greece—Goux's work coincides, here and elsewhere, with the ideas of the philosopher Alfred Sohn-Rethel (1978), a parallel which I have not the space here to pursue further—and inaugurated a "fundamental configuration" of symbolic practices and value-economies which has persisted and reached its most developed expression in our present societies. This fundamental configuration Goux summarizes thus:

> the accession of the father to the rank of privileged subject, controlling the conflict of identification; the elevation of the phallus to the place of centralized standard of objects of drive...the privileged position of language as a phonic signifier potentially equivalent to all other signifiers through the operation of verbal expression—all these appear to be promotions of a general equivalent. In each case, a hierarchy is instituted between an excluded, idealized element and the other elements, which measure their value in it. In short…the *Father* becomes the general equivalent of [psychoanalytic] subjects, *Language* the general equivalent of signs, and the *Phallus* the general equivalent of [psychoanalytic] objects, in a way that is structurally and genetically homologous to the accession of a unique element (let us say *Gold*, for the sake of simplicity) to the rank of the general equivalent of products. Thus, what had previously been analyzed separately as phallocentrism (Freud, Lacan), as logocentrism (Derrida), and as the rule of exchange by the monetary medium (Marx), [can be seen as] part of a unified process. (Goux 1990, 4)

Goux adds that there is no question here of asserting any particular causal priority (e.g., of "base" over "superstructure"). He further proposes that the emergence of this "fundamental configuration" (the rule of the general equivalent, as we might say), is also identifiable as the historical condition for the birth of philosophy "as we know it"—that is, the *problematic* of equivalence and its logic, as this finds its expression in the "idea of the idea" (an abstraction of identity from perceptible difference) and the law of the excluded middle (here and below I am extending Goux's argument, or at least using a slightly different vocabulary from his).

At this point, we must ask: what precisely is the nature of the equivalence relationship which is constituted between any two terms, the privileged and the non-privileged, that which is the measure of value and that which is measured or equalized? The answer, surely, is that the relationship is *symbolic*, in the anthropological sense that it is the totality of such equivalences[2] which structures the symbolic order of any given society, and which finds its expression both in the language and in the institutions of that society.

Castoriadis (1987) suggests that the realm of the symbolic is not, either in the individual or in the society, one which exists *sui generis* or autonomously: every symbol presupposes an *imaginary* relationship between the symbol and that which is symbolized; conversely, the realm of the imaginary (in the life of both societies and individuals) finds its expression in and through symbolization. As Castoriadis puts it:

> the imaginary has to use the symbolic...to pass from the virtual to anything more than this.... But, conversely, symbolism too presupposes an Imaginary capacity. For it presupposes the capacity to see in a thing what it is not, to see it other than it is. To the extent that the imaginary ultimately stems from the originary faculty of positing or presenting oneself with things and relations which do not exist, we shall speak of a final or radical imaginary as the common root of the actual imaginary and of the symbolic. This is, finally, the elementary and irreducible capacity of evoking images. (Castoriadis 1987, 127)

Castoriadis finds common ground here with cognitive scientists such as George Lakoff (1987) and Mark Johnson (1987), who emphasize the role of image schemas, and the corporeal, in language and cognition (Lakoff and Johnson 1999). They propose that the interpretation of the meanings conveyed or expressed by the symbolic system of natural language depends upon metaphoric projections and identifications which find their "root" meaning in human bodily experience, and in the capacity to evoke and deploy image sche-

2. The term "equivalences" refers in this context to all possible symbolic (classificatory, metaphoric, metonymic) identifications in all possible societies.

mas. Such meanings are held to be universally shared, inasmuch as all language users share a common, embodied, human predicament or being-in-the-world.

However, the social dimension of the imaginary (including its role in language and cognition) extends beyond this "inter-corporeality" (Merleau-Ponty 1964). Castoriadis argues that underlying the symbolic order of any given society, and grounding its system of social significations, is an order or pattern which he refers to as the *social imaginary*, and which he speaks of as providing "the conditions for the representability of everything that the society can give [represent] to itself." Castoriadis continues:

> These patterns do not themselves exist in the form of a representation one could, as a result of analyses, put one's finger on. One cannot talk in this case of "images," however vague and indefinite the sense ascribed to this term. God is perhaps, for each of the faithful, an "image"...but God, as an imaginary social signification, is neither the "sum," nor the "common part," nor the "average" of these images; it is rather their condition of possibility and what makes these images, images "of God".... The imaginary social significations do not exist strictly speaking in the mode of representation; they are of another nature, for which it is of no use to seek an analogy in the other spheres of our experience. Compared to individual imaginary significations, they are infinitely larger than a phantasy (the pattern underlying the Jewish, Greek or Western "image of the world" is of infinite extension) and they have no precise place of existence (if indeed the individual unconscious can be called a precise place of existence.) They can be grasped only indirectly and obliquely...as the invisible cement holding together this endless collection of real, rational and symbolic odds and ends that constitute every society, and as the principle that selects and shapes the bits and pieces that will be accepted there. (Castoriadis 1987, 143)

Castoriadis here seems to be proposing a social variant of Kantianism, using analogous arguments to those of Kant regarding the irreducibility of the schema of "triangle" to any particular mental image of a triangle.[3] I critically evaluate this proposal below. For now, employing Castoriadis' notion of the social imaginary, we can now say that what Goux (1990) identified as the rule of the general equivalent, as a general ordering principle manifesting itself in the sphere of symbolic economies, can be seen as having its roots in the *restructuring* of the social imaginary itself, in such a way that the "space" of the latter is made to conform to an omnipresent "image" of rationality—a rationality that is, as Castoriadis (1987, 156) points out, in its decontextualized self-sufficiency, a second-order Imaginary. In this respect, one can speak of the *colonization* of the social imaginary by the *logos*, such that imagination (radical, figurative, and social) becomes a kind of repressed Other, haunting reason's construction of the world in its own iconoclastic image.

## Metonymy, Writing, and the General Equivalent

The thesis that I shall, speculatively, advance in this section is twofold: (1) that what Lakoff (1987) identifies as the "classical" theory of categorization—a thing either belongs or does not belong, without residue, to a particular category—has as its historical (sociogenetic) condition of existence the emergence, in a complex of sociogenetic economies, of the rule of the general equivalent, and (2) that a major role in this sociogenetic process was played by the invention of the phonetic alphabet.

3. "No image would ever be adequate to the concept of a triangle in general. It would never attain that universality of the concept which renders it valid of all triangles ... it would always be limited to a part only of this sphere" (Kant 1929 [1787], 182).

Before proceeding further, let us explore further what is meant by "sociogenesis." The concept implies some sort of relationship between individual human development (ontogenesis), and the historical, socio-cultural circumstances in which individual human development occurs, and has occurred. In this context, we can refer, in the first place, to the issue of the socio-historical conditions of emergence of the *discourses* in which human cognition and its iconology are given shape and substance as an "object" of philosophy and science. In a later section, I examine theories of human psychological development, with respect to their implicit or explicit iconologies (their stance in iconological matters). There is, however, more to it than this, and I shall also try to step down a level, from the meta-discourse of iconology, to ask (in the final section) the question: in what way can we adequately understand the development, in human beings, of conceptualization as a psycho-social activity, and the role of both the "radical imaginary" and the "social imaginary" in this activity? These two questions can be linked by posing a third question, which is the focus of the present section: What is (or was, and is) the historical process which gave rise to the conceptualization of thought itself in terms of abstract, decontextualized general equivalence?

The difficulty with this question lies in the fact that it *crosses* the two different levels of possible analysis which are referred to above. Goux's analysis of the rule of the general equivalent attempts to show how it is that socially circulating representations (of language, of political economy, of the psyche) have some common point of genetic origin, historically locatable and traceable in the web of discourses and practices reaching down through the centuries. But we might go further, and suggest that the sociogenetic mechanism which (for example) elevates the phallus to the signifier (in the realm of the social imaginary) of the object of power/desire, is the very same mechanism which constitutes not only the subject of psychoanalytic discourse, but also *actual, individual* subjects.

The rule of the general equivalent is, if this is so, a sociogenetic mechanism whose effectivity is at the level both of the description and of the thing described. It is inscribed in our discourses, and also in that which the discourse purports to delineate. Psychoanalytic discourse then "becomes" the *appropriate* discourse for a phallocentrically constituted subject; its "ideal" structure mirrors those sociogenetic mechanisms which place the subject in a specific (patriarchally determined) relationship to power and desire.

The same point can, of course, be made in relation to the "mysterious" commodity-form, as analyzed by Marx (1975–2005 [1887]); it is both an imaginary (ideal) abstraction, and a sociogenetic process of *production* of real social relations. And the same point can also be made in relation to the postulate introduced by Saussure of the "arbitrariness" of the sign: was it meant as a methodological principle, or as a real characterization of a relation between two imaginary terms (signified and signifier), which, on Saussure's own account, cannot in actuality exist without each other?

These mysteries of reflexivity and level of discourse are especially recalcitrant in the case of *thought* (cognition) and its sociogenesis. For, as I have intimated, it is easiest to think of the rule of the general equivalent as constituting, in Western thought, the "idea of the idea": the representation of the representation of an essence which remains unchanged despite the flux of appearance. And this is, in itself, a challenging and bold thesis. But what if we were to make the thesis bolder still? What if we were to hypothesize that this very same sociogenetic mechanism (or complex of mechanisms) constituted or made possible thought itself, not just philosophy (thought about thought), but the *kind of thought* which is philosophy's object?

What if what we call (developed) human cognition is a product of a sociogenetic mechanism (the rule of the general equivalent), that is mediated by ontogenetic processes, and whose entry into human culture can be dated to around about the fifth century BC in the Ionian peninsula (and probably elsewhere, at around the same time)? What if this "developed" human cognition is not a universal of either biology or of culture, but a specific socio-historical acquisition? If this is so, it might help to explain why the "crisis" of the rule of the general equivalent is nowhere so acutely manifest and so laden with implications as in the science of cognition itself, deeply divided between "symbolic" and "post-symbolic" paradigms.

If, however, we are to think of a sociogenetic mechanism (such as the rule of the general equivalent) as having effects at different levels which are at the same time somehow reciprocally linked, we need some reflexive conceptual apparatus, and a vocabulary in which to express it. To attempt an expansion of this: only a certain kind of "thought" can permit a certain kind of "thought about thought," and that "thought about thought" constitutes the thought thought-about as that kind of thought; furthermore, both the thought thought-about and the thought about it are brought into existence by the self-same sociogenetic process.

To make this point more concrete, let us critically examine the thesis advanced by Goux (1990). This is, essentially, that the originary, and still operative, "fundamental configuration" of what we might call the "Western political economy of the social imaginary" is constituted by a constellation of institutions and their associated ideologies: namely, patriarchy, monotheism, monetary exchange, and logocentrism. Now, the problem with this formulation, as it is, is that it leaves unstated and unsolved the problem of the *cognitive* dimension of Objectivist logocentrism; by which I mean the notion of a decontextualised, univocal notion of meaning as residing in word and text, with formal logic as the paradigm against which expressions and meanings are to be analyzed and evaluated (see Rommetveit 1978; Marková 1981; Wertsch 1991).

Goux (1990) does, indeed, attempt to specify the sociogenetic *relationships* which bind together patriarchy, monotheism, the mediation of commodity exchange by money, and Mosaic and Protestant iconoclasm (the latter being, of course, linked with logocentrism), and I shall return to these relationships in the following section, in which I shall analyze the iconoclastic presuppositions to be found in the developmental psychologies of Freud and Piaget. Yet these relationships are best seen in terms of the mutual *motivation* of the different components of the constellation by each other, rather than in terms of specific sociogenetic (generative) processes permitting the historical emergence of any particular one of them. Since my concern here is (primarily) with cognition, and the iconology of "thought about thought," it is upon logocentrism that I wish to focus.

Specifically, I wish to bring into play another social institution and social practice, and suggest that it played a crucial role in making possible the logocentric stance towards meaning and rationality as univocal, determinate and decontextualized. That institution and practice, I suggest, was phonetic-alphabetic writing, and my proposal is that this institution and practice made possible new ways of conceiving of categorization, in terms of an opposition between "essence" and "appearance" and in terms of the law of the excluded middle. In short, I am proposing that the invention of the phonetic alphabet was the primary sociogenetic mechanism making possible the "classical" theory of categorization, and thus in introducing the cognitive prerequisites for the "capture" of the Western social imaginary by the rule of the general equivalent.

As should be clear from the foregoing, I am not suggesting that any single mechanism was solely responsible for the sociogenetic institution of the rule of the general equivalent, which cannot have been a single "event," and which should also in any case be thought of in terms of tendencies and relative dominances, rather than as some absolutely all-governing principle. Rather, what I want to examine here is the role played by the invention of the phonetic alphabet in constituting one of the conditions of possibility—through notational externalization—of the concept of generalized equivalence, with specific cognitive and meta-cognitive properties and consequences.

It is important to note here that it is a particular *kind* or *mode* of writing (phonetic-alphabetic), rather than writing in general, that I am proposing as being implicated in the sociogenesis of the concept of generalized equivalence. However, I shall first very briefly discuss the role of writing in general, in both cognitive and monetary economies.

Olson (1977), Scribner and Cole (1981), and others have drawn attention to the possibility that literacy, in certain particular socio-functional contexts, may induce cognitive effects at the level of metalinguistic and metacognitive processes. These effects may be seen as the inducement of a shift towards "decontextualized" or "abstract" thinking. What I shall suggest is something similar, but with a focus on the congruence between a particular cognitive figural process, and the figurative bias underlying abstract logical reasoning. This bias, as Rotman (1988) and Walkerdine (1988) have pointed out with respect to mathematical and other kinds of formal reasoning, is *metonymic* in character.

The advantage of this formulation is that it recognizes that "abstraction" and the "theoretical attitude" are realized, or made manifest, in both the metaphoric and the metonymic dimensions of language and representation (Sinha 1988). The studies of Olson (1977), Cole (1981), and others have tended to see the issue of "abstraction" as one involving semantic and propositional content: that is, as a process whereby semantic content becomes decontextualized along a *metaphoric* dimension. In contrast, I shall focus here upon the metonymic ("part-whole" or "container-contained") figure, without commitment as to whether writing in general, or alphabetic writing in particular, also induces a kind of textual-semantic "abstract cognitive attitude," of the kind conveyed by Olson's notion of a historic shift, attributable to Protestantism and the rise of mercantile capitalism, from a notion of "meaning as mediated by authority," to a notion of "meaning as solely residing in the text." I shall, however, have something to say about the sociogenetic mechanism whereby alphabetic writing potentiates a notion of abstract or formal *truth*, which is of course intimately connected with the notion of decontextualized propositional or cognitive content. The further exploration of the interactions between metonymic and metaphoric modes of "abstraction," and its relation to writing, lies beyond the scope of this chapter.

There is no question that language-based graphic notation has its roots very far back in human prehistory, perhaps as far back as language itself (Harris 1986). It is also clear that, from the very first, graphic notations of language, that is, writing, played a significant role in ordering and regulating economic transactions, of tribute, trade, and so forth, and in economic activity (e.g., stock-taking and accountancy) in general. The invention of the phonetic alphabet quite definitely post-dates the general use of writing for recording economic information, and I do not wish to suggest that its effects on economic systems (e.g., the emergence of commodity exchange and monetary systems) were direct. Rather, it is more plausible to see the emergence of monetary value-equivalence (monetary token-equivalence), and the emergence of the phonetic alphabet, involving graphic token-equivalence, as being mutually reinforcing processes in the sociogenesis of the rule of the general equivalent.

Let us now turn to the specific role of phonetic-alphabetic writing in the externalization of the notion of generalized equivalence, and the way that this frames meaning and value in such a way that the classical, abstract "concept of the concept" becomes possible. We can begin by distinguishing three general classes of metonymic relations, namely:

*Category – Member*

*Substance – Sample*

*Type – Token*

I want to emphasize that these relations do not map in any straightforward way onto each other, even though each of them does imply some kind of equivalence. Thus, all members of a category are equivalent, inasmuch as they may be named by the term for the category. However, as Rosch (1977), Lakoff (1987) and others have pointed out, this does not mean that all category members are equivalent in their "goodness," "centrality" or "typicality." Category membership is (in natural languages) frequently both vague, and contestable or negotiable in its boundaries.

Similarly, samples of a substance exemplify the substance, and are equivalent in that respect, but some samples may be better, purer, and more "representative" than others. Finally, tokens which instantiate some value are all equivalently tokens of that value (be it a monetary or a semiotic value), but their empirical or perceptible form may vary: the phonetic realization of phonemes being a crucial and well-known case, but we may think also of coins, banknotes, and so forth.

Thus, all these three metonymic relations involve some kind of equivalence relation obtained between the elements comprising that which is "contained," and a metonymic "container." The "container" can be seen as precisely the "abstraction" that "represents" and constitutes the contained as being equivalent in some way. Yet the relations are different: a member of a category is not a sample, since a category is not a substance, and vice-versa. The type-token relationship does, it is true, share properties of each of the other relations. A type can be thought of as a category with an infinite (potential) membership; and tokens may be thought of as sampling either a concrete substance (as in gold coinage) or an abstract substance, as for example the "content" level of Hjelmslev's version of structuralist linguistics (Hjelmslev 1953).

At the same time, the type-token relation exemplifies most clearly the notion of "abstract equivalence," an equivalence which is governed by a *value* which is "beyond" the phenomenal form of the token. This value is imaginary (it is constituted in and as representation); but it has real effects in constituting and structuring the real, and the perception of the real. Value, as constituted in type-token relations, is constituted in and through the social imaginary.

The category-member and the sample-substance relationships also exemplify (or embody) the social imaginary, since categorization is, in a sense, the world as it is structured and composed for any given language and culture. Categorization processes also impose themselves on, and are also partly shaped by, the "substances" which make up the natural (and "spiritual" or "mental") worlds. These two metonymic relations, and their interplay, govern to a large degree the experienced world of any given culture.

A large body of research supports the conclusion that categorization, while being highly culturally variable, is not arbitrary: it is partly determined by properties both of the natural world and of the psychophysiology of human perception. In their ecological situatedness, categorization processes (including those governing conceptions of substance) are *socio-*

*natural* in character; they are Janus-faced, facing both nature and culture, and embodying the co-articulation of nature and culture in specific cultural settings.

The type-token relationship is different. In categorization, "substance" partially lends form to its representation: the category member is an *instantiation* of a category which is, as it were, "marked," or informed, by the properties of its most representative members. In contrast, in the type-token relationship the substance is a *realization* of the abstract form: the token, as "substance," is informed by the purely abstract type. Of course, it is also the case that types do not emerge arbitrarily. The phonological systems of languages find their conditions of possibility in the physiology of the human vocal tract. But this shows only that types can also be considered as categories, subject to the same kind of socio-naturalistic double determination as any other category. What concerns me here is not the type considered as a category, but the type-token relation as such, which is different from the category-member relation, while depending for its operation upon it.

The type-token relation is *signification* in its "purest" form, and is *irreducible* to any naturalistic plane (this irreducibility is constitutive, of course, of precisely the difference between phonology, which is part of linguistics, and phonetics, which is part of the natural sciences.) We can add here, too, that although the type-token relationship depends upon categorization processes, it is also the case that the manifestation of categorization in human language depends upon the type-token relationship. Without language, which operates as a system of types, there can be no linguistic categorization and no linguistic semantics, no metaphor and no cultural meaning.

It is for this reason that I suggest that the type-token relationship is crucial for an understanding of the concept of abstract, general equivalence, and how it came to achieve the status of a dominant structuring principle in the social imaginary. My hypothesis is that the invention of the phonetic alphabet, which *externalized* in a notational system the type-token relationship, made it possible to consider equivalence as an abstract and general "rule." That is to say, whereas before the phonetic alphabet, the type-token relation was merely *immanent* in signification and its products, the phonetic alphabet makes this relationship *manifest* and graspable as a *principle* with diverse and powerful applications. And this principle is precisely the principle of abstract general equivalence, or the concept of the concept. The new departure made possible by the phonetic alphabet is the externalization in notation of the mediation of meaning by abstraction. Once made manifest, the notion of abstract general equivalence can be generalized, over language and over concepts of mind and reality, yielding a radical restructuring of the social imaginary.

The mode of this generalization, I suggest, consists, in the first instance, of the transposition of the concept of general equivalence to the two other metonymic relations which I described above. This transposition yields a transformation of the governing relations between "container" and "contained," such that the "container," rather than being a sign or representation which signifies a "place" for the generality of its "contents," becomes rather a function defining an extensional set. Through this transformation, the (initially distinct) category-member and substance-sample relation become co-articulated and mutually definitive, in the form of the following "abstract" relationships:

*Set – Member of set*

*Concept – Instance of concept*

The co-articulation of these relationships yields the fundamental equivalence mapping which is presupposed by a truth-valued (bivalent) logic:

[set] = {members}

Predicate [Arguments]

Where the first notation expresses an *extensional*, and the second notation expresses an *intensional*, logical semantics. To put it another way, the co-ordination of the transformed metonymic relations stipulates that a category is defined as an extensional set, each of whose members instantiates a property or set of properties expressible as intensional predicates. This co-ordination *constructs* and *defines* the "abstract objectivist" notion of truth.[4]

In summary, I have suggested in this section that the semiotic transformation of metonymic relations, in terms of a concept of general equivalence, derived, via its externalization in the phonetic alphabet, from the type-token relation as "pure signification." This, I suggest, was a specific sociogenetic mechanism enabling the re-constitution of "thought about thought" within an abstract-objectivist, or logocentric, framework. This sociogenetic transformation also constituted a radical restructuring of the social imaginary, by virtue of which the rule of the general equivalent was constructed as extensible over the entire domain of the social, and by means of which, also, its figurative-imaginative "origin" was "repressed" and re-constituted as the Unconscious Other of language, law, and logic.

## Iconology and Developmental Theory

In this section I examine the iconological stances of the most influential classical developmental psychological theories of the twentieth century: those of Sigmund Freud and Jean Piaget. This examination will reveal a striking degree of unanimity: developmental psychology, since its inception, has been unrelentingly iconoclastic and logocentric.

The concept of a *developmental* or *genetic* psychology had, as many commentators have pointed out, its principal roots in evolutionary and sociobiological theories (Sulloway 1979). The evolutionary biological matrix within which developmental psychology took shape was also, from the very beginning, intertwined with the notion of a depth psychology, in which semiotic and iconological transformations mediate between "primitive" and "civilized" psychic layers. The linkage between evolutionary depth, and iconological depth, is expressed in Freudian psychoanalysis in the distinction between "latent" and "manifest" mental content.

For the greater part of both the nineteenth and the twentieth centuries, the dominant theory in experimental psychology was associationism. In an associationist framework, there is no principled *difference* between the "verbal" and the "imagistic" aspects of thought: they are conceived of as simply different kinds of presentation of the same atomistic content. The mechanism of thought—the association of ideas—operates on this content indifferently to its particular mode of presentation in consciousness, and that which remains unconscious is a kind of shadow, which parallels those trains of associations that for one or another reason do break through the threshold of consciousness. Disputes about the possibility of, for example, "imageless thought," did not involve any important departure, on either side, from this model, since what is held to be at question is not whether images and words are essentially different "kinds," but whether the one can be present to consciousness without the other.

---

4. Proposals that natural kind terms should be seen as paradigmatic of meaning (e.g., Kripke 1980) represent an attempt to resolve the current "crisis" in the theory of meaning by formulating the substance-sample relation as foundational: see also Putnam (1981).

Freud's theory of dreams and the unconscious differs profoundly from any such assumption that image and word are merely alternative modes of presentation of an identical content. When a psychoanalyst like Lacan (1966) maintains that the unconscious is "structured like a language," this cannot mean that the latent content of the unconscious somehow obeys a syntax which is akin to that of a natural language, for Freud emphasized continually that the unconscious does not *know* logic, or time, or syntactic ordering. It is not so much the unconscious *itself* that is structured like a language, as the operations by means of which unconscious contents, forced into self-dissimulation by censorships and prohibitions, are articulated and ordered as signifiers at the level of manifest content—the dreamwork and the processes of symptom formation.

The world of the dream may also violate logic and time, but it does so in a way that can be understood as having a "logic" of its own, in which "the symbols and the temporal sequences used are based on contiguity (Freud's metonymic 'displacement' and synecdochic 'condensation') or on [metaphoric] similarity (Freud's 'identification and symbolism')" (Jakobson 1956, 81). The world of the dream is a *partially "logicized"* representative of another world, the actual unconscious, which is entirely other to the logos. The unconscious is not a set of propositions awaiting transformation by the dreamwork, nor even a set of propositions lacking temporal, modal or other inflection or qualification. Nothing like a proposition "exists" in the unconscious, and for this reason it is even misleading to speak of unconscious desires, for desire is (or can be formulated as) a propositional attitude, and as such is distinct from the unconscious "investments" (in a pre-symbolic "economy") that power them.

What then is the content of the unconscious? Freud speaks of *Trieb* (drive, pulsion), but drives must have an object, and the "place" of this object is the *imagination.* The unconscious is the realm of the imaginary, of fantasy, and of the *body imagined.* And it is here that Goux's remarkable analysis of Freud's text on Michelangelo's statue of Moses becomes relevant. Freud, he says, "was unable to interpret the Moses completely without analyzing analysis, that is, without analyzing the cultural presuppositions of his discovery and, somewhere along the line, denouncing, perhaps we had better say, 'disclosing' the roots of psychoanalysis in Judaism" (Goux 1990, 144); and specifically in the link "between the Judaic prohibition against worshipping images and the prohibition against incest with the mother" (Goux 1990, 137). Goux continues:

> What Moses condemns when he castigates the idolaters is actually, in a historical perspective, the cult of female, maternal deities, and the fertile incest rites accompanying this cult. The law given to the Jewish people by Moses...is first of all a radical prohibition not only against incest itself but against any incest with the mother in the *imaginary*.... With his written law, Moses institutes a blindness with regard to image, imagination and phantasy —all of which are associated with sensuality and love for the mother. Thus he brings to completion the exodus from Egypt—not merely out of the land that bears this name, but also out of the inner Egypt, the land of sphinxes, tombs, and hieroglyphics, land of the imagination and of its invasive icons.... From this perspective, we might say that Freud's psychoanalysis constitutes a second exodus from Egypt, for it proclaims that image and phantasy are delusion and that behind the seductive draperies of the imaginary, an enduring symbolic structure homologous to a language must be discovered. Psychoanalysis seeks not to interpret the image with another image but to dissolve it, to reduce it, to repress its welling forth by translating it into language. (1990, 138–139)

But there is something strange here. For, as I have already said, we cannot seriously maintain that "beyond" the manifest level of dream-content—the "Egypt" to which we can travel in our imagination—is a *language*. No, "beyond" that is only the unconscious, another Egypt, which is governed by an imagination (not a law) that *precedes* "subjection in" language and submission to the patriarchal law. The only sense in which a language, or anything like it, lies "beyond" the imaginary is the sense in which the journey from latent to manifest content can be described as a *kind of* translation, subject to certain modes of quasi-linguistic operations, whose dynamics can be understood only on condition that certain prohibitions have themselves *taken seat in the unconscious*. And *this* condition, the condition of *repression*, is also the precondition for psychoanalysis itself, the double condition which makes it possible for Goux to speak, apparently so strangely, of psychoanalysis itself "repressing" the imaginary, by repressing its own condition for existence as Freud's "science of the talking cure."

What then is *repressed*? Not just the imaginary, perhaps not even the imaginary as such, because despite Freud's deep suspicion of it, he of course recognizes it as a place or *topos*, while seeking to subject it to language, in practice as well as in theory. What is repressed is rather the recognition that the imaginary is the other, not just to language, but also to the law of the father: it is the place of the mother. As Goux says, in Judaism:

> The prohibition against figural representation amounts to a violent, radical rejection of the maternal. The Jewish tradition thus develops as a systematic exclusion of all metaphysical images of the mother. Conversely, the corollary of this rejection is the importance attributed to the letter, to writing, to alphabetic signifiers. Moses' scribes cannot use hieroglyphics without breaking the law. Rid of its iconic dimension, the symbolic will tend to be reduced to the articulation of scriptural strokes. The utter extirpation of the prolific imaginary, in favor of the agency of the letter, effected by those who followed in Freud's footsteps, is without doubt fully in keeping with this violent devaluation of the imaginary by a writing that decrees law and name, beginning with Mosaic iconoclasm. (1990, 145)

Let us now attempt a summary of the above. Freudian psychoanalysis is a depth psychology in which the imaginary is a *topos* other to language: images do not accompany words or constitute a complementary mode of presentation of the same content as that expressed in language, but are representatives of a deeper, more "primitive" layer of the psyche. The imaginary does not ground language, but subverts it, and must be checked by *rendering it in* and *subordinating it to* language. Psychoanalytic theory is fundamentally iconoclastic, and this iconoclasm is sociogenetically related to the rejection, expulsion, and repression of the maternal and the material. The devaluation of the image of the mother, and of sensuous materiality, is coupled with the domination of patriarchal law and language as representatives of both spirituality and reason.

The subject of psychoanalysis (both the actual, individual subject/patient and the topic or *topos*) is thus constituted not only *in* language, but also *against* the imaginary, from whose bondage science must release him/her/it. This emancipation (or exodus) is simultaneously a recognition of, and subjection to, the law, which must (on pain of neurosis or of the loss of one's identity) be inscribed or introjected in the structures of the Unconscious itself, as operations which simultaneously enable repression and render that which is repressed intelligible. Freudian psychoanalysis is a *rationalization* (in every sense, including the psychoanalytic one) of a sociogenetic process of the production of a specific type of subjectivity, in specific historical circumstances.

Furthermore, the iconoclasm of psychoanalysis mirrors the iconoclasm accompanying the "exodus" from the land of the glyph to the land of the alphabet. According to Harris, "the modern consensus view [of the origin of alphabetic writing] favours the North Semitic alphabet as the earliest known form and dates its appearance to the first half of the second millenium BC" (1986, 31). Alphabetic writing, then, may have been not only a cognitive technology favoring and permitting the elaboration of an abstract-objectivist notion of truth, coupled with the concept of the concept, but also a means of preserving a written record whose form of expression, in being stripped of iconicity, makes it uniquely appropriate for the revelation of the One Truth of the One God in a sacred text.

Alphabetic writing was, if this speculative, sociogenetic story has any validity, the cognitive-mechanical link between two different aspects of the notion of univocal meaning, which has played such a central role in Western thought: meaning as *logically* connected to truth conditions on single sentences, and meaning as *authoritatively* inhering in textually-connected, written sentences. The alphabet *represents* the binding together of the different strands of sociogenetic process—patriarchy, monotheism, the monetary economy, and writing—eventuating in the rule of the general equivalent; and the elevation of a logocentric, abstract-objectivist Reason to a position of dominance and exclusion over and against Imagination. As we have seen, the opposition between Reason and Imagination (the latter seen as primitive, dangerous, unruly, and feminine) is enshrined in the theory and practice of Freudian psychoanalysis.

I want now to suggest that the same iconoclasm is also to be discerned in the developmental theory of Jean Piaget, albeit in a different manner. In Piaget's work (e.g. Piaget 1945) we encounter what seems at first sight a paradox: although Piaget is dismissive of, and accords an essentially conservative (non-progressive) role to imagination and to what he terms figurative thought, he does not, as we might then expect, valorize language over figuration. On the contrary, both language and figurative thought are considered to be instances of a unitary "semiotic function," which can be understood as that dimension of cognition that comes into being with the advent of *mental representation*. Representation is a literally ambivalent (double-valued) concept in Piagetian theory. First, it denotes the ability to mentally evoke absent realities, that is to say, imagination. Second, it denotes *symbolization*, by which Piaget understood not a relation between a sign and an object, but a relation between a (figurative) signifier and a schema; and a schema, while having a figurative (indexical/iconic) component, is principally defined by (or grounded in) actions and (later) logico-mathematical operations.

Representation, for Piaget, is certainly a crucial aspect of human cognition, its advent being coterminous with the acquisition of full object permanence, and constituting what Piaget terms (echoing, no doubt consciously, Kant's phraseology) a "Copernican revolution" in the thought of the child. However, although representation is for Piaget a necessary condition for the development from pre-operational to operational thought, it is neither in itself a sufficient condition, nor the principal cause and motor of that development. This latter role is fulfilled by *action*, and the co-ordination of actions through their government by operative structures. These structures lie "beyond" figuration, and constitute the conditions of possibility for the "logicization" of representation.

In the structure of Piaget's theory of representation, we can discern striking similarities to the structure of psychoanalytic theory. Piaget's theory, like Freud's, is a "depth" psychology, in which the "manifest" content of representation (figuration and symbolization) stands in a relation of rule-governed (language-like) dependence to a deeper structure which is *not* language, but which can only be *rendered* in a language-like, symbolic notation. This deep structure, for Piaget, is the *logos* in its purified form, stripped of the figurative and imagistic contaminations of actual, natural languages.

Natural languages, for Piaget, are doubly removed from the "operative" locus of pure reason: first, in that they depend upon sensible figures (the signifiers), and second in that these signifiers, being genetically related to the imagination, are mediated by the imagination in their relation to the operative structures. For this reason, natural languages can only imperfectly reflect and approximate to the "ideal" notation of a logical language (a notation that Piaget variously sought in the work of "Bourbaki," in the propositional calculus, and in the theory of groups). In virtue of this imperfection and contamination, natural languages cannot be accorded more than a peripheral role in the developmental unfolding of the Logos immanent in the relation between the epistemic subject and the real-rational world that it constructs.

The quest for an ideal, logical, universal language is, of course, a familiar trope of Western philosophy, from Leibniz through to logical positivism. But I want to suggest that it is also one manifestation of an equally deep-seated iconoclasm which can be traced back at least as far as the Protestant (and especially English Puritan) concern to expunge rhetoric and "fanciful" imagery from scientific and scholarly discourse. Piaget was, as is evident from his early writings, a thinker with a deeply religious view of the mission of science and philosophy, a view suffused with an idealistic and humanist Protestantism: not so much a harsh, predestinarian Calvinism, as a vision of the harmonious and dialectical unfolding of a nature which is ultimately and immanently identified with reason.

Like Freud with Judaism, Piaget inherited from his own intellectual and religious background a profound iconoclasm, an iconoclasm that extended to *language itself*, which he saw as an imperfect vessel of reason. And, like Freud, Piaget can be seen as the constructor of a psychology from which the Mother, and desire for the Mother, has been expelled—not just literally, in Piaget's neglect of emotion and of the early socio-communicative environment of the infant, but also metaphorically, in his mistrust and neglect of the sensuous *materiality* of both representation, and the reality that representation informs (see Sinha 1988, on the notion of the "materiality of representation"). Piaget's patriarchal, masculinist, Apollonian psychology is based upon a denial and a refusal (itself a characteristically Protestant theme) of *mediation*; his epistemic subject unfolds in monadic self-sufficiency towards a rationality immanent in, and guaranteed by, a law that is beyond imagination. For Piaget, as for Freud, this "beyond" of imagination is also "above," commanding imagination and engraving its authority within it.

In summary, Freud and Piaget both view the imagination as being other to rationality: the imaginary is identified with primitivity, with the (desire for the) maternal and the feminine, and with egocentrism, and rationality is identified with the reality principle, with the law of the father, and with operational thinking. What I have elsewhere (Sinha 1988) called the "phylocultural complex," the mapping of notions of childhood and primitivity onto each other within an evolutionary framework, can in many respects be read as the principle theoretical expression of an iconoclasm that has marked developmental psychological theory since its inception.

## Ontogenesis and the Social Imaginary

The present article is by no means the first to link Freudian and Piagetian developmental theories by employing Castoriadis' notions of the social and the radical imaginary. Hans

Furth (1990) identifies the ontogenetic origin of the radical imaginary with the emergence, at around 2 years of age, of the capacity for symbolic ("pretend") play and mental imagery.

Cognitively (and following Piaget), Furth proposes that this capacity is due to the emergence of a representational ability, predicated upon the differentiation of the object from specific sensori-motor action structures. Affectively (and following Freud), he suggests that this same capacity involves a libidinal investment in the object, dissociated from any biologically-given reproductive function.

> These two ruptures in biological adaptation form the indispensable core elements of the radical imaginary and explain why adaptation is an inadequate conception for the open-ended constructiveness of human psychology. This constructive power requires the logical comprehension of the separated object and...the motivational freedom of a libido that is not instinctually tied to its biological purpose. From this perspective, the meaning of an image (or of pretending) is the mental object to which the child's libido is attached: "want-my-object," where object stands for "my world, my society." (Furth 1990, 208–209).

Like Castoriadis, Furth (1990) sees the radical imaginary as the source of the social imaginary, inasmuch as the cognitive-affective significations that constitute the play of radical imagination are *social* significations: the ultimate object of imaginary desire and knowledge is the social world or surroundings that lend meaning to the child's actions, even as the child's imagination projects personal meaning onto the world. The dialectic of development involves a co-constructive relationship between the radical imaginary (the source of intellectual and social transformation), and the specific, historically given social significations (or symbolic order) that constrain the field of play of the radical imaginary.

> The radical imaginary, as a particular developmental creation, is peculiar to the one individual, while as a particular historical creation, it is peculiar to the particular society. In the first case we see society within the individual, in the second the individual within society. (Furth 1990, 212)

It seems at first sight, then, that Furth is departing from the iconoclasm that I have argued to be central to both Piagetian and Freudian developmental theories, and that, like Lakoff (1987) and Johnson (1987), he is arguing that the imagination should be seen as being at the core of human mental life. However, Furth explicitly reintroduces a logocentric and iconoclastic dimension to his theory, in his discussion of the child's development from the age of about 7 years (that is, the age period characterized by Piaget in terms of "concrete operations," and by Freud in terms of "latency"). Furth writes: "children must first construct a play reality before awareness of social, logical and physical constraints leads them to make the adult distinction between pretend and serious...around age 7, at which time the play reality of children is safely conserved in their 'unconscious.'" (207). The telling notion of "safe repression" recurs in Furth's text:

> After [7 years], logical operations reach a critical closure.... Faced then with the internal demands of conscious logic and the external pressures of social reality, children by this time have completed the psychological repression of their pretend world...in the early years serious learning is secondary and is in fact constantly distorted in the service of the child's imaginary. But after its safe repression and unconscious conservation, learning and cooperation take on an increasingly serious, nonplayful character...having more or less willingly repressed their imaginary worlds, [children] are open to the impact of the "real"

> social world...progress in logical and societal understanding in the few years following age 7 is truly phenomenal and easily observable in all cultures. (Furth 1990, 210–211)

Given our previous discussion of the sociogenesis of the rule of the general equivalent, it is natural to read Furth's theoretical approach as involving a translation of the processes which I have designated as a historical transformation of the structure of the social imaginary, onto the timescale of ontogenetic development (at least within Western cultures). Just as (I have proposed) the imagination is positioned by logocentric, univocal reason as an excluded and repressed Other, so within the (Western) individual (Furth proposes), the radical imaginary is "safely" repressed as an "indelible core of human society and each person within it" (1990, 210). Behind all adult achievements are unconscious forces which have been gathered and structured in childhood. Because "These forces are pregnant with what is considered the best [creative imagination, etc.] as well the worst [ambivalence, frustration, delusion, etc.]," (1990, 212) their "safe" socio-psychological repression is a *necessary* condition for the constructive growth of the social individual.

There several critical remarks that I wish to address to Furth's developmental theory. The first concerns his apparent obliviousness to the culture-specific nature of his developmental story. Gustav Jahoda, in his critical commentary appended to Furth's article, notes that "the picture drawn by Furth of the life of the young child ... is EurAmerico-centric" (215). The elevation of the culturally specific to the status of universal has been criticized in the work of both Freud and Piaget, and Furth's theoretical synthesis of the two fails to break with this weakness. Secondly, Furth's account, far from abandoning the traditional iconoclasm of developmental theory, merely repeats its central presupposition: Imagination represents, once again, a "primitive" stage which must be simultaneously abandoned and (unconsciously) preserved, in the growth towards mature reason. Logic, and awareness of its objectivizing constraints, are equated with both social stability and psychic health.

In this respect, Furth's account is a normative generalization of Western views of the relations between reason and imagination, over all places and times. This is not to deny that this normative account does indeed express a truth of our culture's social practices of education and socialization of children, inasmuch as these practices are themselves expressive and reproductive of the symbolic order and social imaginary of modern Western society. Furthermore, in so far as this society "functions properly," Furth's account also expresses a truth of child development itself. Yet it is, at best, a half-truth, since it denies and excludes the *constitutive* and continuing role of embodiment and imagination in human cognition.

In a sense, because Furth wishes, on the one hand, to *recognize* the centrality of the imaginary in human development, and on the other hand, to *preserve* the iconoclastic premises upon which Piaget and Freud erected their theories, his own text can be seen as a deeply ambivalent expression of the iconological crisis of the contemporary cognitive sciences, riven by the split between a "Classical", rules-and-symbols based, logocentric paradigm and a "New Romantic" (Sinha 1993), image-schematic, connectionist, situated embodied paradigm (Sinha 2007; Zlatev 1997).

At a more specific level of ontogenetic theorizing, Furth counterposes (as did Piaget) pretend or symbolic play (considered by Piaget to be dominated by assimilation) to logico-mathematical reasoning; and the pretend or imaginary social world to the "real" social world. In contrast with this dualistic view, Sinha (2005) argues that pretend play depends on the ontogenetic development of the cognitive capacity for what Fauconnier and Turner (2002) call conceptual integration, or blending. Far from being a "primitive" stage repressed in later development, Fauconnier and Turner demonstrate the centrality and ubiquity of conceptual blending in the everyday construction of novelty in language and cognition. Sinha (2005) also argues (following Vygotsky 1986 [1934]), that pretend play, far from involving a negation of the real social world, provides a tool for its exploration and, in particular, for the exploration of the negotiability of norms and the construction of identity.

Despite their dialectical emphasis on the mutuality of the "radical" and the "social" imaginaries, both Castoriadis and Furth seem to reproduce in their accounts the dualism, fundamental to the Western intellectual tradition, between the individual and the social, seeing the former as both the condition and the antithesis of the latter. I remarked above on the neo-Kantian structure of Castoriadis' argument against the reducibility of the social imaginary to individual representations. Perhaps this argument appeals to Furth, an eminent Piagetian, precisely because it is analogous to Piaget's distinction between the Kantian-inspired "epistemic subject," with its necessarily universal developmental stages, and the actual, historical, culturally and inter-individually variable psychological subject. Castoriadis, Furth and Piaget all follow Kant in considering the proper object of science to be not the messy, empirical and historically specific subject, but the transcendent subject that grounds all possible cultural modes of subjectivity.

This is not the place to explore an alternative account of human cognitive, linguistic, and communicative development. It is possible, however, to sketch some key features of such an alternative. First, the notion of "object" needs to be recovered from the anonymity of abstraction, whether in its role as the "general equivalent" of traditional epistemology, or in its Freudian guise as a mere screen for the projection of libidinous impulses. A more adequate conception of the object, and its role in development, addresses the social significations that are materially instantiated in its use-values (Sinha 1988, 2005; Sinha and Rodriguez 2008). Second, the constitution of subjectivity in language, discourse and narrative needs to be conceptualized as a *part of*, and not merely a consequence of, the process of acquisition and development of language (Bruner 1990; Sinha 2000). Third, and perhaps most fundamentally in relation to the theories of Castoriadis and Furth, the separation between the "primitive," "individual" radical or root imaginary, and the "developed" social imaginary, needs to be challenged with the recognition that Piaget's "semiotic function" is social from the very beginning.

The development of the semiotic function is not, at any stage, a purely intra-individual affair. The discovery and exploration of meaning (and of the acts that constitute meaningful communication) involves participation in normatively framed intersubjective and "inter-objective" engagements. It is this framing, not an "in itself" quality of signs, or an intrinsic distinction between signifier and signified, that lends to certain actions the status of "counting as" acts of meaning (Bruner 1990; Searle 1995). Thus, the radical imaginary *is* the social imaginary, the foundational source not only of the Vygotskian semiotic mediation of individual cognition, but also of the construction of all social institutions and social objects.

Such an alternative account of human development is both necessary and possible, and its construction could contribute to both a theoretical and a practical emancipation of imaginative reason from the Objectivist straitjacket of logocentric, univocal rationality that currently constrains and discursively "represses" it. Furth, as we have seen, regards "repression" (that is, the repression of the imaginary) as being both a psychological and a social process. He remarks cryptically:

> Personal repression was described in Freudian terms as the conservation of the radical imaginary in a person's psychological structure. But insofar as this becomes the structure to which adult society is presently assimilated, it is clear that social repression in the Marxian sense is already at work in that primary repression. Personal and social unconscious are in fact two sides of the same human situation. (1990, 212)

Furth goes on to reject the possibility of "a utopian social situation that would have no repression, no alienation nor false consciousness, even as there is no person without these characteristics" (1990, 212). No doubt this latter, as a generality, is true. No doubt, either, that it is a sentiment in tune with our time, when we have become wary of Utopias as a result of historical experience of where attempts to realize them have actually led to.

The problem, however, is that Furth seems to locate the "danger" of Utopias exactly where iconoclasts have always seen danger: in the imagination, and (presumably) particularly in its destructive and aggressive aspect (perhaps here Furth is taking inspiration from Melanie Klein as well as from Freud). Yet it is arguable (and has often been argued) that the totalitarian nightmares that motivate Furth's rejection of the utopian imagination are grounded, not in the imaginary as such, but in the *rationalization* (bureaucratic, logocentric, and univocal) of the social imaginary that so largely characterizes modern, Western societies—including their dystopian nightmares as well as their utopian dreams.

If it is true (and I certainly don't wish to quarrel with this) that reason is a bulwark of sanity against the not-so-imaginary dangers posed by the resurgence of both nationalist and religious ideologies of "imagined community" (Anderson 1991), it is also true that Objectivist reason and its socio-technological concomitants have, while professing universality, systematically excluded, denied, and repressed the claims of human communities that "fail" to live up to its own image of what it is to be rational and civilized. None of us, either, can fail to be aware of the danger in which the unremitting exploitation of the natural world places our, and other, species.

There is a temptation to say that "rationality" is part of the problem, not the solution. We need not, however succumb to this temptation, if we are able to recognize that the problem with the dominant mode of rationality lies not so much in its internal normative machinery, as in the *exclusions* that, in the guise of Objectivism, it exercises on the claims of imagination, empathy, embodiment and community. To reassert the ethical, social and political primacy of these claims is not to indulge a Utopian fantasy. It is a matter of survival in our current, global historical predicament.

## References


Anderson, B. 1991. *Imagined Communities*. London: Verso.

Bruner, J.S. 1990. *Acts of Meaning*. Cambridge. MA: Harvard University Press.

Castoriadis, C. 1987. *The Imaginary Institution of Society*. Cambridge: Polity Press.

Fauconnier, G. and M. Turner. 2002. *The Way We Think*. New York: Basic Books.

Furth, H.G. 1990. "The 'Radical Imaginary' Underlying Social Institutions: Its Developmental Base." *Human Development* 33: 202–213.

Goux, J.J. 1990. *Symbolic Economies*. Ithaca, NY: Cornell University Press.

Harris, R. 1986. *The Origin of Writing*. London: Duckworth.

Hjelmslev, L. 1953. *Prolegomena to a Theory of Language*. Bloomington: Indiana University Press.

Jakobson, R. 1956. "Two Aspects of Language and Two Types of Aphasic Disturbance." In *Funda-*

*mentals of Language*, by R. Jakobson and M. Halle, 115–133. The Hague: Mouton.

Johnson, M. 1987. *The Body in the Mind: The Bodily Basis of Meaning, Imagination and Reason*. Chicago, IL: Chicago University Press.

Kant, I. 1929. *Critique of Pure Reason*. London: Macmillan.

Kripke, S.A. 1980. *Naming and Necessity*. Oxford: Blackwell.

Lacan, J. 1966. *Ecrits*. Paris: Editions due Seuil.

Lakoff, G. 1987. *Women, Fire and Dangerous Things: What Categories Reveal about the Mind*. Chicago, IL: Chicago University Press.

Lakoff, G. and M. Johnson. 1999. *Philosophy in the Flesh: The Embodied Mind and Its Challenge to Western Thought*. New York: Basic Books.

Marková, I. 1982. *Paradigms, Thought and Language*. Chichester: Wiley.

Marx, K. 1975–2005. *Capital*. Vol. 1. Collected Works of Karl Marx and Friedrich Engels, vol. 35. London: Lawrence and Wishart.

Merleau-Ponty, M. 1964. *Signs*. Evanston, IL: Northwestern University Press.

Mitchell, W.J.T. 1986. *Iconology: Image, Text, Ideology*. Chicago, IL: Chicago University Press.

Olson, D. 1977. "From Utterance to Text: The Bias of Language in Speech and Writing." *Harvard Educational Review* 47: 257–281.

Panofsky, E. 1939. *Studies in Iconology*. Oxford: Oxford University Press.

Piaget, J. 1945. *La Formation du symbole chez l'enfant*. Neuchatel: Delachaux and Niestlé.

Putnam, H. 1981. *Reason, Truth and History*. Cambridge: Cambridge University Press.

Rosch, E. 1977. "Classification of Real-World Objects: Origins and Representations in Cognition." In *Thinking: Readings in Cognitive Science*, edited by P.N. Johnson-Laird and P.C. Wason, 212–222. Cambridge: Cambridge University Press.

Rommetveit, R. 1978. "On Piagetian Cognitive Operations, Semantic Competence, and Message Structure in Adult-Child Communication." In *The Social Context of Language*, edited by I. Marková, 113–150. Chichester: John Wiley and Sons.

Rotman, B. 1988. "Towards a Semiotics of Mathematics." *Semiotica* 72 (1/2): 1–35.

Saussure, F. de. 1966. *Course in General Linguistics*. New York: McGraw-Hill.

Searle, J. 1995. *The Construction of Social Reality*. London: Allen Lane.

Scribner, S. and M. Cole. 1981. *The Psychological Consequences of Literacy*. Cambridge, MA: Harvard University Press.

Sinha, C. 1988. *Language and Representation: A Socio-Naturalistic Approach to Human Development*. London: Harvester-Wheatsheaf.

———. 1993. "On Representing and Referring." In *Conceptualizations and Mental Processing in Language*, edited by R. Geiger and B. Rudzka-Ostyn, 227–246. New York, Mouton de Gruyter.

———. 2000. "Culture, Language and the Emergence of Subjectivity." *Culture and Psychology* 6: 197–207.

———. 2005. "Blending Out of the Background: Play, Props and Staging in the Material World." *Journal of Pragmatics* 37: 1537–1554.

———. 2007. "Cognitive Linguistics, Psychology and Cognitive Science." In *Handbook of Cognitive Linguistics*, edited by D. Geeraerts and H. Cuyckens, 1266–1294. Oxford, Oxford University Press.

Sinha, C. and C. Rodriguez. 2008. "Language and the Signifying Object: From Convention to Imagination," in *The Shared Mind: Perspectives on Intersubjectivity*, edited by J. Zlatev, T. Racine, C. Sinha and E. Itkonen, 358–378. Amsterdam: John Benjamins.

Sohn-Rethel, A. 1978. *Intellectual and Manual Labour*. London: Routledge.

Sulloway, F. 1979. *Freud: Biologist of the Mind.* New York: Basic Books.
Vygotsky, L.S. 1986. *Thought and Language.* Cambridge, MA: MIT Press.
Walkerdine, V. 1988. *The Mastery of Reason.* London: Methuen.
Wertsch, J. 1991. *Voices of the Mind: A Socio-Cultural Approach to Mediated Action.* London: Harvester-Wheatsheaf.
Zlatev, J. 1997. *Situated Embodiment: Studies in Spatial Semantics.* Ph.D. dissertation, University of Stockholm. Stockholm: Gotab.

# — 6 —

# The Nature of Narrative: Schemes, Genes, Memes, Dreams and Screams!

Rukmini Bhaya Nair

## Introduction

Perhaps appropriately for a paper that has religious narrative cognition as its theme, but inappropriately for an academic presentation, I want to risk introducing my subject here with a confession—*peccavi.*[1] The confession I have is simple. It is that, in addition to its irritatingly long title, marked by such an awful excess of rhyme, my paper sins against what may be thought of as a fundamental ethical imperative in narrative. It fails to present a story with a clearly demarcated beginning and a quasi-moral ending that would satisfactorily clinch my own argument. In this respect, it appears to contradict the very thesis that I earlier outlined with naïve conviction in my book *Narrative Gravity* (2002)—namely, that narrative constitutes a well-attested species of "proto-" or "natural theory." There, I argued that:

> Cognitively, narrative seems to have been designed as a evolutionary mechanism to probe experience and create competing theories about the world, especially the world of emotional experience, which is so crucial to cultural survival. What narratives do is provide us with relatively low-cost "theoretical" means of taking mental risks. Consider, for example, our physiological reactions when we read ghost-stories or adventure fiction. Our palms sweat, our pulse-rates shoot up and our throats go dry, even as we "know" that the words on the page are pure make-believe. We don't actually have to go and climb that cliff or love that bounder in order to realize the dangers and aches inherent in these situations. Just imagine, then, how infinitely useful the "lazy machine" of fiction is: it extracts intellectual labour from us while giving us the illusion of enjoying our leisure. (2002, 343–344)[2]

1. This Latin tag has a particular resonance in the colonial/postcolonial Indian context owing to a widely disseminated historical narrative that attaches to it. It is reported that when Sir Charles Napier conquered the territory of Sind in 1843, he sent a one-word message to headquarters: "Peccavi" which translates as "I have sinned." While sources tell us that the pun in question only appeared in the magazine *Punch* a full three years after Napier's conquest of Sind, this fact has not prevented the Peccavi story circulating vigorously as a "true" meme in the imperial narrative ever since (Morehouse 1983, 16–17).

2. The phrase "lazy machine" is from Umberto Eco's book *Six Walks in the Fictional Woods* (1994, 49).

In other words—in my examination of a host of putative narrative schema presented by linguists from Labov onwards, through the structuralist story-grammars of Rumelhart, Mandler and Johnson to Speech Act and Relevance theory—I came in *Narrative Gravity* to a rather unsurprising but perhaps plausible conclusion. The universal presence and power of narrative across all known human cultures derived from the way in which this genre uniquely combined emotional gratification with the possibilities of rational speculation, leading to definite conclusions—one might call them quasi-religious certainties—that a listener could carry away with from the text.

That is why I described narrative as a dynamic structure, the main function of which is to convert *talk into text.* One of the main structural problems in narrative, I suggested, has to do with something that John Searle might call a problem of "fit." By this I meant that "fitting"[3] the monological, linear A-B-C-D-E-F… structure of narrative into the dialogical, alternating, A-B-A-B-A-B… structure of conversation constitutes a chief problem for any teller of a narrative. A "good" teller of stories in any culture provides his hearers with a narrative that manages to resolve this problem by giving the listener a tale that can be taken away from a specific interaction, stored in memory, inserted into other conversations and infinitely retold. We find such divinely gifted or "demented" storytellers in everyday life, amongst the elite of fiction-writers such as Salman Rushdie for example, as well as among the great inventors of religious parables from a Gautama Buddha to a Jesus Christ. In short, the narratives constructed by these tellers grant them that most transcendent of qualities—cultural immortality. Their narratives comprise flexible "meme-holders" of sorts, facilitating the crossing of cultural boundaries and becoming vehicles for ideas to travel across time and space. Any such narratives, then, would have, so to speak, a sort of funnel-effect: they would convince you, at least for the moment, of a certain theory—a religious theory, a scientific theory, a sociological theory—about the phenomenal world. That is why I wrote in *Narrative Gravity*:

> Narrative…works by introducing us to "primitive" but foundational versions of biological theory (how the leopard got its spots); political theory (Robin Hood was a revolutionary who challenged the oppressive class-structures of a feudal society); moral theory (Cordelia's "sin" was to plainly tell the truth and Lear's to stubbornly refuse to hear it); aesthetic theory (the mirror on the wall in "Snow-white" proved itself a dispassionate judge of beauty)—and so forth. That these one-line summaries are crude reductions of exquisite myths is not the point. The point is that the stories to which they refer perform, in essence, a function not dissimilar to the cultural "work" done by, say, Darwinian theory for biology or Marxist theory for political science or Christianity for ethics or Platonic theory for aesthetics. That is, they present hypotheses about phenomena in the world and a paradigm to explain them. (2002, 343)

This paper, however, seems to lack precisely this narrative drive. Partly, this may be because it is not a narrative at all but an attempt at that conversational oddity—an academic paper. Even so, I have to admit that it does not proffer, even within its own terms, a worked-out "explanatory paradigm." Instead, what it does is to loosely bring together a number of observations and research questions that have arisen for me *after* my book was

3. Note that the notion of "fit" is used here somewhat differently from Searle's well-known discussion of "direction of fit" but takes inspiration from it. Searle's suggestion is that there is either a "words-to-world" or "world-to-words" "direction of fit [that] is always a consequence of illocutionary point" (Searle 1976, 1). However, in this essay and in *Narrative Gravity* (84–88), I extend this seminal idea to the *structural* difficulty of fitting narrative-to-conversation or conversation-to-narrative as distinctive speech modes with quite different structural characteristics.

published—but which I have not yet managed to answer satisfactorily.

In this sense, the structure of the paper is conversational rather than narrative. It is full of lacunae and interstices that just scream out for objections to be raised and suggestions to be made by readers, until a final deconstruction, or disintegration, results. That is, this paper has—to put the best possible construction on it—an opening-out fan-like, rather than that inverted funnel-like, structure that could be said to characterize a prototypical narrative. The elliptical scheme it pursues, insofar as it has a scheme, is simply to try and follow through on some of the themes indicated in its over-the-top subtitle, namely:

*Genes and Emotions*: In this section, I attempt to reconsider a basic set of questions, raised in Charles Darwin's seminal text the *Expression of the Emotions in Man and Animals* (1872), in relation to recent work on primate and human cognition. In particular, I refer to the programmatic paper by Hauser, Chomsky and Fitch (2002) outlining possible research links between linguistics and evolutionary biology—and then discuss the implications of this program for narrative theory and religious cognition.

*Memes and Impliculture*: In this section, I return to the idea of "impliculture" that I initially presented in *Narrative Gravity* as a supplement to Paul Grice's influential Theory of Implicature.[4] I begin by providing a working definition of this new term and outlining some of the advantages that accrue, in my view, from using it to analyze issues in narrative production. Illustrating my argument with a couple of modern "Western" novels that deal, broadly speaking, with the "religious" themes of spiritual anxiety and cognitive doubt, namely, Thomas Pynchon's *Gravity's Rainbow* (1973) and David Lodge's *Thinks* (2001), I then consider the role of implicultural constraints on the production of narrative "memes." This move enables me to dwell in my final section on the specific cultural milieus in which memes, and the "memes of religion" in particular, acquire their emotional connotations and their "death-defying" resonance.

*Dreams, Screams and the Theme of Religion in Rushdie's Novels*: In *Narrative Gravity*, I considered in detail the "danger-of-death" narrative believed by Labov to be central to narrative epistemology (1972, 370–371). There, I debated Labov's contention that the notion of death—of all species happenings, the most inevitable as well as the most unimaginable—seems essential to the construction of the "most tellable" stories across cultures. My conclusion was that thoughts of death, like dreams and nightmares, are prototype *emotion-boosters* in two basic ways. They arouse in us feelings of both speculative curiosity (prompting questions like: *what lies beyond this earthly compass?*) and of knee-jerk fear, panic, love, etc. (encouraging the query: *what's most precious—present delights or the promise of afterlife?*).

In this last section, I offer a brief case-study of the contemporary narrative of religion as it affects the emotive subjects of "life and death" as well as "meme-production" in a context that might be deemed stereotypically "non-Western." I review here certain universal soul-and-body conundrums as they are tackled in Salman Rushdie's works, especially *The Satanic Verses*, via a reading of Dawkins' 1993 paper "Viruses of the Mind." Insofar as they have at their core the matter of intellectual survival in a culturally hostile environment, I argue that an analysis of Rushdie's novels could throw some light on a very basic question: *What is the evolutionary value of narrative as a linguistic form?* My aim in this final section is to show how controversial religious narratives of the high-pitched, dream-based, magic-realist sort that Rushdie constructs are in fact very pertinent to evolutionary biological accounts of narrative.

4. Grice, "Logic and Conversation" (1989, 22–40), first published in Cole and Morgan (1975).

## Genes and Emotions

A classic text recently republished—Charles Darwin's voluminous *The Expression of the Emotions in Man and Animals* (1872)—contains a detailed list of questions which Darwin "circulated early in the year 1867" to missionaries and others in the far-flung outposts of empire (1872, 22–23). This original list appears on page 121.

To begin with the obvious, it seems apparent that these questions of Darwin's were intended to seek evidence for Darwin's thesis that there was an essential continuity between the basic emotional repertoire of humans and the higher primates. Studying the body language and the facial expressions of human beings across cultures could, in short, indicate how some kind of universal "emotional language" was shared, in evolutionary terms, by the entire human species.

Not so obviously, and far more debatably, it could be said that I have a parallel agenda with regard to the evolutionary function of narrative in the present paper, roughly involving the following questions:

1. What can we understand about the narrative/culture/biology intersection by examining the generic "body language" of a narrative text so as to discover clues to its emotive/emotional content?
2. Are there specific emotional repertoires associated with different genres of narrative—for example, religious narrative, detective stories, romance, etc.—and could *these* be said to constitute the "body language" of specialized narrative texts?
3. If the "best" narratives across cultures obey that hoary dictum of creative writing to "show, not tell," then how are emotions best displayed in particular genres of narrative, so they *at once* enable the reader/listener to experience the appropriate feelings *and* leave ample place for freedom of interpretation and speculation?

Note that the above questions rely on a strong assumption *linking a specific set of emotions to particular narrative genres*. That is, it is presumed that, while emotions such as curiosity, excitement and fear might be appropriate to the genre of the thriller, for instance, they would be less suited to a religious tale, where humility, awe, respect and so forth might prove more relevant emotions. On the face of it, such a move appears plausible enough, but I shall return to a closer examination of this premise by looking at the especial case of the religious narrative as it is "subverted" in the novels of Salman Rushdie. I shall argue that this novelist specializes in challenging generic distinctions and in provocatively crossing emotional boundaries, so as to create a narrative "stance," *pace* Dennett, that turns out to be quite dangerous. Before I do this, however, I want to connect those primary Darwinian questions about the expression of emotions in everyday life to current research in primate cognition and linguistics and its implications for narrative theory. Consider just a few of these perspectives:

The Indian researcher V.S. Ramachandran (2000) has maintained that "mirror neuron" studies in the F6 areas of monkeys lead to fundamental questions about the acquisition of culture in human communities. This is so, he believes, because "their emergence and further development in hominids was a decisive step."[5] Bold statements like this show that there is certainly some current speculation in the cognitive sciences that "mirror-neurons" or similar devices could provide a brain architecture that might help explain why we can "feel what others feel," i.e. the sources of our empathy with others.

5. Ramachandran 2000. Other studies on mirror neurons and language include: Rizzolatti and Arbib 1998; Gallese and Goldman 1998; Arbib 2002.

**Darwin's Questionnaire on Body Language and the Emotions**

1. Is astonishment expressed by the eyes and mouth being opened wide, and by the eyebrows being raised?
2. Does shame excite a blush when the colour of the skin allows it to be visible? And especially how low down the body does the blush extend?
3. When a man is indignant or defiant does he frown, hold his body and head erect, square his shoulders and clench his fist?
4. When considering deeply on any subject or trying to understand any puzzle does he frown and wrinkle the skin beneath the lower eyelids?
5. When in low spirits are the corners of the mouth depressed and the inner corner of the eyebrows raised by that muscle which the French called the "grief muscle"? The eyebrow in this state becomes slightly oblique with a slight swelling at the inner end; and the forehead is transversely wrinkled in the middle part but not across the whole breadth as when the eyebrows are raised in surprise.
6. When in good spirits do the eyes sparkle with the skin a little wrinkled round and under them and with the mouth a little drawn back at the corners?
7. When a man sneers or snarls at another, is the corner of the upper lip over the canine or eyetooth raised on the side facing the man whom he addresses?
8. Can a dogged or obstinate expression be recognized which is chiefly shown by the mouth being firmly closed, a lowering brow and a slight frown?
9. Is contempt expressed by a slight protrusion of the lip and by turning up the nose with a slight expiration?
10. Is disgust shown by the lower lip being turned down, the upper lip slightly raised with a sudden expiration, something like incipient vomiting or like something spit out of the mouth?
11. Is extreme fear expressed in the same general manner as with Europeans?
12. Is laughter ever carried to such an extreme as to bring tears into the eyes?
13. When a man wishes to show that he cannot prevent something being done or cannot himself do something, does he shrug his shoulders, turn inwards his elbows, extend outwards his hands and open the palms; with the eyebrows raised?
14. Do the children when sulky pout or greatly protrude the lips?
15. Can guilty or sly or jealous expressions be recognized though I know not how these can be defined?
16. Is the head nodded vertically in affirmation and shaken laterally in negation?

Such a conjecture is obviously in consonance with Dennett's idea that humans as a species are essentially storytelling beings who emotionally "centre" their individual and communal senses of self via narrative performance. Moreover, such "centering" is achieved mainly through the projective and teleological linguistic structure of narrative embedded in a lifetime of community conversations.

In keeping with my own hypothesis about "narrative as natural theory," my further supposition would be that such conversational and co-operative tendencies, wherever they are, or are not, located in the brain, also enable us to construct mental representations of ourselves as "intentional agents."

Surprisingly, one detects such a neo-Gricean strain today even in "hard" brain research conducted by neuroscientists like Rilling *et al.* (2002), which could also be significant in exploring narrative cognition. Using MRI scans to show that players in the game "Prisoner's Dilemma" seemed to prefer to cooperate even when there were no obvious pay-offs from such an "altruistic" strategy, Rilling *et al.* report that:

> Despite the fact that defecting is the most rational option, when both players were given the freedom to choose as they wished...out of a total of 20 rounds, players chose to cooperate an average of 11 times, compared to 2 to 3 times for other decision patterns.[6]

6. Rilling *et al.* 2002. See also the related report in the *New York Times* (20.7.2002).

From the above evidence, these scientists conjecture that "social co-operation is intrinsically more rewarding to our brains" and that for some reason biology privileged this mode among human primates by flooding the brain with "feel-good" neurotransmitters when they gave up and gave in rather than grabbed. Commenting on this intriguing piece of research, evolutionary sociologist Richerson contrasts the tendency towards "mutualism" rather than "me-ism" among humans with the behaviour of other advanced primates. "If you put 50 male and 50 female chimpanzees that don't know each other in a lecture hall there would be a social explosion."[7] Apparently, no such obvious "explosion" results even when human beings are put into a social space where their express purpose is to *compete* against each other. Such emotional altruism, centrally expressed in the genre of religious discourse, is apparently characteristic of human negotiations.

In a slightly different vein, a number of other neuroscientists such as Gazzaniga (1994, 2006), Damasio (2000, 2005) and LeDoux (1998, 2003) have also independently underscored the value of emotion in guiding rational decision-making. Emotions, from the perspective of these researchers, are not peripheral to our sense of self but essential in shaping our identities. As the psychologist George Mandler puts it:

> Riding on a roller coaster produces serious disturbances and discrepancies between our expectations and our current feelings of balance and bodily support. Whether the ride is seen as joyful or dreadful depends on what we expect about the ride, who accompanies us, what we are told to expect and whether we feel in command of the situation. Some love it, others hate it.... The modern view answers the ancient concern about emotions. They are not necessarily remnants of our pre-sapient past but rather they are important characteristics of an active, searching, thinking human being. Novelty, discrepancy and interruption generate visceral responses while our cognitive system interprets the world as threatening, elating, frightening or joyful.[8]

Listening to a "good" narrative could be compared to Mandler's roller-coaster ride. It offers us an emotionally exciting means of interpreting the world via "novelty, discrepancy and interruption," but unlike the roller-coaster, it is pretty much a universal nominal kind. Hence, my contribution to this line of research was to present in detail in *Narrative Gravity* the "paradoxical" thesis that stories might be the main vehicle in all cultures that train us emotionally or "non-rationally," while at the same time coaching us in rationality and hypothesis formation.

We may, for instance, think of ourselves as "calm" people or as "quick to anger" or "easy to please." Such self-reflexive images are the result of a lifetime of narrative roller-coasting, beginning with the childhood training we all receive in reading the barometer of community emotions. Consider again that "chaotic chart" of the emotions that I presented in *Narrative Gravity*, based on my observation that when people spoke of their "feelings" in any narrative situation, they willy-nilly incorporated a great many different parameters, including the following:

7. This remark is attributed to Peter Richerson by Nathalie Angier ("Wired to Cooperate," *New York Times*, 23.7.2002). It is worth noticing, however, that Rilling *et al.*, in line with previous work by Robin Dunbar (1998) and others on the evolutionary import of "social grooming" by the female of the species, used all and only women in their "Prisoner's Dilemma" experiment. Richerson, on the contrary, envisages a scenario that includes an equal number of male and female chimps, a situation that would surely have very different social consequences, both among humans and chimps! But the general point made here about "altruism" and "accommodation" as organizing principles in human cultures through which potential interpersonal "problems" can be resolved and "explosions" contained, stands.

8. Mandler in *The Oxford Companion to the Mind* (1984), 220.

**Preliminary Classfication of Emotional Registers**

| TRAITS | FEELINGS | EMOTIONS |
|---|---|---|
| Patience | Love | Anger |
| Selfishness | Affection | Jealousy |
| Stupidity | Hatred | Irritation |
| Kindness | Sorrow | Boredom |
| Generosity | Friendship | Shame |
| **MOODS** | **SENSATIONS** | **QUALIA** |
| Depressed | Pain | Seeing red |
| Energetic | Pleasure | Hearing thunder |
| Worried | Ticklishness | Tasting sweetness |
| Solemn | Warmth | Feeling water |
| Ecstatic | Thirst | Smelling eucalyptus |
| **INTUITIONS** | **IMPRESSIONS** | **INFERENCES** |
| About danger | From art, films | From linguistic usage |
| About truth | From music | From dress or appearance |
| About grammaticality | From the weather | From professional conduct |
| About logic | From expressions | From gender or age |
| About right and wrong | From the scenery | From race or colour |
| **NEEDS** | **DESIRES** | **INTENTIONS/AMBITIONS** |
| For food | For wealth | To visit your aunt |
| For water | For approval | To be rich |
| For shelter | For progeny | To be a cinema star |
| For sleep | For immortality | To study management |
| For sex | For consumer goods | To solve an equation |

Chart from Rukmini Bhaya Nair, *Narrative Gravity: Conversation, Cognition, Culture* (Oxford University Press, Delhi 2002; and Routledge, London and New York, 2003).

Various combinations of the features listed in my neo-Gallian chart above would inevitably yield what in folk-psychological terms might be called the "character and personality" of an individual. Since the permutations and combinations—or recursive possibilities, as Chomsky might say—thus engendered are endless, we can have very large or small variations of personality from individual to individual. Moreover, we can have extremely rich and teeming interior worlds or relatively limited and impoverished ones according to the specific ways in which we make inter-connections between all these cognitive parameters—moods, instincts, feelings, desires, etc.

The above argument has a bearing on studies of a general 'Darwinian' nature insofar as we can try and investigate whether the affective inner worlds of our primate cousins might possess some telling similarities to our own, even if qualitatively different from ours given that they do not have brains specialized for symbolic/linguistic representations. Other studies in "hard" primate cognition appear broadly to support this perspective. Hurford for example, has argued that in the areas of vision and hearing, there is evidence of "two independent neural pathways; one locates objects in *body-centered* space, the other attributes properties, such as colour to *objects*. In vision, there are dorsal and ventral pathways. In audition, similar separable "where" and "what" pathways exist" (2003, 261). The implication of this sort of primate research for human narrative cognition seems to be that "mental scene descriptions are necessary for practical tasks among primates and pre-exist language phylogenetically" (2003, 261).

Given that the ability to connect "scenes" or "episodes" with their "meaning or "point" is central to narrative processing, Hurford's claim that the connections that primates need to make between "self" and "surrounding" formed an evolutionary platform which could be "reused for more complex, ultimately linguistic purposes" (2003, 261) makes basic sense (see also Hurford 2004). The scenarios in religious narrative and the role they play in enabling members of a community to fuse a notion of "self" with the comforting presence of the familiar thus seem to be broadly anticipated in current research on primate "teleology."

Likewise, in a programmatic article partly devoted to predicting trends in future research, Hauser, Chomsky and Fitch maintain that:

> ...an understanding of the faculty of language requires substantial interdisciplinary cooperation. We suggest how current developments in linguistics can be profitably welded to work in evolutionary biology, anthropology, psychology and neuroscience. (2002, 1569)

To this long list of possible areas of alliance in the cognitive sciences, I would suggest that the field of "narrative theory" be added. In the research mentioned earlier, comparisons between primate and human faculties are obviously central to the research trends that Hauser et al. forecast, and they present some sensible perspectives on carrying out such experiments—for example, testing differences between chimps and humans on numerical ability and navigational skill. In a similar vein, researchers might attempt to devise tests that investigate whether primates possess elementary skills at *narrative recounting* as well as explore the *environments* in which these skills thrive best.

On the specific hypothesis concerning "universal grammar" (UG) that underpins the research programme advocated by Hauser *et al.*, there are likely to be many who have qualms, but few, I think, would deny their general contention that *interdisciplinary work on language* is likely to pay rich future dividends. Tedious as it may be to repeat the point, it is this general emphasis which unites a bewildering variety of approaches to "cognition" today—and which has direct implications for cognitive studies of narrative.[9]

It is true that in the important paper authored with Hauser and Fitch, Chomsky still attempts to "save syntax" by proposing a new distinction between human language in the narrow sense (FLN) and human language broadly defined (FLB). FLN, in Chomsky's view,

9. See for example, Mieke Bal who firmly states "Our approach is synthetic, recognizing the importance of both cognitive and sociological approaches to narrative and their implications for each other" (2004, 168). Or see the reviews of *Narrative Gravity* in *Language* (Volume 80/2, 2004) by Chaoquan Xie or Jonathan Goodwin in *Modernism/Modernity* (Volune 11/2, April 2004), both of which engage with the interdisciplinary approaches. Xie writes: "In this thought-provoking and stimulating book, the author integrates various theoretical frameworks and perspectives, such as structural linguistics, speech act theory, the cooperative principle, relevance theory, and conversation analysis...help[ing] to deepen our understanding of narrative and a variety of issues pertaining to it. This excellent text will be of much interest to those concerned with cognitive, anthropological, and sociological approaches to the study of narrative"; while Goodwin remarks: "The intersections between cognitive science and narrative theory are gaining increasing critical interest, and Rukmini Bhaya Nair's detailed and creative study applies this synthetic methodology to the rhetorical analysis of postcolonialism in the discourse of disaster. *Narrative Gravity* argues that narrative should be the basic analytic unit of pragmatics and other forms of discourse analysis. Nair engages sociolinguistics, speech act theory, deconstruction, Paul Grice and David Lewis's semantics, and Daniel Dennett's arguments about the narrative illusion of intentionality. She then uses this interdisciplinary theoretical apparatus to uncover constraints in official administrative accounts and resistance in villagers' oral narratives of the West Bengal." Thus, interdisciplinary approaches seem to be clearly recognized as valuable today across a range of disciplines and especially in the fields of linguistics, narratology and cognitive studies.

"only includes recursion and is the only uniquely human component of the faculty of language" (2002, 1569), but of course this leaves open the logical possibility that "recursion" is a feature of other human faculties such as vision, hearing, etc. We could, for instance, have an FLN vision component involving only recursion and an FLB that involves other interrelated computations, and so forth. But then my question is: would recursion in this case *uniquely* identify the faculty of language at all? This remains an open question, but more crucially for my own argument, *storytelling*—as I have argued throughout *Narrative Gravity*—does seem fundamentally to involve structural repetition or "recursion" (2002, 30–31, 212–229).

Would Chomsky then be prepared to see *narrative production* as an intrinsic part of FLN? In the paper under discussion, the implicit suggestion seems to be that genres like narrative rightly belong to FLB which "includes a sensory motor-system, a conceptual intentional system, *and* the computational mechanisms for recursion" (2002, 1569). In other words, Chomsky would probably not deny that narratives employ recursion but would contend that such recursion is not totally *insulated* in the sense that syntax is; rather, it involves essential interactions with a "conceptual intentional system" as well as "a sensory motor system." Now, since my view is that narratives offer natural tools for conceptualizing "scenes" and "selves" and thus afford relatively low-cost, iterative "pathways" for emotional exploration, I would not deny that the feature of recursion in stories is in fact tied in with these parallel systems. However, I would tend to support Jackendoff's more systemic proposal that "the proper approach is [to envisage] a parallel architecture…[which] directly involve the rules of grammar in processing…[and] finally leads to a natural account of the incremental evolution of the language capacity" (2003, 651).

A late and sophisticated stage in that "incremental evolution of language," I suggest, is the embedding of "monologic" narratives in "dialogic" cooperative conversations as a means of transferring memes, exchanging emotions and negotiating the boundaries of the self. And it is here that we might return for some initial, indirect, answers to the first of the questions I earlier posed:

> What can our current understanding of the narrative/culture/biology intersection tell us about the "body language" of a generic narrative text and its emotive/emotional content?

In brief, current research in primate cognition shows that the impulse towards reaping the emotional rewards of bonding and social cooperation and combining various modalities is basic to the way in which we deal with our surroundings and organize our communities around our "selves" (Rilling *et al.* 2002). While the chimpanzee ability to make such connections between self and environment may indeed provide an "evolutionary platform" for human behavior (Hurford 2003), this tendency towards "mutualism" is much more marked among humans (Richerson 2002), who shape their communities through language. This faculty of language narrowly defined (FLN) is uniquely marked by the feature of "recursion" which enables certain basic elements of language to be infinitely combined and recombined (Chomsky 2003). At the level of discourse and interaction, we find the same recursive tendencies both in narrative and conversation, with their specialized ABCDE and ABABAB structures. In this connection Dan Everett, embroiled in an explosive debate with Chomsky on the Pirahã language and the possible "absence" of syntactic recursion in this language, makes the following interesting point:

> If you go back to the Pirahã language, and you look at the stories that they tell, you do find recursion. You find that ideas are built inside of other ideas, and one part of the story is

> subordinate to another part of the story. That's not part of the grammar per se, that's part of the way that they tell their stories. So my idea is that recursion is absolutely essential to the human brain, and it's a part of the fact that humans have larger brains than other species... [but] it's certainly not clear [at the moment] that recursion is part of language as opposed to part of the brain's general processing." (2007)

Thus, it could be important to study how stories "as part of the brain's general processing" constitute one of the most basic forms of uniquely human recursive communication creating "imagined communities" (Anderson 1983) that go far beyond immediate face-to-face interaction (Nair 2002).

The cross-cultural generic discourse of religion is particularly fertile in this sense of vaulting over space-time barriers. It typically involves several imaginative transactions in multiple interactional voices—through prayer, ritual interchange, priestly intervention, etc.—and thus, almost recklessly, opens up the boundaries of the self to the metaphysical. The "body-language"—or perhaps one should call it the "linguistic body"of the religious narrative allows for the most extravagant of emotional gestures, such as throwing up one's hands and "abandoning" the corporeal self up to a transcendent God or an abstract ideal of "goodness." The altruistic "feel-good" factor that Rilling *et al.* identify as typical of human reactions to competitive situations appears to be even more heightened in religious discourse where one can mentally cast off the petty shackles of everyday rivalries and bond without constraint. Or as Durkheim put it in his seminal *The Elementary Forms of Religious Life*, referring to the "collective effervescence" that religion promoted:

> We have seen...that if collective life awakens religious thought when it rises to a certain intensity, that is so because it brings about a state of effervescence that alters the conditions of psychic activity. The vital energies become hyper-excited, the passions more intense, the sensations more powerful.... Man...feels somehow transformed and in consequence transforms his surroundings. (1912/1995, 424).

How narrative in general and religious or spiritual narratives in particular effect emotional transformations and thereby affect "psychic activity" or cognition is the subject of the next section of this paper.

## Memes and Impliculture

The issue of the connection between altruism and the cognitive pleasure mentioned in the previous section is central to any analysis of the narrative impulse. Hence, one of my first questions in *Narrative Gravity* was: one can understand why a story pleases, or might please a listener, but *what's in it for the teller*? In brief, my answer was that the altruism displayed by storytellers results in emotional satisfaction. Such satisfaction is derived from making what I call *implicultural gains*, that is, the storyteller's expenditure of time and effort, provided he delivers a "good" story, is rewarded by exactly that "psychic effervescence" to which Durkheim referred. Not only are tellers able to investigate and hold competing theories in mind and explore difficult emotional terrain via the structures of narrative, they are crucially enabled to *assert their membership—and even leadership—within a specific community* through the device of narrative.

Typically, the excitement or interest of all communication depends on tellers and audiences jointly negotiating a maze of inferential clues, using the cultural cues at their disposal. My idea of "impli*culture*" is meant to supplement Grice's theory of generalized "impli*ca-*

*ture*" in this respect. So what does the notion of "impliculture" buy us? Briefly, we could say that it allows us to focus in more detail on:

1. the other modalities that complement "words" such as gesture, gaze, etc.
2. the things left unsaid within a cultural community, including taboo areas
3. the community as a holistic, polyphonic conversational group of interpreters

Because the Gricean system strongly encourages us to think in terms of universal deductive implicatures available to all and any set of conversationalists, it tends to minimise attention to cultural differences. My attempt at redressing this imbalance, as I've just suggested, has been to develop the concept of *impliculture,* or *cultural* implicativeness, where we can theorise particular cultural communities as people perpetually engaged in a large, polyphonic conversation. Since there are so many potential speakers all vying for "turns" in such a model, a great deal depends on efficiently "conveying the unstated" through the use of modalities other than speech, as well as strategies that allow for joint statements of community opinion. And it is here that the generic "body-language" of narrative plays an incalculably important role particularly in the case of religious discourse whose business it is to convey experiences that often involve the "unsayable." In any large, ongoing group conversation, *insider status* or "member-shipping" (Sacks 1989) would obviously come from listeners' picking up on particular *implicultural* meanings, rather than simply deducing generalized Gricean meanings.

Implicultural indicators subtly differ, in this sense, from such "non-natural" Gricean meaning implicatures. To consider a workaday example, in a business organization or a university department that, say, has a regular staff meeting every Tuesday, the phrase "so let's all talk about it next week" specifically indicates Tuesday and not any day of the week, as an "outsider" might presume.[10] Many such small pieces of information including "in-jokes" and yarns constitute an "insider's knowledge" of what is often called "the culture of the organization." Such knowledge may be of little use beyond the organization but is extremely useful in marking institutional boundaries and defining hierarchies. It permits sub-cultures to keep out "freeloaders" and gives those who are familiar with the culture's internalized narrative a "warm feeling" of belonging and of sharing. Unspoken knowledge of implicultural conventions is a means of creating trust and emotional bonding and, as a consequence, productively and cooperatively "working together." They therefore make "rational" sense even though they might seem mere arbitrary injunctions if considered in isolation from a context of social use.

Rilling *et al.*'s experimental findings from brain scans, reported in the previous section, thus seem supported by earlier Gricean theory as well as its extension into the realm of narrative impliculture. At the same time, Grice's fundamental idea of conversation as a form of rational cooperation and deductive theorizing gains new vigour from such results in cognitive neuroscience and their putative implications for a culturally rich notion of "self."

One absolutely fundamental way in which our internal representations of the "self" differ from those of animals, as I'd begun to argue in the last section, is that we constantly *make connections between artistically created worlds and the mental universe of the emotions.* Even if we maintain that animals show some capacity to correlate symbols as well as to deceive, they do not, as far as we know, *habitually* create fictions or stories. And if they do

10. I am grateful to Robert Eaglestone for suggesting this example to me of the implicultural inferences that have to be regularly made in office environments where professionals are in everyday "competition" with each other.

not, it follows, animals *do not use stories as an important learning or cultural device*, whereas humans consistently do. There is, after all, no society in the world that has failed to develop its own characteristic forms of artistic expression and literary representation—which are inducted in us from infancy on.

Here, I would once again concur with Jackendoff in the previous section, when he concedes that Chomskyan generative grammar was "correct to focus on the child's acquisition of language as its central problem" (2003, 651). When a child acquires stories in particular, I have further proposed that she learns to be emotionally resilient, theoretically sophisticated and culturally savvy. While *Narrative Gravity* did not consider stories to or by children in any depth, this is undoubtedly one of most fascinating fields for further research on narrative. As recent works by Toolan (2001) and Ochs and Capps (2001) demonstrate, there are other revealing insights to be gathered from how children gradually stabilize their centres of narrative gravity via "bed-time" stories and other cultural rituals. Here is a quote from an interview with the Indian novelist, Shashi Tharoor, who retells the story of the *Mahābhārata*, one of the two great Indian religious epics, as a "parable" for modern India:

> In my first novel, *The Great Indian Novel*, I took the stories of the great epic, and I reinvented these stories as an account of the political history of India, from the British days to the present. The tales of the *Mahābhārata* are learned by practically every child at his grandmother's knee, and the tales of the nationalist struggle, the "freedom movement" as we call it, are part of the stuff of what we are brought up on in independent India. In the process, I tried to illuminate some aspects of the Indian condition. (1999)

The use of a phrase like "illuminating the Indian condition" by Tharoor in this passage clearly echoes the "greater" theme of providing enlightenment about "the human condition" in general that animates the complex religious "danger of death" tale recounted in the *Mahābhārata*.

It is this re-usability or, loosely speaking, "recursive" property that in my view characterizes the "good" story. Returning to theory at this point, the sort of cultural resources that I have just commented on, as well as their cognitive impact have been the impetus for a radical account of the fundamental "literariness" of all language mooted by Mark Turner. Turner's provocative thesis is articulated with panache in the passage below:

> Parable creates structure for voice by projecting structure from story. The structure it creates is grammar. Grammar results from the projection of story structure…in a community that already had parable…. The dynamic *processes of parable* are basic to the construction of meaning and the construction of language. Story precedes grammar…. Cognitive mechanisms whose existence we must grant independently of any analysis of grammar can account for the origins of grammar. *The linguistic mind is a consequence and subcategory of the literary mind.* (1996, 141–142, 168, italics mine)

The idea that it is not grammar that shapes narrative patterns but narrative that structures grammar obviously turns conventional wisdom on its head, but this non-conformist school of thought has already attracted considerable attention from literary theorists, psychologists and philosophers, amongst others. It has also developed a substantial theoretical vocabulary of "conceptual blending," "mental leaps," "emergent structure," "primary metaphors" and so on which serves to knit the theory together.

An urgent task before narratologists today is to link insights from new research paradigms such as Turner's to other theoretical perspectives and experimental results from studies of the brain. For example, Turner uses the term "parable" to indicate a mechanism which "projects" from story-structure into grammatical structure and gives it "voice." How would this notion of the relation between parable and "voice" be translated into the terms of evolutionary biology or neurological experiment? Alternatively, how would it cohere with my own attempts to integrate the ABABAB interplay of "voices" in conversation with the ABCDE linear lattice of narrative that continuously links temporal past with present and "projects" into the future? Then again, would literary or religious parables of the "Shahrazad" kind that Turner examines in his very first chapter, and that I mention in my Introduction, be obvious candidates for Dawkins' cultural "meme" selection and transmission, as I have suggested?

The answers to these questions depend on how we construe the idea of a "meme." Dennett bases his explanation on Richard Dawkins' revolutionary ideas:

> The gene, the DNA molecule, happens to be the replicating entity which prevails on our own planet.... But do we have to go to distant worlds to find other kinds of replication and other, consequent, kinds of evolution? I think a new kind of replicator has recently emerged on this very planet.... It is still in its infancy, still drifting clumsily about in its primeval soup, but already it is achieving evolutionary change at a rate that leaves the old gene panting far behind. (Dawkins 1976, 206)

The new kind of replicator that Dawkins is referring to is the "meme," which obeys the laws of natural selection:

> Just as genes propagate themselves in the gene pool by leaping from body to body via sperms or eggs, so memes propagate themselves in the meme pool by leaping from brain to brain via a process which, in a broad sense, can be called imitation. (Dawkins 1976, 206)[11]

Ten examples of memes that Dennett cites are: *arch*, *vendetta*, *wheel*, *clothes*, *alphabet*, *calendar*, *the Odyssey*, *chess*, *calculus*, *impressionism*, while candidate "Indian" memes might include:

1. the zero and the Hindu/Arabic numeral system
2. the idea of ahimsa or non-violent action
3. yoga
4. the sari
5. hot pickles
6. rasa theory
7. the *Rāmāyaṇa* and the *Mahābhārata*
8. the idea of reincarnation
9. ayurveda and herbal health products
10. miniature paintings

Sharing "memes" implies sharing presuppositions about the structure of the world, a common culture. The Muslim Ummah or brotherhood, the Hindu caste-system, the members of various Christian denominations are all examples of communities sharing implicultural systems and sets of memes which have tremendous "symbolic value" à la Durkheim. But at this point one must also ask: wherein lies the relation between memes and narrativity?

My own speculation, as I've mentioned, is that narratives can be seen as offering a kind of natural harbour, a "safe house" for the memes produced by a culture. If our chances of cultural survival are increased by going in for meme-imitation, then surely our propensity for narrativization aids this survival instinct. For narratives give us our admittedly illusory but

11. See also Dennett 1991, 201–202 and 2006.

nevertheless crucial sense of "self"; they offer moulds for memes to "house themselves."

Dennett himself links his enthusiasm for memes to his central "Paradox of the Authorless Narrator." In Dennett's scheme of things, there *is* no "true self," no real centre of "selfhood." Paradoxically, though, our stories help us *imagine* an agentive centre, and often times a cultural centre. And what do we see with our illusion-making internal eyes? Well, flowing, rushing, gushing into the human cognitive machine at all times are a variety of visual, linguistic, aural and somatic sensations. The task of the mind's eye is to read narrative coherence into this jumble and to make such inferences as render this chaotic material relevant to the mind's "I."

The ordinary pragmatics of everyday living requires us, in short, even if we *know* otherwise, to subscribe to the illusion of a unitary self and a unitary culture, where memes have to be constantly shared and meaningfully interchanged. To use an analogy at this point, our lives are a lot more like television serials than like classical plays. They straggle along, lurching from episode to episode, often dull and slow and occasionally coming up to speed. But what narratives do is to enable us to organize this chaotic text-in-progress of our lives, imposing on it retrospective and prospective temporal patterns and, occasionally, even allowing for a classic memetic elegance. They bring, as Bakhtin might have phrased it, the "chronotope" of fictive order to bear upon the perturbed "carnival" of real life. The novel was for Bakhtin the most challenging of modern literary forms because it best captured this psychological conjunction between chronotope and carnival. Or, what in terms of this paper may be thought of as religious "explanation" on the one hand, relating to temporal and causal order ("and on the seventh day he rested") and religious "ecstasy" (an emotional release) on the other.

Contrary to "expressive" aesthetics, form is not *pure* expression of the hero and his life, but an expression which, in giving expression to the hero, also expresses the author's relationship to the hero: aesthetic form cannot be founded or validated from within the hero, out of his own directedness to objects and meaning.... Aesthetic form is founded and validated from within the *other*—the author, as the author's *creative* reaction to the hero and his life. (1990; see also Dentith 1995 and Todorov 1984).

Dennett, who comes from a very different disciplinary background than Bakhtin, makes a similar observation with regard to the interpretative role of the "other" in fictional texts when he explains his idea of the self as a gravitational centre for cognitive activity:

> A center of gravity is just an abstractum. It's just a fictional object.... [Similarly] a self is also an abstract object, a theorist's fiction. The theory is not particle physics but what we might call a branch of people-physics; it is more soberly known as a phenomenology or hermeneutics, or soul-science (*Geisteswissenschaft*). The physicist does an interpretation, if you like, of a chair and its behavior, and comes up with the theoretical abstraction of a center of gravity, which is then very useful in characterizing the behaviour of the chair in the future, under a wide variety of conditions. The hermeneuticist or phenomenologist—or anthropologist—sees some rather more complicated things moving about in the world—human beings and animals—and is faced with a similar problem of interpretation. It turns out to be theoretically perspicuous to organize the interpretation around a central abstraction: each person has a self (in addition to a center of gravity). In fact we have to posit selves for ourselves as well. The theoretical problem of self-interpretation is at least as difficult and important as the problem of other-interpretation.... Now how does a self differ from a center of gravity? It is a much more complicated concept. I will try to elucidate it via an analogy with another sort of fictional object: fictional characters in literature. Pick

> up *Moby Dick* and open it up to page one. It says, "Call me Ishmael." Call whom Ishmael? Call Melville Ishmael? No. Call Ishmael Ishmael. Melville has created a fictional character named Ishmael. As you read the book you learn about Ishmael, about his life, about his beliefs and desires, his acts and attitudes. You learn a lot more about Ishmael than Melville ever explicitly tells you. Some of it you can read in by *implication.* Some of it you can read in by *extrapolation.* (1992)[12]

In this cautionary passage, Dennett makes four points that reinforce connections between literary and cognitive research on narrative from Bakhtin to Turner. The first, by now familiar, reminder he has for us is that there is *no location* where the "self" is to be found. The self does not "reside" exclusively in brain or in culture, although it certainly is an epiphenomenal creation of brain processes, produced by a meat-machine in which millions of chemical transmissions across synapses together contrive to give us a strong "illusion" of self. This illusion of "self" is important in humans because it then serves as a *focus of interpretation*—which brings us to Dennett's second idea, namely, the juxtaposition of "self-interpretation" with "other" interpretation. Dennett's third argument is that "theories" about the self help "explain" the future behaviour of selves, that is, they have *predictive* power—and this remark is obviously related to my own ideas about narratives as proto-theory. Most interestingly, Dennett emphasizes, as does Bakhtin, the interpretive dimension of *literary, including religious, narrative.*

Literary narratives, as I see them, strikingly bring together the "narcissistic" desire of an author to reflect on his own life and the "altruistic" urge of a storyteller to gift his readers/listeners with pleasure and self-knowledge. When Dennett reminds us that literary narratives work by "implication and extrapolation," I interpret this as the suggestion that even the simplest narratives need the gravitational pull of a "moral" to render them acceptable and reusable. A primary task of narrative is to underwrite these shared premises. Yet, because narratives are intrinsically indeterminate, there is always infinite scope for further interpretation, and one could further surmise that it is exactly into this ambiguous space that "culture" and its conflicts manage to insert themselves.

We could perhaps recall with profit at this point two literary texts which seek to grapple with the sorts of metaphysical problems typically assigned to the domain of the religious: *Gravity's Rainbow* by Thomas Pynchon (which shares a keyword in common with my book) and *Thinks* by David Lodge. The first of these novels sports a hero caught in an impossible quasi-religious danger of death situation. Each time Tyrone Slothrop has an erection, a destructive V-2 bomb hits the earth in a mimic version of the "sign" God sent Noah after the floods, but it turns out that those Biblical floods are replaced by lethal fire raining down from the sky in Thomas Pynchon's strange and estranging work. So, we might argue that the question that confronts us today, in the wake of 9/11 and other horrifying world events, inevitably concerns the predictive power of fictional narrative. *Exactly how prophetic was* Gravity's Rainbow *as a species of literary/quasi-religious theorizing?*[13]

12. Dennett adds: "This article is based on my summary remarks at the 1983 Houston Symposium on the nature of the self and consciousness. It also draws fairly heavily, and without specific citation, on several recent papers of mine, in which the topics discussed here are developed in more detail. References to these can be found in my book *The Intentional Stance* (MIT Press, 1987), x.

13. Prophecy is an integral part of religious discourse in many cultures, from the Greek oracles to the Scandinavian casting of runes to Biblical prognostications. See Hogue (1997) for extensive coverage attempting to trace the roots of prophecy back to 8000 BCE. In particular, the dream narratives of religious leaders or "seers" are important cultural texts. For a detailed study that employs the elements of

Moving on, if we consider the "long" quarter century that has elapsed between Pynchon's novel, published in 1973 and David Lodge's *Thinks* written in 2001, at the head of this millennium, we notice that new research debates on the nature of "mind-machines" have self-consciously entered the area of literary-religious fiction. Pynchon's intricately metaphoric text on the dialogue of "body" (Slothrop's sexual arousal) with "soul" (the V-2 fire from the sky, signalling heavenly wrath) relies, however unconventionally, on the genre of sci-fi. In Lodge's fiction, these sci-fi imaginings have grown much closer to everyday "reality" and metaphor can be blithely cast aside.

Few would contest the fact that Lodge and Pynchon are writers whose "voices" contrast in every possible way, yet my speculation would be that there are not just "stylistic" preferences at play here but, additionally, what one could think of as a change in *zeitgeist*.

It has just grown considerably easier today to exercise the preference for prosaic conversations on brain mechanics, even in a love story—albeit between academics. "Straight" conversations in a realist mode on the complex Dennettian "illusion of selfhood" fostered by culture and the brain acting in collusion are by now entirely plausible in literary fiction. The conversation I cite below from *Thinks* is between the two main protagonists in Lodge's novel, Helen, a writer, and Ralph, who is exactly that "new man for a new century," a cognitive scientist.

> "I just think that we shouldn't confuse what we would like to be the case with what *is* the case," says Ralph.
> "But you admit that we have thoughts that are private, secret, known only to ourselves.'
> "Oh, yes."
> "You admit that my experience of this moment, lolling here in hot water, under the stars, is not exactly the same as yours?"
> "I can see where this argument is going," he says. "You're saying, there's something it is like to be you, or to be me, some quality of experience that is unique to you or to me, that can't be described objectively or explained in purely physical terms. So one might as well call it an immaterial self or soul."
> "I suppose so, yes."
> "And I'm saying it's still a machine—a virtual machine in a biological machine."
> "So everything's a machine?"
> "Everything that processes information, yes."
> "I think that's a horrifying idea."
> He shrugs, and smiles. "You're a machine that's been programmed by culture not to recognize that it's a machine." (2001, 102)

Lodge and Pynchon's novels are in this respect fine illustrations of Bakhtin's idea that the novel creates a self-reflexive space where an author's life-story and that of his protagonist are in "dialogue" with each other. For example, we cannot help but notice that the Helen who "questions" Ralph the cognitive scientist's stance is herself a writer like Lodge. The metaphysical "mind-matter" concerns in both books are of course age-old, but what each establishes—to my mind, convincingly—is the recurrent tendency in literary fiction to boldly use the lexicon of its time to "theorize" the effects of culture on mentality and vice-versa.

That conversations in literature are so contiguous with those in academia once again reinforces the view that our narratives, whether extremely sophisticated as in the novels of

Freudian analysis (condensation, displacement, dramatization and secondary elaboration or revision) in order to engage with Rushdie's use of the dream-narrative form and its prophetic potential in his novels, see Nair (1995, 103–109).

Pynchon or Lodge, or apparently artless as in oral stories, appear always to build in our "language instinct"[14] to theorize. At this point we may turn to the second question posed at the beginning of this essay, namely:

> Are there specific emotional repertoires associated with different genres of narrative—for example, religious narrative, detective stories, romance, etc.—and could these be said to constitute the "body language" of these narrative texts?

In reply, we might now cautiously answer that each genre of narrative specializes in "theorizing" a particular set of emotions. In the case of religious narrative, we may further surmise that such theorization includes some of the following factors:

1. The religious narrative, consisting of a network of "parables," in Turner's sense, places the human being within a community of believers, who prototypically share their faith in some particular divine or supra-human agency, or set of agencies, governing their lives. This invisible, or intermittently visible, divine "body" is thus a foundational concept in almost all religious stories, even though its presence may only be indirectly indicated through *implicultural* indicators (e.g. the Holy Ghost, the Holy Koran, a dancing Shiva or a Padmapani Buddha or other such "memes").
2. Such putative "memes" representing divinity in a religious narrative have special emotional appeal. They are counters designed to "reveal" the divine to human beings. The memes in a religious narrative thus typically encapsulate the human pursuit of moral codes, as laid down in community narratives. Brahman, angels, priests, errant souls, devils, etc. may also be candidate examples of such memes. The evolutionary task of the memes that a religious narrative generates is to keep its own community emotionally bonded and also to rhetorically persuade other communities of "outsiders" to accept the superiority of its "story" about the human relationship to the divine.
3. The goal of every narrative is, in Dennett's terms, to help us imagine a centre for the human "self" to provide a focus for interpreting the self. In a religious narrative, this centre could be, and typically is, an exo-centric one, residing in a realm "beyond the human" or in "the body of the divine." Emotional turbulence is curbed and "peace" is found, according to the meta-narrative of religion, when a seeker manages to discover for himself, by charting out a path guided by the various memes in a received narrative, this "still centre" of the self. Doubt, sorrow, angst and despair followed by ultimate happiness, peace, consolation, victory over ordinary human desires, transcendent love and final freedom from fear, thus constitute the emotional repertoire of religious narrative, *ceteris paribus.*
4. Like all narratives, religious narratives make hypotheses about the nature of "reality" and predictions about the future. When religious narratives are transmuted into modern literary forms such as the novel, they must be "embodied" in the language and idiom of their own time if they are to effectively communicate some of the emotions listed above. As we observed in the writings of Pynchon and Lodge, writers can deploy language and the spirit of the age in very different ways to describe the present moral dilemmas of the human race and to "prophesy" its future. In each case, however, the purpose of the author, or *intentio auctoris* as Umberto Eco called it (1992, 25), is to set up an independent "dialogue with the divine" which encodes in Bakhtin's words, "the author's creative reaction to the hero and his life."

The dangers and advantages of conducting such an "independent" literary conversation with "God" about the sacred is the subject of my next and final section where I examine

14. This phrase is from the title of Steven Pinker's book of the same name: *The Language Instinct* (1995).

the issues of religious conviction, the fear of the "other" and the evolution of community in relation to the provocative novelist oeuvre of Salman Rushdie.

## Dreams, Screams and the Theme of Religion in Rushdie's Novels

Who's afraid of Salman Rushdie and who's Salman Rushdie afraid of? These questions intersect at a defining moment in the history of the postcolonial, post-cold-war world. In this scenario, if satanic figures like Salman the Apostate did not exist, it would certainly be necessary to invent them. Indeed, the reinvention of Satan appears to be one of the snappiest name-games in geo-politics today, and we have had a long line of Satans paraded before our eyes in the new millennium: Osama bin Laden, Colonel Gaddafi or Saddam Hussain emerging disgraced from his devil's "hole" underground. One has just to open a newspaper practically anywhere in the world to observe how this satanic metaphor has taken hold in the ordinary discourse of geo-politics, as the following lament from an Indian source bears witness:

> Having taken close to 30,000 lives, and utterly destroyed what was left of the Iraqi state, the Americans will withdraw to their fortress on the other side of the Atlantic, claiming both virtue and victory. But where will it leave Iraq and where will it leave us? Al-Qaeda and every other Islamic jehadi fanatic will claim another victory over the "Great Satan" and turn their reinforced and revived attention to "little Satans" in other parts of the world.... Mussaraf could be their first target for having become a traitor. India and its leaders could be their second choice.[15]

Perspectives such as this make it clear that, given a certain vocabulary apparently derived from religious discourse, Satans can be manufactured at will, whether it is the "Christian" West, "Hindu" India or "Islamic" Pakistan that happens to be seen as a focal threat. In the first instance, it is this emotive notion of threat—both morally and physically destructive—and the instinctive fear that it engenders that may be related to questions in evolutionary biology, as well as, ultimately, to an explanation of Salman Rushdie's religiously provocative fiction.

Those interested in "the religious experience" among evolutionary theorists seldom attribute the intense emotions associated with it to any direct genetic influence. Instead, they argue that religious narratives seem to have provided an ancient cultural resource whereby counterintuitive and counter-factual ideas could be explored as "revealed" truth without fear or debilitating anxiety, since they were told under the protective umbrella of priestly sanction.

Another way to make the same point would be to say that religious narratives tell us about the "security" arrangements adopted by early populations in order to cope with the less than tractable phenomenon of "faith." Religious stories provided essential community resources that could be used to both protect and/or punish potential renegades. Thus, the extreme reactions to Rushdie's subversion of Islamic texts might partly be explained in evolutionary terms as a self-protective drive kicking into action when a community perceives itself confronted by a free-loader who uses its assets without sharing its beliefs.

Two recent papers in evolutionary theory approach the issue of religion in slightly different ways. The first, by Scott Atran and Ara Norenzayan, argues that:

> A key feature of the supernatural agent concepts common to all religions is the triggering of an "Innate Releasing Mechanism," or "agency detector," whose proper (naturally-select-

15. Prem Shankar Jha, "Let's get outta here!" editorial page, *The Hindustan Times*, November 20, 2003.

> ed) domain encompasses animate objects relevant to hominid survival—such as predators, protectors and prey—but which actually extends to moving dots on computer screens, voices in wind, faces on clouds. (2004, 713)

According to this theory, an "agency detector" originally intended to spot predators in the course of hominid survival is also set in motion during a religious experience in which a subject "sees," for example, an angel in the trees or "hears" a thunderous God speaking to him. As I understand it, this sort of evolutionary explanation is relevant to our understanding of literary and sociological productions in three ways. *First,* a seer's experiences are never independently verifiable but depend on individual testimony. *Second,* no community can afford too many seers because then it would have to contend with conflicting versions of fundamental "truths," resulting in cognitive chaos. *Third,* once a seer's views are accepted by the community, their cognitive cost is so high that they must be held as basic articles of faith to be trusted and defended at all costs.

To take the above points a little further, it seems clear that "agency detection" by any one individual seer "chosen" from among the mass of the community, must involve a largely unique and original experience. A logical consequence of this is that recovering this experience depends on the kinds of layers of interpretative "work" within the community that we typically associate with an implicultural reading of a poem or a piece of fiction. So, in the first instance, a set of inferences is made by the seer, based on his private interpretation of the phenomenon he thinks he has observed, and then groups of others have somehow to be persuaded to accept this personal revelation as a "higher truth." The obvious danger, of course, is that the community, or powerful factions within it, will not readily agree to such a willing suspension of disbelief, for it is the part of the logic of a seer's situation that he has no proof, indeed, *cannot* have proof. In such cases, the conclusion must be that he has either been duped by "Satan" or some equivalent force, or is deliberately deceiving his fellow beings of his own free will; but, in both instances, as Atran and Norenzayan put it, he "threatens the social order"(2004, 713).

Thus, the "false prophet's" freeloading deserves the severest punishment, for it tends to produce communally costly mental confusion and consequent social disintegration. On the other hand, as these researchers point out, "these same metacognitive capacities provide the hope and promise of open-ended solutions" (2004, 713) to fundamental existential problems, so no social group can afford to weed out the religious experience and the seer altogether. For example, a seer's "vision" may keep the flock together in times of crisis or offer a version of a "life beyond life" that is a crucial psychological counter to the near universal fear of death in all human communities. So while too many seers may constitute a problem, *some* seers seem necessary to the survival of the community as a cohesive unit. That is why the role of the evolutionary "agency detector" is so crucial; its existence among the human species enables them to *build communities that share foundational beliefs without question and visualise a common future for themselves.* For the only species on the planet which possesses the linguistic gift of "displacement"—that is, the ability to summon up scenarios that do not exist in the here and now—such a device could well prove invaluable to survival in a competitive world.

Rushdie's novels, especially *The Satanic Verses*, I suggest, specifically provoke the defensive reactions that come about when a seer's vision, long taken on trust, is challenged in a special way. That is, it is challenged from *within* and through what could be construed as deliberately hurtful techniques of deception. This is because Rushdie's novels invoke familiar patterns of reference that draw attention to themselves as texts of religion, repeatedly referring to Islamic holy days like the feast of Ibrahim, pilgrimages like the Haj and holy sites like the mountain in the Hijaz. Not only does this author boldly make the angel Gibreel and the Prophet Mahomet central to his discourse, but he also incorporates figures like Khalid, Bilal and Hamza—each one of them a Koranic reference point. This move immediately confers on him "insider" status, from which vantage point he then proceeds to further display his intimate acquaintance with the roles played by these great champions of Islam.

Consider his masterly description of Bilal, for example. It echoes the Koran with absolute accuracy in rendering Bilal as the first muezzin, an Abyssinian slave out of Africa chosen by the Prophet himself, but in Rushdie's story this traditional figure undergoes a radical metamorphosis. Paradoxically, confusingly, in the text of *The Satanic Verses*, Bilal becomes at once a mere mouthpiece, an ersatz incantor on "certain surreptitious radio waves" and at the same time a revolutionary folk-hero—Bilal X. The passage in which Rushdie achieves this transformation is worth quoting at length.

> [O]n certain surreptitious radio waves...the voice of the American convert Bilal is singing the Iman's holy song. Bilal the muezzin: his voice enters a ham radio in Kensington and emerges in dreamed-of Desh, transmuted into the thunderous speech of the Iman himself.... The voice is rich and authoritative, a voice in the habit of being listened to; well nourished, highly trained, the voice of American confidence, a weapon of the West turned against its makers, whose might upholds the Empress and her tyranny. In the early days Bilal X protested at such a description of his voice. He, too, belonged to an oppressed people, he insisted so that it was unjust to equate him with the Yankee imperialist. The Imam answered, not without gentleness: Bilal, your suffering is ours as well. But to be raised in the house of power is to learn its ways, to soak them up through that very skin that is the cause of your oppression. The habit of power...is a disease, Bilal, infecting all who come through it. If the powerful trample over you, you are infected by the soles of their feet. Bilal continues to address the darkness. "Death to the tyranny of the Empress Ayesha, of calendars, of America, of time! We seek the Eternity, the timelessness of God...." Burn the books and trust the Book; shred the papers and hear the Word, as it was revealed by the Angel Gibreel to the Messenger Mahound and explicated by your interpreter and Imam. *Ameen*, Bilal said, concluding the night's proceedings, which in his sanctum the Iman sends a message of his own: and summons, conjures up, the Archangel, Gibreel. (1988, 211)

The enormous range of references in this passage, its literary ambiguity and its liberal morality are typical of Rushdie. Bilal remains black and a convert, but now the slave history of his people takes in Malcolm X's conversion to Islam, symbolising a heartfelt rebellion against "Yankee imperialism." On offer is both an empathetic reading of the theme of power and the way in which being "raised in the house of power" corrupts because those underfoot tend to mimic the violent ways of their oppressors *and*, astonishingly, of Ayatollah Khomeini's or "the Imam's" real comprehension of such oppression. The whole familiar vocabulary of "black power" is being reassessed here. Simultaneously, this historical perspective, studded with a succession of real-time, worldly events is contrasted with the eternalist gaze of religion, which envisions a pure world freed from the bondage of the temporal. What's more, Rushdie's presentation of these opposing viewpoints seems remarkably even-handed, at least at first brush. It's only when one takes a closer look at the mimic-hermeneutic commentary positioned just outside the quotation marks that enclose Bilal's own impassioned radio commentary that one begins to suspect where the author's sympathies lie. "Burn the books and trust the Book; shred the papers and hear the Word"—how do we read these words? Even here, we have to work at recuperating the ironic modernist subtext that lies under the simple echo-effect of these words. Nothing is straightforward. We find the great overarching themes of revelation, symbolised by the ethically challenged figure of the Archangel Gibreel; prophecy, embodied in the Prophet's dubious doppelganger, Mahound; and interpretation, exemplified by the Iman, placed alongside what seems like the most frivolous of puns. "*Ameen*, Bilal said"—another echo-effect, where the slight change of spelling from the familiar "Amen" phonetically copying the Arabic pronunciation of the word also mimics the deep tones of the black English "Ah mean" or "I mean." How much more complex can Rushdie's meanings get?[16]

The point is really that little of this engaging literary complexity is available to the mullahs who defend the faith against Rushdie's apostate writings. In the unlikely event that they did find the time and energy to sit and pore over Rushdie's highly recondite texts, they would, according to my argument, probably dismiss his wordplay as frivolous and his views as deceptive. In any case, it is self-evident that, in actual fact, neither the mullahs nor the general public, nor the media, go in for any such complicated readings. As Rushdie puts it himself, in answer to an interview question about whether he was surprised by the fatwa, the forces that condemned him could not, for the most part, be bothered to read him.

> Clearly, there are more people in the world who have not read me than have. Unfortunately!… Well, I wasn't surprised that a bunch of mullahs didn't like it, because they never liked anything I wrote anyway. And I didn't particularly write for them. I thought there would probably be a bit of an argument about it, you know? But I thought, that's fine. It's actually one of the good things that books can do, is to start interesting arguments…. So I thought that would happen.... [But] who would have thought this kind of thing? That the leader of a foreign power would suddenly instruct his minions to have me killed? It would never really happen to a writer. And, frankly, if I'd have told you on the publication date of *Satanic Verses* that that would happen, you'd have thought I was mad.[17]

Mullahs are not literary critics and neither are they mad. It is just that they seem to prefer a much simpler, time-tested, means to suss out right from wrong. This is the test of what Rushdie would call "the Word" and I call "the Name."

Names with "memetic" value, like Gibreel and Satan, heavy with incantatory significance, are simply *not* to be fooled about with, however wonderful the literary flair with which they are claimed. At the very least they have to be taken seriously, even when they are criticised. For example, in the wake of one of Rushdie's "private" visits to India in 2004, the youthful leader of a small Islamic group in Delhi, who, a-typically, claimed to have read *The Satanic Verses*, appeared on an India-wide television channel. In that highly publicized interview, he declared the book, without mincing words, to be "shit." As a follower of the Koran, he opined that there was little doubt that the author deserved to be liquidated.

16. While my question here is rhetorical, the issue of the ambiguity of meaning raised by Rushdie's text is fundamental to our understanding of modern linguistics from Chomsky's pathbreaking analysis of the differences between syntactic and semantic ambiguities in *Syntactic Structures* (1957) to current work (2009) in post-Gricean pragmatics on "salient meanings" by Keith Allan, Katazyna Jaszczolt and others (see, for instance, the papers presented at the International Pragmatics Association Conference, Melbourne, July 2009, abstracts at http://ipra.ua.ac/be).

17. Interview with Linda Richards in the Canadian magazine *January*, 2002. http://januarymagazine. com/profiles/rushdie2002.html

However, as a good citizen of the India, he had himself chosen to follow "the laws of the land." This created a conflict within him, yet torn though he was, he was persuaded in his heart that the death of Rushdie was a not undesirable resolution to the conflict.

Within such a world-view, it is unlikely that one would accept that the great, iconic "Names" of religion can be utilized merely to "start interesting arguments," as Rushdie would have it. Such names carry great political and ideological charge and even the most refined of academics recognise this. For example, when Rushdie spoke in 2003 at the University of Michigan, Ashutosh Varshney, then director of the Center for South Asian Studies, introduced him with the question: "Why should we talk of the political Rushdie at all? Why should we not entirely abandon Salman to the literary scholars? Because Salman is an intensely political being" (Herwitz and Varshney 2009, 1).

The truth is that "abandoning Salman to the literary scholars" has not been an option ever since the fatwa was issued by "the Imam," Ayatollah Khomeini, as long ago as 1989. It's not that Rushdie is not "an intensely literary being"; indeed he has often argued passionately for literature, and especially the genre of the novel, as one of the most sacred spaces within the paradigm of modernity because it stands for free speech and an unfettered imagination. Moreover, passages like the one above on Bilal, illustrate the superb dexterity with which he handles language and thought in his own novels—but this is not the issue.

Rather, I would argue that Rushdie has over the past fifteen years been propelled along a path that has in fact *made* him ever more "an intensely political being," but that this has very little to do with the way he has been "read" by conservative Islam. While it is undeniable that "chance" events like the fatwa and, on a larger scale, the increasingly violent global profile of Islam, have influenced Rushdie to become far more overtly political, there is perhaps a more fundamental explanation for the configuration of events around his texts. The shock of 9/11 and continuing suicide bombings across the world (Palestine, Indonesia, Pakistan, Iraq, Mumbai) may also have played their part in making him to speak out in what he calls "this strange, unattractive historical moment." However, my contention is that a more basic, if seemingly remote, question needs to be asked if we are to really understand the face-off between Rushdie's complex fictions and the "fundamentalist" Islam that they oppose.

What are the deep sources of the cultural angst that drive a suicide bomber to self-destruct in the face of the deep evolutionary instinct that cues each one of us to survive as long as we possibly can? What underlying cognitive processes go into the selection of a "satanic enemy" and then animate religious "believers" to hunt him down at all costs? What combination of external political circumstances and internal psychic fears caused the panic buttons to sound off in the Ayatollah's head (or those of his advisers) when they were first informed of the advent of Rushdie's "insulting" text?

Answers to unlikely queries like these could go some way towards explaining not only why the Ayatollah, but also those with strong communitarian convictions such as President George Bush and Prime Minister Tony Blair react in a pathologically aggressive way when they perceive the presence of "Satan." Similarly, they might explain Rushdie's "coming out" so strongly against his own demon of dictatorial Islam in ever more direct and confrontational terms during the last decade. Witness the following statement, for example:

> For a vast number of "believing" Muslim men, "Islam" stands, in a jumbled, half-examined way, not only for the fear of God—the fear more than the love, one suspects—but also for a cluster of customs, opinions and prejudices that include their dietary practices; the

> sequestration or near-sequestration of "their" women; the sermons delivered by their mullahs of choice; a loathing of modern society in general, riddled as it is with music, godlessness and sex.... (2002, 395)

Rushdie's emphasis in this article on the emotion of fear is significant from the point of view of the argument that I am making—for fear is the most basic of evolutionary triggers. Together with its affective cousins—terror, awe, anxiety, panic—it could well be responsible for activating various pathologically "rational" narratives, including religious ones, that seek to "explain" why we feel and act the way we do. In a recent article, Ronald Pies, a clinical professor at the Tufts University School of Medicine, analyzing narratives by depressives, writes:

> The inherent problem in positing a depressive trigger is that we humans are famous for constructing explanatory narratives, even when the facts of the situation are not clear. Indeed, in her book, *Narrative Gravity*, linguistics professor Rukmini Nair argues that human beings have a genetic drive to fabricate narratives that serve our emotional needs. (2009, 21)

And it is here that we return to the notion of the religious "agency detectors" propounded by Atran and Norenzayan. My thesis in this essay is that when memetic religious names such as those in the Koran suddenly appear in "new" texts at times of grave political crisis or turmoil where, in Pies words, "the facts of the situation are not clear," they tend to reflect a situation of wide-spread social anxiety and fear. It is at such times of crisis that "seers" are most needed but "false prophets" equally abound. Just as a seer might once have interpreted "voices in wind, faces on clouds," iconic names in religious texts, I suggest, are phenomena that urgently require interpretation. Since they constitute *abstract semantic patterns*, exactly like "moving dots on computer screens," they activate, in the contemporary social context, those very primal evolutionary triggers that "once animated objects relevant to hominid survival—such as predators, protectors and prey."

Quick and unsparing reactions are needed in such times; there is no time to cogitate and dither, especially if it is thought that the vulpine "enemy"—in this case, Rushdie—is one already ensconced in the fold. This is a chief reason why it was not necessary for the Ayatollah Khomeini to read *The Satanic Verses* before issuing his fatwa; his move was an instinctive yet shrewd act, both defensive and aggressive. Indeed, it would be "mad" or irrational for him to have settled down in a leisurely fashion to read Rushdie's text and deliberate on its merits and demerits in the light of the immediate danger that it seemed to represent to the ruling Iranian dispensation at the time. As usual, Rushdie states the case most succinctly himself: "Iranian officials have admitted that Khomeini never so much as saw a copy of the novel."

In the same way that an atavistic "fight or flee" reaction is set in motion when one is confronted by what appears to be a serpent—or what one is told is a serpent—one cannot afford to pause and reflect in such a situation. There is no "mental space"[18] at such a time to carefully sift evidence that could indicate that the dreaded snake might be a harmless rope instead. The mere fact that the "animate object" before one *might* be a snake is good enough to set one's "agency detectors" going. Consequently, it is a little disingenuous for Rushdie to plead, as he has done so often on and off, that he should be *read* before being condemned by Islamic believers. For, Rushdie's predicament arises more basically out of the way in which communities, and especially religious communities, have evolved from the earliest times.

18. This phrase echoes the title of Gilles Fauconnier's book *Mental Spaces* (1985).

As the evolutionists contend, once certain perceptual designs, such as iconic "name-triggers" appear in a text, the default presumption is that the text in question will have a certain pattern, follow a certain trajectory, or as Rushdie phrases the same thought in a different context: "We all have a kind of description of the world and out of that come our instincts and how to behave and what is appropriate and what is inappropriate and so on."

It is this basic "description of the world," as he accurately puts it, "out of which come our instincts"—instincts that we tend to take on trust and, for the sake of our communal "sanity" *and* "sanitization" not to question. That is why it continues to be the case that it is not in the least required for Islamic scholars to have read Rushdie's erudite multi-layered fictions to "dislike him." It is also why many liberal Muslims in India, for instance, who are completely against any sort of ban on works of literature, continue to "personally somehow suspect Rushdie's novels" even as they earnestly defend his right to freedom of speech.

Culturally, as products of the liberal ideal of the Muslim Ummah or community of brotherhood, Indian Muslims seem to continue to feel a deep unease with the way Rushdie has apparently "distorted" or "used" his privileged access as a Muslim to the texts and symbols of Islam. These belong to the discourse of faith or "the Word," yet here they are in a text which, although superficially *patterned* like the "original" is not about faith at all, but about its satanic antitheses—doubt, rebellion and pride. The "choices of the faithful" are deliberately mocked in such writings. As he puts it in *Fury*, in his usual acutely self-reflexive way: "When a man without faith mimicked the choices of the faithful, the results were likely to be both vulgar and inept" (2002, 81).

Few can accuse Rushdie of ineptitude as an author, but vulgarity or "showing off" is a different matter. Rushdie has been charged time and again, especially in India, of staging high voltage performances for the "natives" of the West, with little regard for sensibilities of the poor and the faithful natives of the subcontinent. Of course he at once produces an explanation for such thoughtless behaviour on his part, and it is an interesting one in terms of evolutionary theory. A few sentences down the line from *Fury* just quoted, he describes the immigrant hero of his novel in New York, city of cities wherein is lived the "brilliant, brittle, gold-hatted, exemplary American life":

> [He] had landed himself in a city in which he was not completely unknown, which was not completely unknown to him, in which he could speak the language…and understand, up to a point the customs of the natives. He had acted without thinking…before he allowed himself to reflect; then he'd simply accepted the imperfect choice his reflexes had made, agreed to proceed down the unlikely road on to which his feet, without prompting had turned. A *sanyasi* in New York…. (2002, 82)

Rushdie's claim here, in effect, is that his decisions, his turn towards the glittering West was instinctive. He had "acted without thinking" because he was familiar with this world and not the backwoods world of the Indian *sanyasi* (or the small-time Muslim cleric, for that matter), turning towards it as naturally as a sunflower turns towards the sun. Just as with the "blind" followers of Islam in a different cultural context, his literary and intellectual "reflexes" had *made* his choices for him.

From the point-of-view of his co-religionists, on the contrary, such betraying reflexes would be quite "unnatural." They could be "worms," as Richard Dawkins, whose views on the evolution of religion we're to consider shortly, calls them. Exactly like William Blake's "invisible worm" at the heart of the rose,[19] the burrowing worms of doubt and self-love are

19. "The Sick Rose" from "Songs of Experience" in *The Complete Poetry and Prose of William Blake* edited

known to sicken and destroy the community from within, which accounts for the sharp animus that they cause. Rushdie, indeed, "instinctively" seems to know this, and places the matter of false documentation—the perspective of the doubter—plumb at the centre of *The Satanic Verses.* Focussing on the figure of his namesake Salman the Persian, Rushdie's sophisticated narrative shows this character *deliberately altering* the text that the Prophet is dictating to him. But what psychological gains accrue to Salman the Persian from this extraordinary "instinctive" rebellion of his?

Or, to place this textual question in a broader psychological context, one might ask the following question. What is the significance of the ubiquitous figure of the apostate, the prodigal son, and other "rebellious" figures within the paradigm of modern religions like Christianity and Islam? It is this question that the other perspective from evolutionary theory that I've mentioned, namely, Dawkins', can be said to address.

Religion persists, Dawkins argues, in his book *The God Delusion* (2003), because children have a deep-seated psychological tendency to take on their parents' beliefs. Differing from Atran and Norenzayan on the role played by "agency detectors" in religious evolution, Dawkins fixes not on the unique psychology of the seer, but instead on what he thinks of as a capacity just as "unique." This, according to him, is the universal orientation towards obedience among human children, which makes them highly vulnerable to the "mind-viruses" and "worms" that could constitute a rebellion against the prevailing parental order.

Society, in turn, encourages such absolute loyalty by labelling children with the religion of their parents. Or, as Dawkins more colourfully puts it: "A child is not [naturally] a Christian child, not a Muslim child, but a child of Christian parents or a child of Muslim parents" (2006, 382). Thus conditioned by their religious labels, children from deeply religious backgrounds face constraints in making independent decisions as adults with regard to their beliefs. They regard other religions with fear as possible competitors and tend to react by fiercely defending their own religious terrain. So, citing the example of Protestant fundamentalists in Belfast spitting at young Catholic girls merely because their parents labelled them Catholic, Dawkins contends:

> Though it is universal, religion is not widely beneficial. [It]...not only provides false comfort—it is actively divisive and harmful. Society provides a breeding ground for the "virus" of religion by labelling children with the religion of their parents. Designated as Christians or Muslims by their parents, children are apt to face the discrimination associ-

---

by David V. Erdman, with a Commentary by Harold Bloom, p. 23. Note that William Blake presents us with the case of a relatively recent literary figure who seriously admitted to being a "seer." As a child, for instance, the poet claimed he saw God "put his head to the window and set him a-screaming," and at age ten, he tried to convince his father that he had seen hosts of angels in a tree in Peckham Rye. Later, at his brother Robert's death-bed as an adult, he observed casually that he saw his brother's soul "ascend heavenward clapping its hands for joy." What is noteworthy from an evolutionary perspective is: a) that Blake retained this strong faith in the spirit world throughout his life, affirming that he often spoke with the apparitions that populate his work, and b) that his writing continues to be valued by the community two centuries later as a "faith" resource. The visionary psychological propensities of the man who wrote, "I can look at the knot in a piece of wood until it frightens me," thus seem in some way important for wider community affirmation of an unconventional yet deeply-felt "Christian" faith. One of many differences between Rushdie and Blake is that though both have produced unconventional literary texts steeped in religious experience, Rushdie's works are decidedly *not* those of a seer, and he has repeatedly refused to reconfirm "his" community's most fundamental reasons for being, namely, their allegiance to Islam. This is the move that appears to confirm him as a freeloader not to be communally tolerated.

ated with these labels.[20]

A similar line of thought is also to be detected in an earlier essay by Dawkins called "Viruses of the Mind" (1993). Here, he had begun by remarking that "a human child is shaped by evolution to soak up the culture of her people." Pursuing an epidemiological metaphor in that preliminary paper, he took the view that the institutional hotbeds of religion and science in particular offer fertile breeding grounds for "mind viruses" among the impressionable, such as children. These fields, he argued, take in the gullible but also draw in the most profound of thinkers.

More generally, I would suggest that what is common between these institutional fields (science and religion) is that they offer specialized, doctrinal and yet capacious and socially legitimate "niches" for a) community member-shipping, and b) pattern-recognition. As a result, emotional, ethical *and* aesthetic needs are simultaneously satisfied when any individual hosts mental viruses from these areas. Since religious narratives comfortingly repeat semantic motifs and syntactic units, creating sets of iterated patterns, they are foundational to our sense of aesthetics and at the same time memetically serve to strengthen our belief in the ethical norms that guide us.

The upshot of all this for an analysis of Rushdie's texts is that he seems to be caught in an evolutionary cleft stick. By Dawkin's yardstick, he is a prototypical "bad boy" or "disobedient child" and thus calls down upon himself the patriarchal wrath of the community. Panic is therefore automatically activated by his literary gestures. At the same time, there is always a special place for the Prodigal Son and the Satanic Verse in all religious narratives. This is because they serve a two-fold purpose: they highlight the tolerance of the father figure, as when the Prophet forgives Salman for producing false testimonies in his Name. Simultaneously, they create a pattern or meme-stencil that serves to identify and reinforce the boundaries of the community, reminding the faithful not to "stray." Rushdie, of course, is the prototypical "stray"[21] and as such a study of his texts provides an indirect but very

20. Report in *The Harvard Crimson*, "Oxford Scientist Launches Sharp Critique of Religion," by Asya Troychansky, November 20, 2003. "Rejecting the theory of many of his contemporaries, Dawkins argued that religion has not helped people to adapt or to survive. Beyond acting as a source of solace, religion provides no protection against diseases or physical threats. 'A person who is faced with a lion is not put at ease when he's told that it's a rabbit,' he said. Children absorb their parents' religious beliefs because they are conditioned to do so. 'Religion should be something for children to choose or not when they become old enough to do so,' Dawkins said. He did, however, concede that 'The sense of transcendence is something that is shared even by those who don't call themselves religious [such as scientists in pursuit of theoretical truths].' Following Dawkins' speech, Johnstone Professor of Psychology Steven Pinker proposed two amendments to Dawkins' theory, both of which Dawkins accepted. Pinker argued that peers, as much as parents, affect children's beliefs. He also attributed the spread of beliefs to the active faculty of the human brain, which he referred to as 'the intricate activity of neurons.'"

21. In both his texts and his essays, Rushdie in fact fashions himself not only as a "stray" but as a stray *dog*. He explicitly likens himself to the dog in Saul Bellow's novel *The Dean's December* in which a character "hears in the distance the…insistent, unstopping, barking of a dog... And he imagines what the dog is saying in his barking is 'For God's sake, open the universe a little more,' and I've often thought that the job [of the writer]. That's the job, open the universe a little more" (Herwitz and Varshney 2009, 38–39). Yet, for all his enthusiastic identification with a dog's duty to "open up the universe a little," Rushdie must recognize the irony of the fact that dogs in Islam are generally accounted "unclean animals; for according to a tradition by Abu Hurairah, Muhammad said that when a dog drinks in a vessel, it must be washed seven times, and that the first cleansing must be with earth" (*Mishkat*, book iii, c. ii pt 1. Thomas Patrick Hughes, *Dictionary of Islam*, 1885, 91).

complex and detailed answer to the third and last question that I'd raised at the beginning of this essay:

> If the "best" narratives across cultures obey that hoary dictum of creative writing to "show, not tell" then how are emotions best displayed in particular genres of narrative, so they at once enable the reader/listener to experience the appropriate feelings and leave ample place for freedom of interpretation and subversive speculation?

Current biology tells us that all human beings share 99% of their genetic material. So what makes each one of us so "different"? How are renegades like Rushdie to be identified and punished because of their intentionally projected difference from others in their community? In my view, the role of a powerful cultural device like narrative is to function like a natural, internalized, magnifying glass. It is a tool which "blows up" these minute one-part-in-thousand differences so that they can be made the basis of individual character traits and cultural distinctiveness. Literary fictions, with their great scope for interpretative expansion, serve this deep cultural purpose. Referring to my own work, Marianne Horsdahl, for example, writes:

> We construct and listen to narratives to make sense of our experiences and the experience of others. Stories are told and shared as a reservoir of vicarious experience. Stories are mental time travels in imaginary contexts, in different worlds, narrated from different perspectives. Rukmini Bhaya Nair, an Indian researcher, describes the human possibility of narrative cognition as a great evolutionary progress, increasing our chances of survival. Our knowledge of environment is not limited to our own direct experiences but extended by the means of narratives, which give us a wider range of possible interaction.[22]

Rushdie, Pynchon and Lodge each create this "wider space" for interaction and in the process generate considerable, but in the final analysis, possibly healthy, social perturbation. Insofar as they infuse the received narratives of religious discourse, and their time-worn memes, with a new literary energy that speaks directly to the times and engages with contemporary intellectual and social dilemmas, they may thus be said to contribute, in evolutionary terms, to keeping the religious conversation alive. By "re-theorizing" the narrative of religion, they do for religion what contentious papers in the academic arena do, let's say, for sociology or history or political theory, as I argued at the beginning of this essay.

Should we now turn, at the end of this essay, to a last commentator, this time located within the strict confines of the cognitive sciences, we seem to find an echo of just such a move towards "complexifying" the discourse of mind *à la* Rushdie in literature. Intervening between the programmatic essay by Hauser, Chomsky and Fitch in 2002 discussed earlier in my paper and one by Stephen J. Gould appositely entitled "Sociobiology: The Art of Storytelling" written in 1978, is a rough quarter of a century. The subject of Gould's paper is the contrast between two stories or theories which "make all the difference in the world." The first of these takes the position that "human behaviors develop and stabilize by cultural evolution" while the second argues for "direct Darwinian selection for genes influencing special adaptive action" (Gould 1999, 84).[23] But while these opposing stances may

22. Marianne Horsdahl, "Narrative Cognition and Construction—the Implications for Biographical Narrative Research," in the ESREA Biography and Life History Network (European perspectives on Life History research: theory and practice of biographical narratives, Geneva, 2002).

23. For an exhaustive literary reading of Darwin and his cultural context, see also Gillian Beer 2000. This book is of some interest in terms of the larger argument of this essay since, as the Preface by George Levine to Beer's book points out, "Darwin's language shows him to be as much a believer in cooperation

have been well defined back in 1978, the situation seems to have grown messier in 2008, owing perhaps to a packed phase of invention and discovery in the area of the biological sciences. Consequently, the implications to be drawn for narrative studies from developments in allied disciplines like evolutionary theory have also changed dramatically. There appears to be a shift of research emphasis towards an integrative "thick description" (Geertz 1973) of cognition. The rather impressionistic account of current research initiatives I have outlined in this paper seems to suggest that at least four major areas of research are likely to contribute to future "meta-retellings" of "the story of narrative." These consist in:

1. Neurobiological and psychological experiments that seek to discover how the brain reacts, emotionally and otherwise, to narrative input;
2. Studies in evolutionary biology and genetics that could throw light on why narrative structures have developed as they have and on possible constraints on these structures, (such as the brain's capacity for mnemonic retention, etc.);
3. Work in linguistics and especially on children's acquisition of narrative skills, including research on story-telling abilities among groups such as autistic children who might suffer from putative "theory of mind" impairments;
4. Interpretive moves in literary and philosophical theory which enable a better understanding of narrative as a key element in cultural process and cognition.

Certain attitudes, however, have remained reassuringly stable. Gould, for instance, very properly warns his audience against the sweeping "Just So" stories that socio-biologists—and by extension, their academic cousins, the cognitive scientists—sometimes choose to tell about evolution, but the odd thing is that in the end he too succumbs to the overwhelming allure of narrative interpretation.

> There is no such thing in nature as a self-evident and unambiguous story.... [T]he world of human behaviour is too complex and multifarious to be unlocked by any simple key.... [T]his richness is both our hope and our essence. (Gould 1999, 82)

As one anticipates the myriad schemes, genes, memes, dreams and screams that are bound to be as inevitable a part of our unknown future as they were of our genetic past, one cannot help but respond with a resounding "Amen" to Gould's upbeat vision of a species "religiously" committed to narrative invention and interpretation for its very survival.[24]

## References

Anderson, Benedict. 1983. *Imagined Communities: Reflections on the Origin and Spread of Nationalism.* London: Verso.

Arbib, M.A. 2002. "The Mirror System, Imitation, and the Evolution of Language." In *Imitation in*

and what Kropotkin called 'mutual aid' as ruthless competition."

24. In my most recent book *Poetry in a Time of Terror: Essays in the Postcolonial Preternatural* (2009), I attempt to include other literary forms like poetry within the ambit of an evolutionary explanation. Among other things, I suggest in this book that poetry can be seen to function as "a form of secular prayer" and poems as "texts of crisis" that constitute an affective response to private as well public emergencies. Despite their apparent "uselessness," I argue that song and story survive in all known human cultures because they constitute deep evolutionary safeguards against that fatal terror of "the other" that has, paradoxically, always caused our species to wage war against itself. By engendering multiple languages, genres and discourses that ensure a stubbornly indelible space for interpretative license, poetry, narrative and other literary genres teach us, generation after generation, to delight in our imaginative capabilities, and thereby eschew self-hatred, fear and despair.

*Animals and Artifacts*, edited by C. Nehaniv and K. Dautenhahn, 229–280. Cambridge, MA: MIT Press.
Atran, Scott and Ara Norenzayan. 2004. "Religion's Evolutionary Landscape: Counterintuition, Commitment, Compassion, Communion." *Behavioral and Brain Sciences* 27: 713–770.
Bakhtin, M.M. 1990. *Art and Answerabilty: Early Philosophical Essays*. Austin: University of Texas.
Bal, Mieke, ed. 2004. *Narrative Theory: Critical Concepts in Literary and Cultural Studies*. London: Routledge.
Beer, Gillian. 2000. *Darwin's Plots: Evolutionary Narrative in Darwin, George Eliot and Nineteenth Century Fiction*. Cambridge: Cambridge University Press.
Cole, P. and J. Morgan, eds. 1975. *Syntax and Semantics*. Volume 3. New York: Academic Books.
Dahlbom, Bo, ed. 1993. *Dennett and his Critics: Demystifying Mind*. Oxford: Blackwell.
Damasio, Antonio. 2000. *The Feeling of What Happens: Body and Emotion in the Making of Consciousness*. New York: Harvest Books.
———. 2005. *Descartes' Error: Emotion, Reason, and the Human Brain*. New York: Avon Books.
Darwin, Charles. 1872. *The Expression of the Emotions in Man and Animals*. London: John Murray, (London: Harper Collins, corrected Third Edition, 1989).
Dawkins, Richard. 1976. *The Selfish Gene*. Oxford: Oxford University Press.
———. 1993. "Viruses of the Mind." In *Dennett and his Critics: Demystifying Mind*, edited by Bo Dahlbom, 13–27. Oxford: Blackwell.
———. 2006. *The God Delusion*. London: Bantam Press.
Dennett, Daniel C. 1991. *Consciousness Explained*. Boston and New York: Little Brown.
———. 1992. "The Self as a Center of Narrative Gravity." In *Self and Consciousness: Multiple Perspectives*, edited by F. Kessel, P. Cole and D. Johnson, 103–114. Hillsdale, NJ: Erlbaum.
———. 2006. *Breaking the Spell: Religion as a Natural Phenomenon*. New York: Viking Adult.
Dentith, Simon, ed. 1995. *Bakhtinian Thought*. London: Routledge.
Dunbar, Robin. 1998. *Grooming, Gossip and the Evolution of Language*. Cambridge, MA: Harvard University Press.
Durkheim, Émile. 1995 [1912]. *The Elementary Forms of Religious Life*. Translated by Karen E. Fields. New York: The Free Press.
Eco, Umberto. 1992. *Interpretation and Overinterpretation*. Cambridge: Cambridge University Press.
———. 1994. *Six Walks in the Fictional Woods*. Cambridge, MA: Harvard University Press.
Everett, Daniel L. 2007. "Recursion and Human Thought: Why the Pirahã Don't Have Numbers." *Edge: the Third Culture*. http://www.edge.org/3rd_culture/everett07/everett07_index.html
Fauconnier, Gilles, 1985. *Mental Spaces*. Cambridge, MA: MIT Press.
Gallese, V. and A. Goldman. 1998. "Mirror Neurons and the Simulation Theory of Mind-Reading." *Trends in Cognitive Sciences* 2(12): 493–501.
Gazzaniga, Michael A. 1994. *Nature's Mind: Biological Roots of Thinking, Emotions, Sexuality, Language, and Intelligence*. New York: Basic Books.
———. 2006. *The Ethical Brain: The Science of Our Moral Dilemmas*. New York: Harper Perennial.
Geertz, Clifford. 1973. *The Interpretation of Cultures*. New York: Basic Books.
Gould, Stephen J. 1999. "Sociobiology: the Art of Storytelling." In *The Biological Basis of Human Behavior: A Critical Review*, edited by Robert W. Sussman. New Jersey: Prentice Hall. Originally published in *New Scientist* 80 (1978): 530–533.
Grice, Herbert Paul. 1989. *Studies in the Way of Words*. Cambridge: Cambridge University Press.
Hauser, Marc. D., Noam Chomsky and W. Tecumseh Fitch. 2002. "The Faculty of Language: What

Is It, Who Has It, and How Did It Evolve?" *Science* 298: 1569–1579.
Herwitz, Daniel and Ashutosh Varshney. 2009. *Midnight's Diaspora: Encounters with Salman Rushdie*. New Delhi: Viking Penguin.
Hogue, John, ed. 1997. *The Millennium Book of Prophecy*. New York: HarperOne.
Hurford, James. 2003. "The Neural Basis of Predicate-Argument Structure." *Brain and Behavioral Sciences* 26 (3): 261–283.
———. 2004. "Language Beyond Our Grasp: What Mirror Neurons Can, and Cannot, Do for Language Evolution." In *Evolution of Communication Systems: A Comparative Approach*, edited by D.K. Oller and U. Griebel, 297–313. Cambridge, MA: MIT Press.
Jackendoff, Ray. 2003. "Précis of *Foundations of Language: Brain, Meaning, Grammar, Evolution*." *Brain and Behavioral Sciences* 26(6): 651–665.
Labov, William. 1972. *Language in the Inner City*. Philadelphia: University of Pennsylvania Press.
LeDoux, Joseph. 1998. *The Emotional Brain: The Mysterious Underpinnings of Emotional Life*. New York: Simon and Schuster.
———. 2003. *Synaptic Self: How Our Brains Become Who We Are*. New York: Penguin.
Lodge, David. 2001. *Thinks*. Harmondsworth: Penguin Books.
Morehouse, Geoffrey. 1983. *India Brittanica*. London: Harvill Books.
Nair, Rukmini Bhaya. 1995. "Why Does Rushdie Offend?" *Seminar* 437, Annual Issue, December, 103–109.
———. 2002. *Narrative Gravity: Conversation, Cognition, Culture*. New Delhi: Oxford University Press.
———. 2009. *Poetry in a Time of Terror: Essays in the Postcolonial Preternatural*. New Delhi and New York: Oxford University Press.
Ochs, Elinor and Lisa Capps. 2001. *Living Narrative*. Cambridge, MA: Harvard University Press.
Pies, Ronald. 2009. "Narratives, Normality, and Diagnostic Neutrality." *Psychiatric News* 44(5): 21.
Pinker, Steven. 1995. *The Language Instinct*. New York: Harper Collins Perennial.
Ramachandran, V.S. 2000. "Mirror Neurons and Imitation Learning as the Driving Force behind 'The Great Leap Forward' in Human Evolution." *Edge* 69, May 29, 2000. [http://www.edge.org/3rd_culture/ramachandran/ramachandran_p1.html]
Rilling, J.K., D.A. Gutman, T.R. Zeh, G. Pagnoni, G.S. Berns and C.D. Kilts. 2002. "A Neural Basis for Social Cooperation." *Neuron* 35: 395–405.
Rizzolatti, G. and M.A. Arbib. 1998. "Language within Our Grasp." *Trends in Neurosciences* 21(5): 188–194.
Rushdie, Salman. 1988. *The Satanic Verses*. New York: Viking Penguin.
———. 2002a. *Step Across this Line: Collected Non-Fiction 1992-2002*. London: Random House.
———. 2002b. *Fury*. London: Jonathan Cape.
Sacks, Harvey. 1989. *Lectures 1964–1965*. New York: Springer.
Searle, John. 1976. "A Classification of Illocutionary Acts." *Language in Society* 5 (1): 1–23.
Tharoor, Shashi. 1999. "The Art of Writing, the Work of Diplomacy: Conversation with Author and Diplomat Shashi Tharoor," interviewer Harry Kreisler. In *Conversations with History*. Institute of International Studies, UC Berkeley series. http://globetrotter.berkeley.edu/people/Tharoor/tharoor-con3.html
Todorov, Tzvetan. 1984. *Mikhail Bakhtin: The Dialogic Principle*. Minneapolis: University of Minnesota Press.
Toolan, Michael. 2001. *Narrative: A Critical Linguistic Introduction*. London: Routledge.
Turner, Mark. 1996. *The Literary Mind*. Oxford: Oxford University Press.

# — 7 —

# Believing and Doing: How Ritual Action Enhances Religious Belief

Ilkka Pyysiäinen

I outline here the hypothesis that participation in religious rituals contributes to religious belief in a specific way. I first explain what I mean by rituals and religious belief; then follows a cognitive analysis of motor action and agency, followed by an analysis of the nature of shared information. Finally, I try to show in what ways religious rituals can enhance religious belief as an explicit or implicit view that religious concepts refer to something real which is of vital importance.

## Ritual and Belief

Religious rituals consist of collectively sanctioned activities that 1) follow some preexisting plan, 2) involve counterintuitive agents either as actors, patients of action, or as interested witnesses of action, and 3) aim to produce effects or outcomes the satisfaction of which cannot be judged by evidence. This impossibility of evaluating the outcome derives from the fact that the sought effect either is completely transcendent or does not exist apart from the ritual performance itself (see Boyer 1994, 185–198; Pyysiäinen 2001, 88–90).

By religious belief, in the sense of a belief or beliefs (belief$_A$), I mean belief(s) with religious content. Religious belief as an attitude (belief$_B$) is an attitude in which one emotionally feels religious concepts to be relevant. It is not, however, necessary to consciously represent, for example, "I have the belief that God exists." It is enough that one acts and thinks in ways that presuppose the belief$_A$ "God exists" (see Oberlander and Dayan 1996).

Conversely, it is also possible to have the conscious belief "I believe that God exists" without this belief actually serving as a premise in reasoning (Barrett and Keil 1996). I call these two instances of religious belief$_B$ *Boyerian belief* and *reflective belief*, the former so well-described by Boyer (2001), and the latter due to a reflective attitude (see Pyysiäinen 2004b, 113–134). Barrett (2004) makes a similar distinction between intuitive and reflective beliefs. Note, however, that it is also possible to combine Boyerian and reflective belief. In that case, one is consciously aware of belief$_B$ and beliefs$_A$ that influence one's thought and action (see Pyysiäinen 2004a).

It is possible to "have" religious beliefs$_A$ without religious belief$_B$. If I, for example, have the belief$_A$ "Joe believes that 'God exists,'" I do not necessarily myself believe$_B$ that "God exists." I mentally represent "God exists" but I do not have belief$_B$ in this statement. I could, for example, mentally represent "'God exists' is not true" (see Pyysiäinen 2004b, 172–186).

It is relatively easy to explain how we acquire religious beliefs$_A$: we come to have them by hearing or reading about them, or by inferring them from our other beliefs (see Sperber 1997; Pyysiäinen 2004a). It is also possible *in principle* that at least some religious beliefs are literally innate (see Laurence and Margolis 2002), although it is almost as impossible to verify or falsify this claim as it is to verify or falsify the claim that these beliefs are "revealed."

It is more difficult to explain how religious belief$_B$ emerges, or what it actually is. As we will soon see, everyday belief$_B$ has its basis in communication and motor interaction with objects in external reality, thinking being a kind of imagined movement (Lakoff and Johnson 1999; Cruse 2003; Gallese and Metzinger 2003). Reflective beliefs are then abstracted from such immediate experience. We can represent goals, the realization of which requires a long chain of events, without in any sense representing all those intermediary links (Gallese and Metzinger 2003). Such distal goal-states are attained through a series of multipurpose action schemata chained within a highly specific temporal sequence (Gallese and Metzinger 2003). This entails that the tokens we use to refer to objects and events are decoupled from direct token-object relationships.

To follow Deacon's (1998, 70–101) use of a Peircean terminology, referential relationships can be divided into three types: iconic reference is based on similarity (picture of car represents a real car), indexical reference is based on contact (fingerprint represents the person in question), and symbolic reference is based on a convention shared by a number of people. Symbols thus refer only because they have a relationship to other symbols. Semantics is not simply referential but holistic (Quine 1980; Lawson and McCauley 1990, 146–155). There is thus a very important difference between symbolic and non-symbolic cognitive systems.

To learn an iconic or indexical relationship between an image and an object, it is enough to have what Perner (1993, 7–9, 70–75) calls a "situation-theoretical" understanding of referential relationships and which emerges in the second year in the developing child. The child is capable of using secondary representations decoupled from reality, as well as multiple models (see also Sinha 2004). He or she can interpret pictures as pictures of a depicted situation but not as representations: objects *re*presenting something. The child cannot understand that a picture can be given different interpretations, that it has a sense and a reference, and that pictures can also misrepresent. Yet situation theory is sufficient to understand simple iconic and indexical relationships in principle. A younger child with her or his "single updating model" may not be capable of this; for her, pictures are mere boring pieces of paper, objects resembling the depicted, without her necessarily understanding the similarity-relationship. Thus, learning iconic and indexical relationships is learning object-token relationships (see Deacon 1998, 270–271).

Learning symbolic relationships is quite different in the sense that it entails learning a whole symbolic system within which the object-token relationships are established (see Quine 1980). At some point in evolution, the human species crossed the "symbolic threshold" and proceeded from icons and indices to such symbolic reference that characterizes verbal language. It is the prefrontal cortex and its connections with the cerebellum that guides the construction of the distributed mnemonic architecture that supports the sensory-based mental images making symbolic associations possible. In reading fiction, for example, we feel mental images arising incessantly just as we feel sound motor images as we search our memories for the right words to say or the correct names to match with familiar faces. These images are mere neurological tokens of associative paths, like buoys indicating the best course, helping us reconstruct the implicit symbolic reference. The symbolic reference thus emerges from a pattern of virtual links between neurological tokens constituting a kind of parallel realm of associations to those that link these tokens to real sensori-motor experiences and possibilities. The symbols are not located in the brain; they are mere relationships between tokens. The symbolic function is constituted by the virtual set of associations partially sampled in any one instance (Deacon 1998, esp. 73–92, 265-267). The symbolic capacity thus is distributed in a socio-neurocognitive network that makes possible to learn such creative use of words that is not determined by simple object-token relationships.

This, however, is not enough to explain religious belief$_B$ because religious belief$_B$ is an attitude that takes its object-beliefs$_A$ neither as fiction nor as ordinary facts. For the believer, religious concepts refer to real entities but their reference is mysterious (see Edis 2002). The semantic relations of religious concepts are at once suspended and not suspended in the sense that, for example, the Christian Eucharist both means consuming the body and blood of Jesus Christ and yet it is not considered to be a meal in any ordinary sense (see Pyysiäinen 2004b, 113–134). The problem now is to explain how such belief$_B$ can be affirmed by experience. Here I shall restrict myself to outlining a hypothesis of the ways participation in religious rituals might enhance such belief. Rituals need not be the only means of such enhancement, however.

## Motor Action and Agency

The mechanism by which ritual participation enhances religious belief operates at three levels: conscious awareness, cognitive processes, and the brain. The brain and cognitive processes regulate the behavior of the organism as a whole by adjusting its behavior according to the requirements of the external environment and of the organism's own internal states. When any given sensory input can be responded to in differing ways, the system has internal states in the sense of possible motor actions from which it has to choose which one to execute (Cruse 2003). Reactive structures thus can lead to adaptive controllers which can handle tasks with a large number of degrees of freedom. Such a system is able to plan ahead, relying on its internal world model and by temporarily disconnecting the motor output. It is in this sense that thinking is imagined movement.

However, once such organismic systems develop, they will have to react not only to the physical environment but also to other similar systems. In other words, they will have to plan and predict interaction with behavioral systems and not only with causal-mechanical systems. This necessitates that such systems can somehow share information about their internal states. The internal states now must become objects for other internal states. As Sperber (2000) puts it, a cognitive system may associate some sign with some object or event and then use some mental symbol to refer to this relationship. When there is not merely a fixed one-to-one relationship between an internal state as object and another internal state as a token for that other state, full meta-representation takes place. Then mental symbols are not used as simple fixed tokens of relationships, rather they form a system of associations that is as at least as rich as the associations between objects and signs. They form a language-like structure with recursion and compositionality: it is possible to form long chains of recursive embeddings of mental contents. Conscious awareness is an experience of volitional control over meta-representation. Whether this scenario describes evolutionary development or not, these three levels of neural behavior regulation, cognitive structures, and consciousness are present in modern humans.

Following Gallese and Metzinger's (2003) arguments, the brain possesses a set of neurally represented "assumptions" about the elementary building blocks of reality, decomposing target space in certain invariant ways. The ontological "assumptions" of the brain then guide both motor output (behavior, action) and phenomenal output (the conscious mind that arises from such functioning). At the level of personal, phenomenal experience, this is experienced as ontological assumptions in the folk-psychological sense.

An organism capable of motor action has representations of goal-states that consist of representations of the organism's own state, of an aspect of external reality, and of the relation between these two. Organisms aspire to homeostasis and survival; anything that is in this regard relevant for an organism is represented as a goal-state, that is, as possessing some value for the organism. A conscious representation of value makes the represented value globally available in the organism's consciousness. We can also match internally represented goal-states against sensory input (or memories) and evaluate their satisfaction. Goal representations thus can also change the assumptions of the system ontology (Gallese and Metzinger 2003).

Goals thus are fundamental elements of the brain's model of the world. From such functional ontology arises a teleological ontology, the world of behavioral possibilities as represented by the organism. Actional possibilities are abstract possibilities in the sense that they are not experienced from any particular perspective: the brain represents them independent from experienced agency. A goal representation and the corresponding motor output are mediated by the "phenomenal model of the intentionality relation." This is a conscious mental model of an ongoing, episodic subject-object relation. Conscious volition is created when abstract goal representations are integrated into the current model of the phenomenal intentionality relation as object components (Gallese and Metzinger 2003). Ontological assumptions and representations of goals are beliefs in the sense that they are taken to refer to reality in such a way that makes successful action in the world possible for the organism in question (see Boyer and Barrett 2005).

The moment bodily behavior ensues is the moment when the already active motor simulation is integrated into the currently active bodily self-model. At that moment one identifies with a particular action, turning potential action into actual action. The action thus is functionally and phenomenologically embodied. The moment of agency is the moment when the phenomenal model of intentionality relation collapses in the sense that what was previously a possibility distinct from the self now becomes part of it. A possible (form of) self is realized by enacting it. Proprioceptive and kinesthetic feedback allows one to feel the degree to which one has identified with the sequence of bodily movements. A simulated self is realized when it is made the content of a transparent self-model. Agency consists of ongoing representational dynamics that collapse a phenomenal model of the practical intentionality relationship into a new transparent self-model (Gallese and Metzinger 2003).

The empirical basis of this view is the findings on the so-called "mirror neuron system" in the brain dedicated to multi-modal goal representation and involving both a sensory and a motor component. Solid objects are identified and differentiated not just on the basis of their physical appearance as such but from the point of view of motor interaction between the organism and the object. With object representation and action representation thus overlapping, intentionality is brought to the level of the brain's own functional ontology. Neurons with the mirroring function were first found in monkey brains in premotor area F5 and parietal area 7b. These neurons were activated both during the execution of purposeful, goal-related hand actions (grasping, etc.) and during the observation of similar actions performed by conspecifics (see Rizzolatti *et al.*, 1988; Gallese *et al.*, 1996; Rizzolatti *et al.*, 2000). Such sensori-motor integration instantiates an "internal copy" of actions by which goal-related behaviors are generated and controlled. At the preconceptual level, it also provides a meaningful account of behaviors performed by others (Gallese and Metzinger 2003; see Bremmer *et al.*, 2001).

A mirror system has also been found in humans. When a given action is planned, its expected motor consequences are also forecasted. During action observation, there is strong activation of both premotor and parietal areas with inhibition of action taking place in the spinal cord (Rizzolatti and Arbib 1998) much as in dreaming: there is an equivalence of what is being acted and what is perceived. Both are implicit, automatic and unconscious simulation processes that establish a link between an observed agent and the observing agent (Hari *et al.*, 1998; Gallese and Metzinger 2003). Wegner (2002, 44) thus briefly suggests that the existence of the mirror neuron system might explain why willful movement can be experienced by watching *any* body move where one's own body ought to be. When action consistently follows prior thought and when other apparent causes of action are excluded, we have the experience of volitional action.

To give an example, Nielsen (1963) showed in the 1960s that subjects asked to draw a straight line with their hand in a box by looking into the box through a tube, they could be fooled to believe that the experimenter's hand seen in the box through a mirror was actually their own (which they did not see). When both hands followed a straight line, the subjects experienced the alien hand as their own. But when the experimenter drew a curve to the right, the subjects still experienced the hand as their own, although now making involuntary movements.

Similarly, Ramachandran and colleagues arranged an experiment in which a subject with an amputated arm was asked to put his real arm and what was felt as a phantom limb in a box with a vertical mirror in the middle; the reflection of the real arm then appeared in the mirror where the phantom would be if it were real. When the subject moved the real arm, he experienced the phantom as moving voluntarily as well. Then the experimenter put his arm inside the box so that it appeared in the place of the phantom. Observing his real arm and the experimenter's arm in the box made the subject feel as though his phantom was moving (Ramachandran and Blakeslee 1998, 46–48; see Wegner 2002, 40–44). Willful movement is an experience that follows from the convergence of observed movement and one's own thoughts. This shows how the prefrontal system for processing agent causality tends to take over and interfere with folk-mechanical reasoning (see Wegner 2002, 26).

The mirror neuron system codes the relational fact that an observed agent is directed at an object component. When this information becomes globally available in consciousness, it becomes possible to have a representational object component defined by the intentions of *another*. Intersubjectivity follows from the subjective perspective that arises from the goal representation simulation leading to a kind of "mind reading." At the level of conscious experience this means that I can relate my own intentions to the intentions of simulated others. We can now understand social relations from a third-person perspective (Gallese and Goldman 1998; Gallese and Metzinger 2003; Gallese, Keysers and Rizzolatti 2004; Hurley 2008).[1]

We can thus have goal representations that are socially individuated (Gallese and Metzinger 2003): "Marilyn knows that Helen wants that Tom thinks that Allison wishes..." Your goal-representation can be an object for my phenomenal model of the intentionality relation, while at the same time my directedness towards that object can be an object for your phenomenal model of the intentionality relation. I think that you think that I think [something]; you think that I think that you think that I think [something]. This involves the so-called recursion principle, that is, the principle of rules within rules. This feature defines language in particular but it is also involved in many kinds of human thought and action. Pupil A can, for example, watch pupil B watch pupil C watch the teacher. Apes, being unable to use recursion, can only watch each other directly (see Frith 2001; Fitch and Hauser 2004; Premack and Premack 2002, 150–153; Tomasello, Call and Hare 2003). It is in this way that the tendency to attribute agency seems to arise from the representation of action.

## Action and Shared Knowledge

If the outline above is plausible and if religious rituals enhance religious belief$_B$, then this seems to happen through the mechanism I have just sketched. Combining religious concepts and beliefs with motor action in rituals creates such proprioceptive and kinesthetic feedback that enables the subject to feel that religious concepts and beliefs actually guide his or her action and thus are reliable guides for one's orientation in reality. This is what Boyerian belief is all about. This, however, requires that religious concepts actually get associated with some behaviors in rituals. Such associations can be conscious or implicit and be targeted at individual concepts or their meta-representational context.

Because of the complexity and vagueness of religious concepts (Boyer 1994, 48–49), among other things, it is more common that ritual action enhances trust and belief in the general meta-representational context in question. A Christian, for example, may walk away from the service thinking that the pastor had just given a wonderfully persuasive sermon, and yet not remember one sentence from the sermon. In such cases, it is the person's belief in the sermon's meta-representational context ("Christianity") that has been enhanced, not belief$_B$ in the particular beliefs$_A$ expressed in the sermon. If, however, some single religious concept or belief is at the focus of the person's attention while he or she engages in ritual motor action, the person's belief$_B$ in this concept or belief$_A$ may be enhanced. This, however, probably requires that the target concept or belief has special salience for the subject from the start (see Pyysiäinen 2004a).

Wegner (2002, 229–235) lists three things that contribute to something being judged as real: it must seem real (perceptual detail), feel real (emotional impact), and act real (uncontrollability). There is evidence that persons consider their memories to be of real events when there is perceptual detail about the events. Memories that arise from perception convey perceptual information, whereas memories that originate in thinking contain more information about the cognitive operations on which they are based. Thus, a vision of

1. Note, however, that the self is not there first. The self arises from the organism's relation to its environment; "the way on interiority proceeds through the other" as Sartre (1943, 292) once put it.

an angel is regarded as better evidence of the existence of angels than what mere thoughts about angels are. False perceptual details, if rich enough, can cause us to regard even imagined agents as real (see Honko 1962).

This tendency to mistake imagined agents to be real is further enhanced if the agent representations have an emotional impact on us. It is possible to become so emotionally involved in an experience that one no longer cares to remind oneself that the agents are not real. This happens, for example, while watching an emotionally stimulating movie. Conversely, in the Capgras' syndrome, one loses the ability to link emotional reactions to familiar faces and therefore starts to suspect that his or her relatives and intimate friends are mere impostors, not real relatives and friends. Third, if one feels that an imagined agent acts in uncontrollable ways, the agent will likely be regarded as real.

Now, consider religious rituals. McCauley and Lawson (2002) argue that when a ritual is considered to be very important, it is so arranged that it involves a maximum amount of "sensory pageantry": it appeals to all the senses (there are lights, burning of incense, music, and so forth). In other words, there is much perceptual detail. This, in turn, provokes emotions. Third, the participants are assured that it is the counterintuitive agents that act, humans being mere passive objects. Humans cannot control what happens, the locus of control being in the counterintuitive agents. All three of Wegner's criteria thus are met: rituals make religious concepts plausible.

An agent is generally understood to be an organism that seems to act on the basis of its beliefs and desires. Some have argued that humans are "hard-wired" to automatically categorize apparently self-propelled entities as agents (Premack 1990; Leslie 1994, 1996). This seems to be supported by Heider and Simmel's (1944) early work which showed that under certain circumstances we tend to attribute agentive properties even to mere objects and geometric shapes guided by only minimal cues. Agent detection is hyperactive in the sense that even minimal cues can trigger the postulation of agency (Barrett 2000, 31-32). We use agent-explanations whenever possible (see Pyysiäinen 2004b, 1-27).

It is thus a special feature of human psychology that we are capable of an elaborate understanding of the states of mind of conspecifics and thus to predict their behavior creatively, not merely by a mechanical stimulus-response model. Yet there seems to be no unified "theory of mind" system (cf. Carruthers and Smith 1996) with a single evolutionary background (Boyer and Barrett 2005).

Blakemore *et al.* (2003) distinguish between mechanical and intentional contingency, emphasizing that the detection of intentional contingency, or agency, is based either on type of motion or on interaction between objects: movement that is self-propelled and shows unpredictable changes in velocity is perceived as animate even if it is not enacted by human bodies. So too movement is based on reaction at a distance. Mechanical and intentional contingency are also processed in different parts of the brain, detection of mechanical contingency being a low-level process not influenced by higher order processes. The detection of agency thus might be a precursor of the ability to infer other people's mental states (Blakemore *et al.* 2003, 837; see also Gergely and Csibra 2003; Csibra 2003).

There is also empirical evidence to support the conclusion that agency attribution may not be based on seemingly self-propelled movement as the cue, and that it does not necessarily presuppose an understanding of psychological states. Atran (2002, 64-67) argues, on the basis of Csibra *et al.* (1999), that children attribute goal-directedness to certain kinds of "telic" event structures, rather than to entities. What is important here is contingent control over the outcome of the telic situation, not the initiation of action. The importance of control over movement is also supported by the empirical study of Barrett and Johnson (2003).

Infants as young as nine months of age are able to represent the observed behavior of objects in teleological terms; yet they do not necessarily do it in mentalistic terms and are not necessarily committed to any specific ontological categorization of these objects (Csibra *et al.* 1999). There are (at least) two distinct ways of doing this. First, nine-month-olds are capable of taking the "teleological stance" and representing actions as being directed towards a *goal.* Actions are not connected to their antecedent causes but to their consequences, yet without attributing mental states to the acting entities. Second, actions such as pointing and looking at an object can be interpreted as being *referential* in the sense that action is linked to a specific object or an aspect of the environment. The action is *about* something. (Gergely and Csibra 2003; Csibra 2003).

Thus, following Boyer and Barrett (2005), it is advisable to break the "theory of mind" into such separate cognitive operations as animate motion detection, distant reactivity, goal-ascription, intention-ascription, joint attention, and facial cues about emotional states. The detection of animate motion has as its input such things as non-linear changes in direction, sudden acceleration without collision, and change of physical shape that accompanies motion (e.g. caterpillar-like crawling). Animacy can also be attributed on the basis of an object reacting at a distance, rather than on the basis of presupposed self-propelledness. Goal-ascription is a related capacity based on the presupposition that certain trajectories make sense only on the condition that the moving entity is trying to reach or avoid something. A more complicated process is intention-ascription in which the end-result of an action is connected to perceived movement through the link of the intention of the moving organism. Joint attention develops in humans between nine and twelve months, proceeding from joint engagement to communicative gestures and attention following (like gaze following). Lastly, facial expressions are used as cues about emotions as early as in five-month-olds who react differently to displays of different emotions on a familiar face.

In terms of the underlying neural processes, one option is that the prefrontal cortex inhibits the natural tendency to act on mere correlative stimulus relationships, guiding us to search for higher-order associations and recursive embeddings as well as directly helping us process difficult syntactic constructs (see also Fuster 2002). Prefrontal activity thus outperforms other brain processes and we come to rely on its functions whenever possible (Deacon 1998, 265-267).

Another recent suggestion is that "mind-reading" might be based on the mirror neuron system (Gallese, Keysers and Rizzolatti 2004). If "mind reading" has evolved from observed object manipulation and is based on shared intentionality, then religious rituals can be predicted to enhance religious belief$_B$ on the basis of the mirror neuron system. The participants do some things and have some beliefs$_A$; at the same time they watch and imitate others' similar actions, understanding that also the others believe what they believe (and believe that others believe that they believe…). Joint action thus helps coordinate common knowledge in two ways. First, the ritual use of religious beliefs enforces them by combining them with motor action schemata; second, joint action enforces persons to simulate others' doings in their mind, using the mirror neuron system.

Thus, when behavioral systems begin to react to each other's internal states, they come to predict the course of observed action, including their own action, by representations of individuated will and belief$_B$; socially individuated goal representations thus emerge.

Mutual cooperation from then on entails the processing of complex social information. This is what our large cortex most likely was developed for (Dunbar 1993, 2002). When a person needs to act not merely as an individual but also as a member of a defined group, trying to advance or at least not to contradict the norms, values, goals, etc. of the group (see Tuomela 1995, 2003), these common values must be somehow communicated among the members. It is not enough simply to send and receive messages between individuals; an individual also needs to know that others have received the same message, and also others need to know that others have received the same message. In other words, members of the group need to establish "common knowledge." Chwe (2001) argues that this happens in public rituals (ceremonies, rallies, media events, and so forth).

Common knowledge is based on recursion and theory of mind; it refers to events and facts known to everyone in a group when everyone also knows that others know, everyone knows that everyone knows that others know, and so forth. In actual fact, it is enough that one *believes* with something like 90 per cent probability that others also believe, etc. The various layers of belief also do not form an endless chain of independent levels, but rather are recursively embedded in each other. We thus understand them at once, without effort (see Chwe 2001, 76–79). We have no problems, for example, in seeing John looking at Mary looking at Lisa looking at John, although chimpanzees cannot handle such loops. What is entailed here is that one intentional act is about another intentional act which is about another intentional act, and so on.

## Religious Rituals

Religious rituals involve bodily action and experience to varying extent. In a typical Finnish Lutheran service, for example, the participants mostly sit still, only rising a couple of times and possibly participating in the singing of hymns. The Eucharist, which is not celebrated at all services, is an exception: at the Holy Communion one goes to the altar and kneels down to consume the body and blood of Jesus Christ. At the other extreme, the patients of some tribal rituals may be beaten, tortured, mutilated and kept without food and sleep for days or weeks. Their skin may be scarified, a few teeth smashed in, and their genitals mutilated in various ways, and so forth (see, e.g. Barth 1975; Whitehouse 2000).

Also in Christian traditions we find phenomena such as speaking in tongues, baptism by full immersion, singing and dancing with raised hands, falling down at the preacher's touch, or even so-called "snake handling" (Wilson 1970; Williamson and Pollio 1999). In all these practices, one is involved in the ritual through one's own bodily action. Compare, now, the situations of merely listening to a relatively abstract and dry sermon while sitting comfortably (well, relatively so) on the bench, on the one hand, and hearing an emotional sermon which you yourself interrupt every now and then by exclamations like "Hallelujah!", finally going to the altar to confess your willingness to give your life to Jesus and to receive the blessing which eventually knocks you down. Which of these two events would you remember more vividly?

In the first case, information would mostly be saved in semantic memory in the form of abstract ideas and schemata, whereas in the second case episodic memory would be activated: it happened to me there and then. Whitehouse argues that in the first case frequent repetition is needed to keep the tradition alive, whereas in rituals of the second type it is the emotional shocks that engrave the memory in the minds of the participants "one-off" (Whitehouse 1995, 2000, 2001, 2002a, 2002b, 2004; cf. McCauley and Lawson 2002).

Frequent repetition, however, runs the risk of a "tedium effect": people get bored with the dry routine, and various kinds of revivalist movements then emerge (Whitehouse 2000; Pyysiäinen 2004c). Strong emotional stimulation involves the different risk that the affective volume needs to be constantly increased until the whole movement breaks down in chaos. This can only be prevented by inventing emotional rituals in which one can participate in the role of the patient of the ritual (instead of agent) only once in a lifetime (McCauley and Lawson 2002).

We should, however, differentiate between repetition of concepts and verbalized beliefs, on the one hand, and repetition of motor actions, on the other hand (see Whitehouse 2002b). Conceptual repetition involves semantic memory and leads to better conceptual recall and to the formation of unemotional abstract schemata, whereas repeated motor action leads to good procedural memory and to unemotional, nonconscious, and routinized procedures. By combining conceptual information and motor action in infrequent rituals, it is possible to create vivid episodic memories. Most cognitive theorists of religion agree that concepts that are recalled the best get a head start in cultural selection and thus become widespread (Sperber 1996; Boyer 1994; Boyer and Ramble 2001; Barrett and Nyhof 2001).

Although memorability helps explain the spread of beliefs$_A$, to explain belief$_B$ we need something more, as it is possible to remember something without believing it (see Pyysiäinen 2001, 130–139; Atran 2002). Repetition may also enhance belief. There is empirical evidence of the so-called illusory truth effect: a mere repetition of a statement can lead to increases in the strength of one's belief that the statement is true (Begg, Anas and Farinacci 1992; see Schacter and Scarry 2000). "Tedium" thus will not lead persons to neglect their ritual duties if the costs of such neglect are believed to be high enough.

I have previously used Chwe's theory to argue that (at least many) religious rituals are a means of creating common knowledge in the sense that in a preliterate society, information about a baby being born, a man and a wife getting married, and so forth can only be "published" in common rituals (Pyysiäinen 2004b, 135–146). The rituals of baptism and wedding, for example, are a means of disseminating the information that a new member has been born to the group and that two persons are about to establish a new social unit. As this information is important from the point of view of social relationships, it is necessary to make it known to the group. In the absence of modern media, rituals have been an effective way of doing this.

This is not merely a question of distributing information but also of coordinating it in such a way that it becomes common knowledge. Singing, reciting, dancing, and otherwise moving together in a coordinated manner are ways of achieving this. As one moves and speaks while at the same time observing others move in the same way and utter the same phrases, coordination of religious knowledge takes place not only with regard to beliefs$_A$ but also belief$_B$. Ritual action provides an opportunity to actually use religious concepts and beliefs in connection with motor action (and even to guide this action) and thus to create the feeling that these beliefs are true. Although it is possible to use religious concepts in various ways also in everyday life, rituals have two special advantages: first, because ritual action has no external goal apart from the action itself (see below), rituals have the potential to incorporate counterintuitive representations. Second, the coordination of knowledge also has the effect that the belief$_B$ of individual group members cross-strengthens everyone else's belief. Belief$_B$ arises not only from one's own action but also from observing and even imitating others using religious concepts in their action.

Religious concepts and beliefs thus need not remain mere thoughts; they are also related to perceptual and bodily experience and thus convey perceptual information, not just information about the cognitive operations on which they are based. Rituals are often made to appeal to the different senses and thus involve perceptual detail and have emotional impact (see also McCauley and Lawson 2002, 89–123). Therefore, episodic memories of rituals are felt to be about real objects, although the reference of religious concepts is mysterious. The third factor (in addition to perceptual detail and emotional impact) contributing to the feeling of reality is controllability. We are prone to attribute agency to account for movement that is beyond our control and yet does not seem to be merely mechanical. In rituals, certain things that happen are explained with reference to counterintuitive agency. The desired outcome of the rituals is thus understood either as being transcendent or is identified with the ritual performance itself (see Pyysiäinen 2001, 88–90). As we seemingly cannot control what happens, the control is attributed rather to the supposed counterintuitive agents (see Wegner 2002, 221–270).

A Christian baptism, for example, is not considered to be properly performed on the criterion that it has been successful (i.e. the wanted effect is produced). It is considered to be successful if it has been properly performed. There is no way of examining the baby to see whether (s)he actually has been baptized. Rather, we should examine the ritual performance, not the end result. In everyday reality it is the other way around: it is more important that actions (like washing one's car, for example) produce the desired result than that the action is performed in some specified manner. Were someone to insist that the car should absolutely be washed in some specified manner, irrespective of the end result, we would be inclined to say that washing the car has become a ritual for this person. Although baptism can be said to have the goal of removing the guilt related to original sin, this goal is not separate from the performance of the ritual.

Baptism is a typical "rite of passage" (van Gennep 1960). It is meant to move a person from one social position to another (see Honko 1979; Anttonen 1992; Pyysiäinen 2001, 80–84). Rites of passage are "special agent" rituals in the sense that in them, the active agent is a counterintuitive agent. In other types of rituals, counterintuitive agents are the patients, while the human participants are the active agents (see McCauley and Lawson 2002). The so-called crisis rites are one example of this. In their case, the goal can be, for example, healing a sick person, to get the rains to come, and so forth. Yet believers tend to think that even in such cases there is a conceptually necessary relationship between the ritual and the outcome: performing the ritual equals achieving the desired outcome. The fact that the sick person does not get better or that the rains do not come are not regarded as evidence that the ritual does not work, but rather to show that there is something mysterious about this relationship (the rains only came in heaven, etc.) Apparent failure does not destroy belief but rather affirms it.

This is because if the ritual were effective every time, this would create in the participants a feeling that it is their own will causing the outcome: they pray to or otherwise manipulate counterintuitive agents, and what they want from them subsequently happens. In such cases, it would be difficult to resist the feeling that one produces the result directly by oneself. If, however, the sick sometimes recover and sometimes do not and the rains sometimes come and sometimes do not, participants will be prone to feel that what happens is controlled by someone other than by themselves. In such cases, it is natural to ascribe the causal agency to the imagined counterintuitive agents that have been appealed to. When the ritual does not seem to work, this is explained not as a failure in the ritual but rather as a failure of the participants to understand how the will of the counterintuitive agents works.

Rites of passage are different in the sense that their failure is absolutely impossible to assess. The desired outcome is simply dependent on human agreement made manifest in the ritual. Once the ritual is performed, the desired outcome is by the same token realized. Boyer, however, argues that as we from early on have many occasions to witness rituals, we gradually develop the belief that the ritual is in fact necessary for the social change it was arranged to celebrate. We thus develop the belief that it is the ritual that *causes* the change in social position, even if we have no idea of *how* the rituals produce such effects. Boyer calls this the "illusion that the ritual is actually indispensable to its effects" (Boyer 2001, 250–256).

This means that when ritual action receives an aura of magic around it, the representations of counterintuitive agents involved in the ritual also gain plausibility. This is because of three things: the perceptual detail of ritual memories makes these agents seem real, the emotional impact of rituals makes them feel real, and the counterintuitive agents seem to act real because the rituals seem to have a controlled effect that is not yet due to the will of the participants.

Consequently, it is quite natural that in certain types of rituals, believers often feel that they are being acted upon rather than themselves acting. This conviction is also manifest in religious doctrines, such as the Lutheran doctrine of the "justification of the sinner" (e.g. Luther's *De servo arbitrio*) or the Mahayana Buddhist idea that the right effort is non-effort (see Pyysiäinen, 1993). Both emphasize that it is impossible to attain the ultimate goal by one's own effort. We must and we cannot. Ritual and doctrine thus enhance each other.

## Conclusion

The conclusion now should follow that the more elaborate rituals are, *and* the more they include sensory pageantry and all kinds of emotion-provoking elements, *and* the stronger feelings of uncontrollability they evoke, the better they should enhance religious belief. This is a prediction I have come to through relatively complex theoretical reflection. It may sound prima facie convincing, but it can also be easily turned into a testable hypothesis.

The point I have been trying to make here could be empirically tested by a series of experiments in which the subjects would have to participate repeatedly in artificial rituals involving artificial counterintuitive concepts and beliefs. Afterwards they would have to assess how likely the concepts and beliefs are to be true (cf. Begg, Anas and Farinacci 1992). A control group would then be exposed to the same concepts and beliefs but without getting involved in ritual action. Thus we could assess the role of ritual action (as distinct from mere repetition) in enhancing belief in counterintuitive concepts. The next step would then be to apply this knowledge in the study of religion in the wild by studying the actual transmission of religious belief$_{\text{A and B}}$ in various religious movements.

## References

Anttonen, P.J. 1992. "The Rites of Passage Revisited: A New Look at van Gennep's Theory of the Ritual Process and Its Application in the Study of Finnish-Karelian Wedding Rituals." *Temenos* 28: 15–52.

Atran, S. 2002. *In Gods We Trust: The Evolutionary Landscape of Religion*. Oxford: Oxford University Press.

Barrett, J.L. 2000. "Exploring the Natural Foundations of Religion." *Trends in Cognitive Sciences* 4: 29–34.

———. 2004. *Why Would Anyone Believe in God?* Walnut Creek, CA: AltaMira Press.

Barrett, J.L. and Keil, F. 1996. "Conceptualizing a Nonnatural Entity: Anthropomorphism in God Concepts." *Cognitive Psychology* 31: 219–247.

Barrett, J.L. and Nyhof, Melanie A. 2001. "Spreading Non-natural Concepts: The Role of Intuitive Conceptual Structures in Memory and Transmission of Cultural Materials." *Journal of Cognition and Culture* 1(1): 69–100.

Barrett, J.L. and Johnson, A.H. 2003. "The Role of Control in Attributing Intentional Agency to Inanimate Objects." *Journal of Cognition and Culture* 3(3): 208–217.

Barth, F. 1975. *Ritual and Knowledge among the Baktaman of New Guinea.* New Haven, CT: Yale University Press.

Begg, I.M., A. Anas and S. Farinacci. 1992. "Dissociation of Processes in Belief: Source Recollection, Statement Familiarity, and the Illusion of Truth." *Journal of Experimental Psychology: General* 121 (4): 446–458.

Blakemore, S.-J., P. Boyer, M. Pachot-Clouard, A. Meltzoff, C. Segerbarth and J. Decety. 2003. "The Detection of Contingency and Animacy from Simple Animations in the Human Brain." *Cerebral Cortex* 13: 837–844.

Boyer, P. 1994. *The Naturalness of Religious Ideas: A Cognitive Theory of Religion.* Berkeley: University of California Press.

———. 2001. *Religion Explained: The Evolutionary Origins of Religious Thought.* New York: Basic Books.

Boyer, P. and C. Ramble. 2001. "Cognitive Templates for Religious Concepts: Cross-Cultural Evidence for Recall of Counterintuitive Representations." *Cognitive Science* 25: 535–564.

Boyer, P. and C. Barrett. 2005. "Domain-Specificity and Intuitive Ontology." In *The Handbook of Evolutionary Psychology*, edited by David Buss, 96–118. Hoboken, NJ: John Wiley.

Bremmer, F., A. Schlack, N. J. Shah, O. Zafiris, M. Kubischik, K.-P. Hoffmann, K. Zilles and G. R. Fink. 2001. "Polymodal Motion Processing in Posterior Parietal and Premotor Cortex: A Human fMRI Study Strongly Implies Equivalencies between Humans and Monkeys." *Neuron* 29: 287–296.

Carruthers, P. and P. K. Smith, eds. 1996. *Theories of Theories of Mind.* Cambridge: Cambridge University Press.

Chwe, M. S.-Y. 2001. *Rational Ritual: Culture, Coordination, and Common Knowledge.* Princeton, NJ: Princeton University Press.

Cruse, H., 2003. "The Evolution of Cognition: A Hypothesis." *Cognitive Science* 27: 135–155.

Csibra, G. 2003. "Teleological and Referential Understanding of Action in Infancy," *Philosophical Transactions of the Royal Society, London, B* 358: 447–458.

Csibra, G., G. Gergely, S. Bíró, O. Koós and M. Brockbank. 1999. "Goal Attribution without Agency Cues: The Perception of "Pure Reason" in Infancy." *Cognition* 72: 237–267.

Deacon, T. W. 1998/1997. *The Symbolic Species: The Co-Evolution of Language and the Brain.* New York: Norton.

Dunbar, R.I.M. 1993. "Coevolution of Neocortical Size, Group Size and Language in Humans." *Behavioral and Brain Sciences* 11: 681–735.

———. 2002. "The Social Brain Hypothesis." In *Foundations in Social Neuroscience*, edited by John T. Cacioppo *et al.*, 69–87 Cambridge, MA: MIT Press.

Edis, T. 2002. *The Ghost in the Universe: God in Light of Modern Science.* New York: Prometheus Books.

Fitch, W.T. and M.D. Hauser. 2004. "Computational Constraints on Syntactic Processing in a

Nonhuman Primate." *Science* 303: 377–380.

Frith, U. 2001. "Mind Blindness and the Brain in Autism." *Neuron* 32: 969–979.

Fuster, J. 2002. "Frontal Lobe and Cognitive Development." *Journal of Neurocytology* 31(3–5): 373–385.

Gallese, V., C. Keysers and G. Rizzolatti. 2004. "A Unifying View of the Basis of Social Cognition." *Trends in Cognitive Sciences* 8(9): 396–403.

Gallese, V. and T. Metzinger. 2003. "Motor Ontology: The Representational Reality of Goals, Actions and Selves." *Philosophical Psychology* 16(3): 365–388.

Gallese, V., L. Fadiga, L. Fogassi and G. Rizzolatti. 1996. "Action Recognition in the Premotor Cortex." *Brain* 119: 593–609.

Gallese, V. and A. Goldman. 1998. "Mirror Neurons and the Simulation Theory of Mind-Reading." *Trends in Cognitive Sciences* 12: 493–501.

Gennep, A. van. 1960. *The Rites of Passage*. Translated by Monica B. Vizedom and Gabrielle L. Caffee. Chicago, IL: University of Chicago Press.

Gergely, G. and G. Csibra. 2003. "Teleological Reasoning in Infancy: The Naïve Theory of Rational Action." *Trends in Cognitive Sciences* 7: 287–292.

Hari, R., N. Forss, S. Avikainen, S. Kirveskari, S. Salenius and G. Rizzolatti. 1998. "Activation of Human Primary Motor Cortex during Action Observation: A Neuromagnetic Study." *Proceedings of the National Academy of Sciences* 95: 15,061–15,065.

Heider, F. and M. Simmel. 1944. "An Experimental Study of Apparent Behavior." *American Journal of Psychology* 57: 243–249.

Honko, L. 1962. *Geisterglaube in Ingermanland I*. (*Folklore Fellows Communications*; 185). Helsinki.

———. 1979. "Theories Concerning the Ritual Process." In *Science of Religion: Studies in Methodology*, edited by Lauri Honko, 369–390. The Hague: Mouton.

Hurley, S. 2008. "The Shared Circuits Model (SCM): How Control, Mirroring, and Simulation Can Enable Imitation, Deliberation, and Mindreading." *Behavioral and Brain Sciences* 31(1): 1–58.

Lakoff, G. and M. Johnson. 1999. *Philosophy in the Flesh*. New York: Basic Books.

Laurence, S. and E. Margolis. 2002. "Radical Concept Nativism." *Cognition* 86(1): 25–55.

Lawson, E.T. and R.N. McCauley. 1990. *Rethinking Religion: Connecting Cognition and Culture*. Cambridge: Cambridge University Press.

Leslie, A. 1994. "ToMM, ToBy, and Agency: Core Architecture and Domain Specificity." In *Mapping the Mind*, edited by Lawrence Hirschfeld and Susan Gelman, 119–148. Cambridge: Cambridge University Press.

———. 1996/1995. "A Theory of Agency." In *Causal Cognition*, edited by Dan Sperber, David Premack and Ann J. Premack, 121–141. Oxford: Clarendon Press.

Luther, M. 1603. *De servo arbitrio liber*. Edited by J. Kimedoncius. Neustadii. [An English translation available at: http://www.mountainretreatorg.net/classics/bondage_will.html.]

McCauley, R.N. and E.T. Lawson. 2002. *Bringing Ritual to Mind*. Cambridge: Cambridge University Press.

Nielsen, T.I. 1963. "Volition: A New Experimental Approach." *Scandinavian Journal of Psychology* 4: 225–230.

Oberlander, J. and P. Dayan. 1996. "Altered States and Virtual Beliefs." In *Connectionism, Concepts, and Folk Psychology*, edited by Andy Clark and Peter Millican, 101-114. (*The Legacy of Alan Turing* 2). Oxford: Oxford University Press.

Perner, J. 1993/1991. *Understanding the Representational Mind*. Cambridge, MA: MIT Press.

Premack, D. 1990. "The Infant's Theory of Self-Propelled Objects." *Cognition* 36: 1–16.

Premack, D. and A. Premack. 2002. *Original Intelligence*. New York: McGraw-Hill.

Pyysiäinen, I. 1993. *Beyond Language and Reason: Mysticism in Indian Buddhism. Annales Academiae Scientiarum Fennicae, Dissertationes Humanarum Litterarum*, 66. Helsinki: Suomalainen Tiedeakatemia.

———. 2001. *How Religion Works: Towards a New Cognitive Science of Religion*. Leiden: Brill.

———. 2004a. "Intuitive and Explicit in Religious Thought." *Journal of Cognition and Culture* 4(1): 123–150.

———. 2004b. *Magic, Miracles, and Religion: A Scientist's Perspective*. Thousand Oaks, CA: AltaMira Press.

———. 2004c. "Corrupt Doctrine and Doctrinal Revival: On the Nature and Limits of the Modes Theory." In *Theorizing Religions Past: Past: Archaeology, History, and Cognition*, edited by Harvey Whitehouse and Luther H. Martin, 173–194. Walnut Creek, CA: AltaMira Press.

Quine, W.v.O. 1980/1951. "Two Dogmas of Empiricism." In *From a Logical Point of View*, by Willard van Orman Quine, 20–46. Cambridge, MA: Harvard University Press.

Ramachandran, V.S. and S. Blakeslee. 1999/1998. *Phantoms in the Brain*. New York: Quill, William Morrow.

Rizzolatti, G. and M.A. Arbib. 1998. "Language within Our Grasp." *Trends in Neurosciences* 21: 188–194.

Rizzolatti, G., R. Camarda, L. Fogassi, M. Gentilucci, G. Luppino and M. Matelli. 1988. "Functional Organization of Inferior Area 6 in the Macaque Monkey: II: Area F5 and the Control of Distal Movements." *Experimental Brain Research* 71: 491–507.

Rizzolatti, G., L. Fogassi and V. Gallese. 2000. "Cortical Mechanisms Subserving Object Grasping and Action Recognition: A New View on the Cortical Motor Functions." In *The Cognitive Neurosciences*, edited by Michael Gazzaniga, 539–552. Cambridge, MA: MIT Press.

Sartre, J.-P. 1943. *L'être et le néant*. 14th edition. Paris: Gallimard.

Schacter, D.L. and E. Scarry. 2000. "Introduction." In *Memory, Brain, and Belief*, edited by Daniel L. Schacter and Elaine Scarry, 1–10. Cambridge, MA: Harvard University Press.

Sinha, C. 2004. "Biology, Culture and the Emergence and Elaboration of Symbolization." In *In Search of a Language for the Mind-Brain: Can the Multiple Perspectives Be Unified*, edited by Ocke-Schwen Bohn, Albert Gjedde and Anjum Saleemi (*The Dolphin* 33). Aarhus: Aarhus University Press.

Sperber, D. 1996. *Explaining Culture: A Naturalistic Approach*. Oxford: Blackwell.

———. 1997. "Intuitive and Reflective Beliefs." *Mind and Language* 12: 67–83.

———. 2000. "Metarepresentations in an Evolutionary Perspective." In *Metarepresentations: A Multidisciplinary Perspective*, edited by Dan Sperber, 117–137. Oxford: Oxford University Press.

Tomasello, M., J. Call and B. Hare. 2003. "Chimpanzees Understand Psychological States—The Question Is Which Ones and to What Extent." *Trends in Cognitive Sciences* 7(4): 153–156.

Tuomela, R. 1995. *The Importance of Us: A Philosophical Study of Basic Social Notions*. Stanford: Stanford University Press.

———. 2003. "The We-Mode and the I-Mode." In *Socializing Metaphysics: The Nature of Social Reality*, edited by F. Schmitt, 93–127. Lanham, MD: Rowman and Littlefield.

Wegner, D.M. 2002. *The Illusion of Conscious Will*. Cambridge, MA: MIT Press.

Whitehouse, H. 1995. *Inside the Cult: Religious Innovation and Transmission in Papua New Guinea*. Oxford: Clarendon Press.

———. 2000. *Arguments and Icons: Divergent Modes of Religiosity*. Oxford: Oxford University Press.

———. 2001. "Transmissive Frequency, Ritual, and Exegesis." *Journal of Cognition and Culture* 1(2): 167–181.
———. 2002a. "Conjectures, Refutations, and Verification: Towards a Testable Theory of "Modes of Religiosity.'" *Journal of Ritual Studies* 16(2): 44–59.
———. 2002b. "Implicit and Explicit Knowledge in the Domain of Ritual." In *Current Approaches in the Cognitive Science of Religion*, edited by Ilkka Pyysiäinen and Veikko Anttonen, 133–152. London: Continuum.
———. 2002c. "Modes of Religiosity: Towards a Cognitive Explanation of the Sociopolitical Dynamics of Religion." *Method and Theory in the Study of Religion* 14(3-4): 293–315.
———. 2004. *Modes of Religiosity: A Cognitive Theory of Religious Transmission*. Walnut Creek, CA: AltaMira Press.
Williamson, W.P. and H.H. Pollio. 1999. "The Phenomenology of Religious Serpent Handling: A Rationale and Thematic Study of Extemporaneous Sermons." *Journal for the Scientific Study of Religion* 38: 203–218.
Wilson, B.R. 1970. *Religious Sects: A Sociological Study*. London: World University Library.

# — 8 —

# Pumping Intuitions: Religious Narratives and Emotional Communication

Tom Sjöblom

## On Relevance[1]

One of the classical problems in the study of religion is the question of what makes a belief true for someone. For a long time two basic strategies have been used to answer this question. The first strategy, employed mostly by philosophers, has been to search for the logical conditions that make religious statements true. The problem with this approach is that in ordinary circumstances, people do not have religious beliefs because rational argumentation leads them to think that way. Instead, some individuals use rational argumentation to justify religious beliefs that they hold for other reasons (see, e.g. Shermer 2000, 59–88). The second strategy has been to argue, like scholars of culture tend to do, that we hold something to be true because we have been taught to do so by the cultural authorities of our social community (See Geertz 1973; Boyer 1994, ix). Thus, most Europeans believe in the existence of the Christian God but not in the existence of Mickey Mouse because we are coached to do so because of patterns established by our cultural environment. The problem with this approach is that people do not seem to believe in something simply because they are told to do so. Additional and more fundamental cognitive factors seem to be involved (Sperber 1975, 1996; Boyer 1990, 1994).

Cognitive anthropology and psychology have recently come up with a third solution. According to this solution, what makes something to be conceived as true is based on how much relevance it possesses. A highly relevant belief is more likely to be accepted as true than another with little or no relevance (Sperber and Wilson 1995). In other words, people believe in God instead of Mickey Mouse because the latter is not relevant to their lives. This approach has proven to possess more explanatory power than the two classical approaches. It has created a new problem, however, known as the "Mickey Mouse Problem" that has to

1. This paper is based on a paper read at the seminar "Religious Narrative, Cognition and Culture," held at Aarhus University, Denmark on December 2–4, 2004. The writing of this paper was funded by the Academy of Finland (project 1200827).

do with how representations attain relevancy (e.g. Atran 2002, 260).

Three possible answers have been proposed to solve this question. First, it can be argued that most relevance is attached to beliefs that derive from our "deepest desires" (Nichols 2004, 347–371). This solution has its roots in Freud and classical psychology of religion. The cognitive version differs from earlier ones by describing the cognitive processes involved. Thus, the most relevant beliefs appear to be those that in one way or another deal with the survival of the individual either directly by promoting social cohesion or indirectly by offering alternative scenarios to existential questions such as what happens to us after death, etc. (see Nichols 2004, 357; Sosis and Alcorta 2003, 264–274; Bulbulia 2004, 1–32). Proponents of this approach usually think of religious representations as cognitive adaptations.

Two other approaches view religious representations rather as by-products, existing because it is possible, not because they have some essential function (see Gould 1991, 43–46; Pinker 1997, 554–558; Boyer 2001).[2] The first of these by-product approaches argues that once a belief comes into existence, it survives if it contains sufficient strategic information. Beliefs that are invested with strategic information are those that can be used as frames of reference in our everyday interactions. Naturally, the belief-representations with the widest frameworks contain more strategic information. This gives them more relevancy over other representations (Morton 2000, 218–237; Boyer 2001, 150–155; Barrett 2004, 46–51). The second by-product approach argues that instead of, or in addition to, cognitive relevance, what really matters is emotional commitment to some beliefs rather than others. According to this view, people have emotional attitudes towards some representations, and it is this emotional commitment to them that makes them relevant for people (Atran 2002, 260–261; Pyysiäinen 2002, 110–132; 2004, 46, 130–134). The emotional commitment does not have to be due to some inherent connection with our "deepest desires," but can be produced by emotional stimulation (Pyysiäinen 2001, 140).

This chapter is a contribution to this discussion by introducing a new body of data, namely, religious narratives. Narrative intelligence and narrative discourse have been mostly neglected in the cognitive study of religion. This might simply be because narratives are complex representations which are not easily approached on the level of cognition. Or the reason might be that narratives are often connected with written texts and stories. Such extra-somatic evidence is often interpreted to be somehow outside the competence area of cognitive science. Studying narrative intelligence and narrative discourse, however, has a long-standing place in cognitive science, starting with the ground-breaking work of cognitive scientist Roger Schank to the more recent, but even more influential, theories of psychologist Jerome Bruner (see, e.g. Bruner 1986; Schank 1990; Mateas and Sengers 2002, 1–25). It is the findings and approaches of these research programs that I want to link to the theories of relevance mentioned above.

## In the Beginning

The importance of narratives for religious traditions is obvious. One can go so far as to argue that outside the field of experts, religions communicate mainly through narratives. It appears to be a fundamental feature of religious systems that structural beliefs

2. Of course, this does not mean that religious beliefs are without functions. Indeed, many different kinds of functions seem to be attached to them. According to the by-product approach, however, these functions are secondary applications of these beliefs and do not, therefore, explain why they actually exist.

and theological doctrines are supported by narratives and that narratives also integrate different aspects of the religious system into each other and into the self-systems of the believers (Hinde 1999, 101–102).[3] Narrative representation is not, of course, restricted to religious systems and belief environments. On the contrary, it appears to be fundamental for human cognition in general. Our cognition seems to be built for seeking patterns, rhythm and causal connections in our environment (see, e.g. Dennett 1991, 344–356; Guthrie 1993; Thagard 2000). Recent research on these rhythmic patterns of causal connections in human cognition—what I will refer to as the narrative drive—suggests that they are vital for our ability to plan actions and solve problems (Damasio 1994; Greenspan and Shanker 2004, 218).

Given the importance of narratives in human experience and communication, it is not surprising that story-creation and narrative intelligence have been of interest in cognitive science from the beginning. It has, however, proven to be a very elusive object of study, and our understanding of the narrative mode of thought (in contrast to the forms of narrative discourse) has begun to accumulate only during the past decade or two (see Mateas and Sengers 2002, 1–25). One of the leading scholars behind this development has been psychologist Jerome Bruner who argues that there are two modes of cognitive functioning basic to human thought. One of the modes, the *paradigmatic mode*, employs categorization or conceptualisation as its basic devices for ordering human experience, while the second mode, the *narrative mode*, works through creating mental landscapes where present experiences, actions, and conditions are placed into relationship with what has been going on before (and sometimes what is thought will happen in the future)—something I call *the principle of analogy* (Bruner 1986, 11–43; Tulving and Lepage 2000, 208–228; Gentner, Holyoak and Kokinov 2001).

In the original formulation of his argument, Bruner saw these two modes as complementary but irreducible to one another, implying a more or less equal status and independent origins for both of them (Bruner 1986, 11). More recently, however, he has stressed that it is the narrative mode that seems to be cognitively preferred, and it might even turn out to be the fundamental obligatory basis for the paradigmatic mode to exist in the first place (Bruner 2002, 89). There are several lines of evidence that support this argument. For example, the development of narrative intelligence seems to precede that of the paradigmatic mode both historically and psychologically. On the basis of her experiments, Katherine Nelson argues that rudiments of narrative thinking—what she calls *event sequences*—are present in the child's mind around the age of one. However, it takes an additional year or two for full-blown narrative thinking to be in place (Nelson 2003, 25–26; see also Holyoak and Thagard 1997, 35–44). The evolutionary origins of narrativity are much more difficult to investigate, but Michael Carrithers has presented convincing arguments for narrativity as the essential explanatory mechanism for the birth of culture, and the actual evolutionary dynamics involved have been explained in the much more detailed and refined hypothesis of the birth of mythic culture by Merlin Donald (Carrithers 1991, 305–317; Donald 1991, 201–268). Furthermore, Donald argues that the birth of the narrative mind can be equated with the appearance of the so-called archaic *homo sapiens* 160,000 years ago, while the first signs of paradigmatic thought do not appear before the appearance of modern culture starting around 40,000 years ago (Donald 2001, 260–262).

3. According to Robert Hinde, a self-system is the process of how we perceive ourselves in an ongoing interaction with other people and with our environment (Hinde 1997, 23–35). Antonio Damasio describes this as a cognitive narrative consisting of memories of situations in which core consciousness was involved in the knowing of the most invariant characteristics of an organism's life (Damasio 1999, 17).

As discussed by Leda Cosmides and John Tooby through their experiments with the Wason Selection Task, problem solving appears to be easier for us when operating in a narrative mode (Cosmides and Tooby 1992, 163–228). Similarly, our memory also tends to prefer the narrative mode, and a well-known mnemonic technique used by professional mnemonists is to create a mental narrative of the memorized entities that seems to greatly enhance their capabilities to remember even abstract and highly paradigmatic material (see Yates 1966). Experiments by Andrea S. Heberlein and associates seem to indicate that we also prefer to ascribe what we see in terms of social narratives even when we describe changes in relationships of purely abstract entities (Heberlein *et al.* 1998, 1176).[4] The same preference for narrativity is also evident in such prototypical examples of paradigmatic thought as scientific writing, as discussed by Misia Landau and others (Lewin 1987, 30–38; Landau 1991; Stockzkowski 2000, 187–192).

Third and finally, all the above-mentioned studies also suggest that our interactions in the social world are controlled by the narrative mode. We also know that our minds have been ultimately adapted to deal with this same social world, with its challenges and interactions (Chance and Mead 1953, 395–439; Humphrey 1976, 303–317; Byrne and Whiten 1988; Whiten and Byrne 1997; Dunbar 1998, 178–190). We are fundamentally a social species with social minds. Therefore, simply in terms of the evolution of cognition, it would make good sense to think of narrativity as the basic mode of thinking. Indeed, this is basically what Robin Dunbar suggests with his version of the so-called Social Brain Hypothesis, where the origins of language and human communication are understood in terms of emotional communication (Dunbar 1996; 1998 178–190; 2004; Dautenhahn 1999a, 59–66; 1999b, 63–90), to which I will turn in what follows.

## Narratives and Emotional Communication

Narrativity as a cognitive tool should be distinguished from narratives as discourse. If the narrative mode, however, is the cognitive tool used for dealing with social situations, it is not surprising that narratives are the preferred mode of social communication. Dunbar's starting-point is his finding that people use language mainly for gossip, i.e. exchanging information on social matters. Indeed, speaking seems to cover the same functions in human society as grooming does in other primate societies. It establishes and upholds social bonds and mutual relationships between individuals. Thus, exchanging information appears only to be a means for a more fundamental purpose, i.e. to produce social cohesion by creating empathy towards each other in a group of individuals (Dunbar 1996). The problem emerging from this, however, is that telling stories is actually not a very efficient tool for transmitting information. Actually, research in different academic disciplines, such as anthropology, folklore, history, psychology and sociology, all seem to agree that narratives cannot be trusted as sources of information (e.g. Bartlett 1932; Lord 1960; Finlay 1971; Boyer 1990; Hervieu-Léger 2000). In the words of Jerome Bruner, "Stories are surely not innocent: they always have a message, most often so well concealed that even the teller knows not what ax he may be grinding" (Bruner 2002, 5–6).

Why this is so is because narratives are complex representations, and we just do not have the cognitive capacity to keep in mind all the possible interpretations and inferences present in them. We must choose what interests us and later on fill the appearing gaps

4. Heberlein's findings are discussed in Adolphs 1999, 473. More examples of similar experiments with a general discussion are provided by Bloom 2004, 17–19.

with relevant material available to us (see Schank 1990, 56–57). In other words, narratives have the habit of transforming themselves to suit the needs of contextual situations. They do not just inform us about one particular event or type of event but provide material for dealing with social situations on a more general level and even in totally new situations, which makes them powerful cognitive tools in terms of survival. Stories help us to adapt to new conditions and environments (Holyoak and Thagard 1995, 39–73; Fauconnier and Turner 2002, 171–193). The accuracy of the information is, therefore, not the issue—at least not the most important issue—here. A much more important factor for deciding what we think of a narrative is how relevant it is for our personal concerns and thoughts (see Sperber and Wilson 1995).

According to Keith Holyoak and Paul Thagard, narratives achieve this by creating similarities by mapping the source, i.e. the narrative, to the target event. An obvious example of this is mythical stories that people in different cultures conceive as highly relevant and meaningful for their understanding of the world, life and everything. While analogies often are obvious and clear, however, sometimes—as in the case of many myths—they are nothing of the kind, and in those cases, we must ask what made somebody choose a particular narrative as the source analogy to begin with (Holyoak and Thagard 1995, 101). The standard solution to this in anthropology and the study of religions has been to explain this by cultural learning. According to this view, it is the cultural authorities that teach us which narratives are paradigmatic for us and in what way (see, e.g. Barthes 1957; Geertz 1973, 33–54).[5] But this cannot be the whole story. Narratives surround us, and some of them are more widely shared and known than others. Still, we do not find all of them relevant or relevant in the same way just because we have heard them, they are widespread, and/or promoted by cultural authorities. The relevance of a narrative seems to come to us intuitively, as pointed out by philosopher Daniel Dennett. He argues that narratives as a form of communication work by pumping intuitions into our minds, which means that we accept them or not without reasoning or cognitive reflection (Dennett 1984, 12; Damasio 1994, 187).

Dennett discusses intuition pumps in the context of philosophical argumentation and does not really elaborate on what he means by intuitions. The fast and pre-reflective nature of intuitive decision-making, however, suggests that it relies heavily on emotions (Damasio 1994, 2004, 176; LeDoux 1998, 112–114; Evans 2001, 129–134). Indeed, it is the widely accepted view that the power of analogy lies in its reliance on emotional communication, which might even be the most basic and ancient function it has (Holyoak, Gentner and Kokinov 2001, 5). Communicating emotions is not easy. Because they work intuitively, they are hard to fake and are unpredictable (Damasio 1994, 158–164; Evans 2001, 145–176; Ekman 2003, 58–83). Emotions usually communicate therefore by emotional transfer, where emotional signalling by one individual creates an emotional state in another individual, and this all happens outside the level of consciousness (Thagard and Shelly 2001, 344). What appears to be unique to humans is our capacity to manipulate emotions even when emotional signalling is not directly present and, indeed, through different types of extra-somatic aids that make it possible to separate the signaller and the receiver from each other in space and time, a phenomenon called *secondary social sharing of emotions*. Narratives are one of the foremost tools used in creating this sense of sharing (Christophe and Rimé 1997, 37–54; Curci and Bellelli 2004, 881–900).

5. The notion of paradigm narratives is borrowed from Patrick Colm Hogan. According to him, paradigm narratives are those stories that "engage us, the stories we celebrate and repeat" (Hogan 2003, 5).

Narratives achieve this by creating simulations of the real world and what goes on in it. They portray emotions and emotionally consequential events that provoke affective responses in us (Hogan 2003, 5; Curci and Bellelli 2004, 881–900). Indeed, according to psychologist Keith Oatley, the made-up realities of narratives might actually communicate emotions more effectively than any on-line interaction in the real world because interactions in them are usually more controlled and are constructed especially for the purpose of conveying certain emotions (Oatley 1999, 101–117). When narratives evoke strong enough emotional responses from the audience, a somatic-marker is attached to them (Damasio 1994, 173–177). According to Antonio Damasio, this means that the account in question is emotionally marked from then on so that when similar emotions are felt in the future, these activate the memories for the marked out account and bias our minds to use it as a source for solving the present task.

This connection with emotional relevance makes narratives the perfect tool for communicating representations that cannot be approached empirically or logically. For example, if the truth of some account cannot be verified by empirical observations or logical arguments, it can still be accepted as true on the basis of the emotional relevance and meaning it contains for the individual. In many societies, the guardians of the paradigmatic narratives know this and professional storytellers are capable of producing somatic markers in the minds of their audience through their performance (Foley 1995, 275–301; Niles 1999, 212; Sjöblom 2004, 185–208). Unfortunately, while we do know that the use of analogy is the driving technique here, we know much less about exactly how analogies are used to fulfil their purposes. But research is currently being conducted on this issue. For example, Paul Thagard and Cameron Shelly have investigated how analogies create emotions by connecting otherwise neutral representations with others that evoke natural affective responses such as in the phrase: "An associate dean is a mouse training to be a rat" (Thagard and Shelly 2001, 349–350). In some cases, often connected with religious traditions, somatic marker effects can also be caused by arousing strong emotions through non-verbal media during the performance. The initiation rituals of some tribal communities, where the participants are physically and mentally tortured while being told the traditional myths of the community, are extreme examples of the latter strategy (see, e.g. Whitehouse 1995; 2000). While the emotional effects aroused by such non-verbal media cannot be directly and unproblematically connected to the narrative mode of representing the world, I will, in what follows, focus on the use of analogies as a strategy for creating relevance through narrative representations.[6]

## Taboos as a Type of Emotional Communication

The case I will discuss here is that of early Irish taboos. Early Irish tradition covers the period between 600 and 1200 AD (Thurneysen 1946, 1, 673; Nagy 1997, 10). It is accessed basically through literary sources surviving from this period, including fictional narratives, legal tracts, poetry, religious texts and wisdom literature (Findon 1995, 12–17; McCone and Simms 1996). One of the prominent and reappearing themes of the fictional narratives from this period is how leading characters are placed under taboos (OI *gessi*) restricting their powers or possibilities to act. This usually leads them into impossible situations where they are forced to violate their taboos, and this in turn leads ultimately to their deaths (Reinhard 1933; O'Leary 1988, 85–107).

6. But see Sjöblom (2005, 235–254), where it is argued that communicating through non-verbal means is possible only because of cognitive narrativity.

In the context of the present discussion, early Irish taboos have two interesting features. First, in contrast to other types of social rules and injunctions, taboos appear to have been followed intuitively. Early Irish society used different kinds of social and legal contracts, injunctions, and codes of behaviour. Everybody and everything was provided with guidelines on how to behave. Even the natural world was not excluded, and surviving legal manuscripts include, among other details, separate law codes for cats, dogs, wolves, and even bees (See Kelly 1988; 1997; Sjöblom 2000, 78–110). All these different types of rules and restrictions were supervised by different social institutions and experts, and a carefully constructed system of punishments and compensations existed to hold up the society, but not so with taboos. Taboos seem to have had an intuitive hold on the minds of the early Irish, so no external supervision or control was needed or expected (Sjöblom 2000, 111–144). For example, in the tale *Feis Tighe Chonáin* (FTC), Finn mac Cumaill, one of the great heroes of the early Irish narratives, explicitly expresses his dislike of being placed under taboos, but afterwards acts according to them without hesitation or protest (FTC 11.1203). In the same manner, in the tale *Aided Áenfir Aífe* (AAA), Cú Chulainn, the other prototypical hero of the Irish narratives, goes as far as to kill his own son in order not to violate his taboos (AAA §9).

The second interesting feature of early Irish taboos is that while other types of injunctions are carefully codified in the legal tracts and social commentaries by early Irish authors, taboos can be found almost exclusively only from fictional narratives (Greene 1979, 19). It is difficult to see this as a coincidence. Compared to other bodies of early medieval literature, the Irish material is very well preserved and a wide variety of texts in different genres are available to us, as stated above. If taboos were dealt with outside the fictional narratives, at least traces of such texts should have been known to modern scholarship. The few references to taboos found outside narratives are either examples of individual taboos without any explanatory context, or notes on the importance of rulers and heroes to observe their taboos (see Borsje 1996, 220; Sjöblom 2000, 199). If anything, these stray references strengthen the case of approaching early Irish taboos as a cultural practice existing in the real world but communicated and dealt with through fictional narratives. But if emotional communication and the creation of emotional relevance is what narratives are designed to do, then the power of taboos and their relevance for early Irish audiences were based on the emotional power invested in traditional tales.

On a general level, narratives achieve this simply by creating in the audience an emotional urge to share the experiences of the fictional characters (Curci and Bellelli 2004, 881). Early Irish narratives created this urge by making the heroes exaggerate their emotional responses. For example, heroes always act out their emotions; they do not keep them for themselves. Additionally, when they act out their emotions, they do so in a grandiose way, for instance, by starting to sing instead of simply talking, or by rending their clothes, or even through physical changes in their appearance, as when battle-frenzy disfigures the face of Cú Chulainn and colours his hair in the story called *Táin Bó Cúailnge* (TBC):

> Thereupon he became distorted. His hair stood on end so that it seemed as if each separate hair on his head had been hammered into it. You would have thought that there was a spark of fire on each single hair. He closed one eye so that it was no wider than the eye of a needle; he opened the other until it was as large as the mouth of a mead-goblet. He laid bare from his jaw to his ear and opened his mouth rib-wide so that his internal organs were visible. Then champion's light rose above his head. (TBC 11.428–434)

However, I believe it is possible to find additional and more specific strategies for communicating the power of taboos if we take one further step and presume that the emotions of fictional characters are not only related real emotions, but that they actually react to taboos much in the same way as real people would do. Keith Oatley has argued that this is how it actually must be. Fictional characters can show exaggerated expressions of emotions, and they do act in prototypical and schematic ways. But they cannot behave and react in ways that are totally strange to us if the intention is for us to relate ourselves to them and their concerns (Oatley 1999, 101–117). While little attention has been paid to this question, the empirical findings thus far support this idea (see Dunbar, Clark and Hurst 1995, 233–246; Gotschall *et al.* 2004, 102–112).

Thus, if we want to know more precisely how early Irish people felt about taboos, we could do this by focusing on how the fictional characters react to being tabooed in the early Irish tales, and then compare these reactions to similar descriptions from ethnographic data. This is not, however, as simple as it may sound. The problem is that even in narratives only a few descriptions of the actual tabooing events survive. In most cases, taboos are only narrated to the reader in the form of catalogues set out at the beginning of the tale. Still, even the material we have can be enlightening. For instance, in the early Irish tale *Togail Brudne Da Derga* (TBDD), the prototypal mythical king of early Irish tradition, Conáire Mór, is tabooed from killing birds. We are told that this is because he is related to birds and because his father is actually a bird (TBDD §§ 13, 16). In the same vein, in the tale *Tóruigheacht Dhiarmada agus Ghráinne* (TDG), the hero Diarmaid is forbidden to hunt boars, a favourite past-time of early Irish heroes, because one of them, the big boar of Beann Gulban, is his foster-brother and, thus, in effect is related to him (TDG 11.1452–1522). In both of these cases, the emotional force for following the taboo is based on the analogy made between an animal and the close relatives of the tabooed individual. Killing one of your own kin (*fingal*) was one of the gravest crimes in early Irish society and therefore feared by everyone (Kelly 1988, 127–128).

Of course, the fact that killing your own relative is feared and deemed as a horrible crime is not something unique to early Irish society. It is a universal feature of human thinking, probably grounded in our innate instinct for promoting the survival of our genes. This is what makes it so hard to violate and it usually is not done without very strong reasons, such as protecting your own life (Hamilton 1964, 1–52; Dawkins 1976; Dunbar, Clark and Hurst 1995, 233–246). What happens in the context of early Irish taboos is that this innate sense of the fear of kin-slaying is transferred through metaphorical transference of emotions to also include some animals, like birds or boars. Anthropologist Nancy Thornhill found a similar cognitive mechanism at work with the incest taboos of 129 different cultures she collected and analysed (Thornhill 1991, 247–293). Thornhill notes that incest taboos forbidding sexual interaction between relatives are not usually given to close kin but for more distant relatives and the like. Thus, with these taboos our natural disgust for having sexual interaction with close kin is transferred also to cover individuals that do not evoke this natural disgust in us.

## In the End: Religion and Emotional Relevance

In this article I have discussed the place of emotional communication in narrative presentations and how this contributes to making narratives relevant to the audience. How does this contribute to the discussion of religious relevance in general? To begin with, it seems to me that it is quite possible that at least some of the relevance of religious representations is due to their being connected to such basic biological urges as the survival instinct. To my mind, only this kind of biological urge can be thought of as the base of our "deepest desires." However, as discussed above, a connection with such biological urges cannot be the whole story. First, many religious representations have no obvious connection with such urges and, secondly, it is possible to transfer affective and emotional reactions from one realm of experience to another. Thus, the argument that religious relevance is due to some inherent essence of certain representations is clearly insufficient.

To argue that relevance is based on how widely a representation can be referred to and used in everyday social interactions is also obviously accurate to some extent. However, how we decide what is relevant is not purely based on rationality and logic. Emotions play a large role here, and we can safely argue today that without emotions we would probably be unable to make proper rational decisions. Among different possible and working solutions, we decide what to do on the basis of our emotions. Thus, for me, emotional relevance appears to be a prerequisite for choosing what information is deemed to have strategic value to begin with. This also makes it more fundamental to other types of relevancy.

In this chapter, I also argued that emotional relevance can be transferred and created culturally with the aid of emotional stimulation. Narratives are one of the basic tools for achieving this. Emotional commitments to certain representations and their truthfulness becomes essential when other criteria of evaluation, such as empirical testing or logic, cannot be used. This is probably the reason why religious traditions rely so much on narrative discourse for their communication. This also shows that cultural authorities and skilful narrators can manipulate and guide our intuitions and instincts to find relevance in the narratives they promote and relate to us. Thus, a skilful narrator can also enhance and create the strategic value of some narratives over others.

## References

### Sources

AAA. Aided áenfir Aífe. In *Compert Con Culainn and Other Stories.* edited by Anton G. van Hamel), 9–15. Dublin: Dublin Institute for Advanced studies (1933).

FTC. *Feis tighe Chonáin.* Edited by Maud Joynt. Dublin: Dublin Institute for Advanced Studies (1936)

TBC. *Táin bó Cúailgne: Recension I.* Edited by Cecile O'Rahilly. Dublin: Dublin Institute for Advanced Studies (1976).

TBDD. *Togail bruidne Da Derga.* Edited by Eleanor Knott. Dublin: Dublin Institute for Advanced Studies (1936).

TDG. *Tóruigheacht Dhiarmada agus Ghráinne/ The Pursuit of Diarmaid and Gráinne.* Edited by Nessa Ní Shéaghdha. Dublin: Irish Texts Society (1967).

### Literature

Adolphs, Ralph. 1999. "Social Cognition and the Human Brain." *Trends in Cognitive Science* 3 (12), 469–479.

Atran, Scott. 2002. *In Gods We Trust.* Oxford: Oxford University Press.

Barrett, Justin. 2004. *Why Would Anyone Believe in God?* Walnut Creek: AltaMira Press.

Barthes, Roland. 1957. *Mythologies.* Paris: Pierre Vives.

Bartlett, Frederic C. 1932. *Remembering*. Cambridge: Cambridge University Press.

Bloom, Paul. 2004. *Descartes' Baby*. London: William Heinemann.

Borsje, Jacqueline. 1996. *From Chaos to Enemy: Encounters with Monsters in Early Irish Texts*. Turnhout: Brepols Publishers.

Boyer, Pascal. 1990. *Tradition as Truth and Communication*. Cambridge: Cambridge University Press.

———. 1994. *The Naturalness of Religious Ideas*. Berkeley: University of California Press.

———. 2001. *Religion Explained*. New York: Basic Books.

Bruner, Jerome. 1986. *Actual Minds, Possible Worlds*. Cambridge, MA: Harvard University Press.

———. 2002. *Making Stories*. New York: Farrar, Straus and Giroux.

Bulbulia, Joseph. 2004. "The Cognitive and Evolutionary Psychology of Religion." *Biology and Philosophy* 19(5): 655–686.

Byrne, Richard and Andrew Whiten, eds. 1988. *Machiavellian Intelligence*. London: Clarendon Press.

Carrithers, Michael. 1991. "Narrativity: Mindreading and Making Societies." In *Natural Theories of Mind*, edited by A. Whiten, 305–317. Oxford: Basil Blackwell.

Chance, M.R. and A.P. Mead. 1953. "Social Behaviour and Primate Evolution." *Synopsis of the Society for Experimental Biology* 7 (Evolution): 395–439.

Christophe, Véronique and Bernard Rimé. 1997. "Exposure to the Social Sharing of Emotion: Emotional Impact, Listener Responses and Secondary Social Sharing." *European Journal of Social Psychology* 27(1): 37–54.

Cosmides, Leda and John Tooby. 1992. "Cognitive Adaptations for Social Exchange." In *The Adapted Mind*, edited by J. Barkow, L. Cosmides and J. Tooby, 163–228. Oxford: Oxford University Press.

Curci, Antonietta and Guglielmo Bellelli. 2004. "Cognitive and Social Consequences of Exposure to Emotional Narratives: Two Studies on Secondary Social Sharing of Emotions." *Cognition and Emotion* 18(7): 881–900.

Dawkins, Richard. 1976. *The Selfish Gene*. Oxford: Oxford University Press.

Damasio, Antonio. 1994. *Descartes' Error*. London: Papermac.

———. 1999. *The Feeling of What Happens: Body and Emotion in the Making of Cognition*. San Diego: Harcourt.

———. 2003. *Looking for Spinoza*. London: Vintage.

Dautenhahn, Kerstin. 1999. "The Lemur's Tale: Story-Telling in Primates and Other Socially Intelligent Agents." In *Proceedings of the AAAI Fall Symposium 1999 "Narrative intelligence"*, edited by M. Mateas and P. Sengers, 59–66. AAAI Press, Technical Report FS 99-01.

———. 2002. "Stories of Lemurs and Robots: The Social Origin of Story-Telling." In *Narrative Intelligence*, edited by M. Mateas and P. Sengers, 63–90. Philadelphia: John Benjamin.

Dennett, Daniel. 1984. *Elbow Room: The Varieties of Free Will Worth Wanting*. Oxford: Clarendon Press.

———. 1991. *Consciousness Explained*. Harmondsworth: Penguin Classics.

Donald, Merlin. 1991. *Origins of the Modern Mind*. Cambridge, MA: Harvard University Press.

———. 2001. *A Mind So Rare*. New York: W.W. Norton.

Dunbar, Robin. 1996. *Grooming, Gossip and the Evolution of Language*. London: Faber and Faber.

———. 1998. "The Social Brain Hypothesis." *Evolutionary Anthropology* 6: 178–190.

———. 2004. *The Human Story: A New History of Mankind's Evolution*. London: Faber and Faber.

Dunbar, Robin. Amanda Clark and Nicola L. Hurst. 1995. "Conflict and Cooperation among the Vikings: Contingent Behavioural Decisions." *Ethology and Socio-Biology* 16: 233–246.

Ekman, Paul. 2003. *Emotions Revealed.* London: Phoenix.
Evans, Dylan. 2001. *Emotions. The Science of Sentiment.* Oxford: Oxford University Press.
Fauconnier, Gilles and Mark Turner. 2002. *The Way We Think.* London: Basic Books.
Findon, Joanne. 1995. *A Woman's Words.* Toronto: University of Toronto Press.
Finlay, M.I. 1971. *The Use and Abuse of History.* London: Pimlico, 2000.
Foley, John M. 1995. "Word-Power, Performance and Tradition." *Journal of American Folklore* 105: 275–301.
Geertz, Clifford. 1973. *The Interpretation of Cultures.* London: Fontana Press.
Gentner, Dedre, Keith J. Holyoak and Boicho N. Kokinov. 2001. *The Analogical Mind.* Cambridge, MA: Harvard University Press.
Gould, Stephen Jay. 1991. "Exaptation: A Crucial Tool for Evolutionary Psychology." *Journal of Social Issues* 47: 43–46.
Greene, David. 1979. "Tabu in Early Irish Narrative." In *Medieval Narrative: A Symposium*, edited by H. Bekker-Nielsen, 9–19. Odense: Odense University Press.
Greenspan, Stanley and Stuart Shanker. 2004. *The First Idea: How Symbols, Language, and Intelligence Evolved from Our Primate Ancestors to Modern Humans.* Cambridge, MA: Da Capo Press.
Gottschall, Jonathan, Johanna Martin, Hadley Quish and Jon Rea. 2004. "Sex Differences in Mate Choice Criteria Are Reflected in Folk Tales from Around the World and in Historical European Literature." *Evolution and Human Behavior* 25: 102–112.
Guthrie, Stewart. 1993. *Faces in the Clouds: A New Theory of Religion.* Oxford: Oxford University Press.
Hamilton, William. 1964. "The Genetical Evolution of Social Behaviour." *Journal of Theoretical Biology* 7(1): 1–52.
Heberlein, Andrea S. *et al.* 1998. "Impaired Attribution of Social Meanings to Abstract Dynamic Visual Patterns Following Damage to the Amygdala." *Society for Neurosciences Abstracts* 24: 1176.
Hervieu-Léger, Danièle. 2000. *Religion as a Chain of Memory.* London: Polity Press.
Hinde, Robert. 1997. *Relationships: A Dialectical Perspective.* Hove: Psychology Press.
———. 1999. *Why Gods Persist: A Scientific Approach to Religion.* London: Routledge.
Hogan, Patrick Colm. 2003. *The Mind and Its Stories.* Cambridge: Cambridge University Press.
Holyoak, Keith and Paul Thagard. 1995. *Mental Leaps: Analogy in Creative Thought.* Cambridge, MA: MIT Press.
———. 1997. "The Analogical Mind." *American Psychologist* 52(1): 35–44.
Holyoak, Keith, Dedre Gentner and Boicho N. Kokinov. 2001. "Introduction: The Place of Analogy in Cognition." In *The Analogical Mind*, edited by D. Gentner, K.J. Holyoak and B.N. Kokinov, 1–19. Cambridge, MA: MIT Press.
Humphrey, Nicholas. 1976. "The Social Function of Intellect." In *Growing Points in Ethology*, edited by P.P.G. Bateson and R.A. Hinde, 303–317. Cambridge: Cambridge University Press.
Kelly, Fergus. 1988. *A Guide to Early Irish Law.* Dublin: Dublin Institute for Advanced Studies.
———. 1997. *Early Irish Farming.* Dublin: Dublin Institute for Advanced Studies.
Landau, Misia. 1991. *Narratives of Human Evolution.* New Haven, CT: Yale University Press.
LeDoux, Joseph. 1998. *The Emotional Brain.* New York: Phoenix.
Lewin, Roger. 1987. *Bones of Contention.* Chicago, IL: The University of Chicago Press.
Lord, Albert. 1960. *The Singer of Tales.* New York: Athenaeum.
Mateas, Michael and Phoebe Sengers. 2002. "Narrative Intelligence." In *Narrative Intelligence*, edited by M. Mateas and P. Sengers, 1–25. Amsterdam/Philadelphia: John Benjamins Publishing Company.

McCone, Kim and Katherine Simms. 1996. *Progress in Medieval Irish Studies*. Maynooth: Department of Old Irish/Saint Patrick's College.

Morton, Adam. 2000. "The Evolution of Strategic Thinking." In *Evolution and the Human Mind*, edited by P. Carruthers and A. Chamberain, 218–237. Cambridge: Cambridge University Press.

Nagy, Joseph. 1977. *Conversing with Angels and Ancients*. Dublin: Four Courts Press.

Nelson, Katherine. 2003. "Narrative and the Emergence of a Consciousness of Self." In *Narrative and Consciousness*, edited by G.D. Fireman, T.E. McVay Jr. and O.J. Flanagan, 17–36. Oxford: Oxford University Press.

Nichols, Shaun. 2004. "Is Religion What We Want? Motivation and the Cultural Transmission of Religious Representations." *Journal of Cognition and Culture* 4(2): 347–372.

Niles, John D. 1999. *Homo Narrans*. Philadelphia: University of Pennsylvania Press.

Oatley, Keith. 1999. "Why Fiction May Be Twice as True as Fact: Fiction as Cognitive and Emotional Simulation," *Review of General Psychology* 3(2): 101–117.

O'Leary, Philip. 1988. "Honour Bound: The Social Context of Early Irish Literature." *Celtica* 20: 85–107.

Pinker, Steven. 1997. *How the Mind Works*. Harmondsworth: Penguin Books.

Pyysiäinen, Ilkka. 2001. *How Religion Works*. Leiden: E.J. Brill.

———. 2002. "Religion and the Counter-Intuitive." In *Current Approaches in the Cognitive Science of Religion*, edited by I. Pyysiäinen and V. Anttonen, 110–132. London: Continuum.

———. 2004. *Magic, Miracles, and Religion*. Walnut Creek: AltaMira Press.

Reinhard. John Revell. 1933. *The Survival of Geis in Medieval Romance*. Halle: Max Niemeyer Verlag.

Schank, Roger C. 1990. *Tell Me a Story*. Evanston, IL: Northwestern University Press.

Shermer, Michael. 2000. *How We Believe*. New York: A.W.H. Freeman.

Sjöblom, Tom. 2000. *Early Irish Taboos: A Study in Cognitive History*. Helsinki: Department of Comparative Religion/University of Helsinki.

———. 2004. "Wordpower: Narratives, Tradition and Religion." *Temenos* 39–40: 185–208.

———. 2005. "Storytelling: Narratives of the Mind and Modes of Religiosity." *Historical reflections/ Reflexions historique* 31(2): 235–254.

Sosis, Richard and Candance Alcorta. 2003. "Signaling, Solidarity, and the Sacred: The Evolution of Religious Behaviour." *Evolutionary Anthropology* 12: 264–274.

Sperber, Dan. 1975. *Rethinking Symbolism*. Cambridge: Cambridge University Press.

Sperber, Dan and Deirdre Wilson. 1995. *Relevance: Communication and Cognition*. 2nd edition. Oxford: Blackwell.

Stockzkowski, Wiktor. 2002. *Explaining Human Origins*. Cambridge: Cambridge University Press.

Thagard, Paul. 2000. *Coherence in Thought and Action*. Cambridge, MA: MIT Press.

Thagard, Paul and Cameron Shelly. 2001. "Emotional Analogies and Analogical Inferences." In *The Analogical Mind*, edited by D. Gentner, K.J. Holyoak and B.N. Kokinov, 335–362. Cambridge, MA: MIT Press.

Thornhill, Nancy. 1991. "An Evolutionary Analysis of Rules Regulating Human Inbreeding and Marriage." *Behavioral and Brain Sciences* 14: 247–293.

Thurneysen, Rudolf. 1980 [1946]. *A Grammar of Old Irish*. Dublin: Dublin Institute for Advanced Studies.

Tulving, Endel and Martin Lepage. 2000. "Where in the Brain Is the Awareness of One's Past?" In *Memory, Brain and Belief*, edited by D.L.Schacter and E. Scarry, 208–228. Cambridge, MA: Harvard University Press.

Whitehouse, Harvey. 1995. *Inside the Cult: Religious Innovation and Transmission in Papua New Guinea*. Oxford: Oxford University Press.

———. 2000. *Arguments and Icons: Divergent Modes of Religiosity*. Oxford: Oxford University press.

Whiten, Andrew and Richard Byrne, eds. 1997. *Machiavellian Intelligence II: Extensions and Evaluations.* Cambridge: Cambridge University Press.

Yates, Frances A. 1966. *The Art of Memory*. London: Pimlico.

— 9 —

# Moses and the Invasion of the Body Snatchers: Durkheim, Cognition and Deuteronomy[1]

Hans J. Lundager Jensen

## Introduction

The aim of this paper is to explore possible connections between three theoretical positions relevant to our understanding of religion: (1) Durkheim's classical theory of religion as presented in his *Elementary Forms*; (2) certain potentially fruitful perspectives in actual cognitive studies (meme-theory and Merlin Donald's evolutionary model of the development of human cognition); and (3) a central point in the most "theological" part of the Hebrew Bible, the theology in the book of Deuteronomy. I maintain that Durkheim's theory is fundamentally "cognitive" in so far as the concept of "society" in his use of the term basically means a reservoir of representations shared by a number of individual consciousnesses, the acquisition, maintenance, dissemination, and disappearance of which the theory of religion wishes to explain. I point out similarities between Durkheim's basic religious scenario, basic scenarios in cognitive theories and central Deuteronomic themes in its normative reflection on its own religion.

## Durkheim: Society Is a Set of Representations

In its formative period, before WW I, the new science of sociology was fascinated by religion. The last and greatest work by Émile Durkheim, the founder of modern French sociology, was his *Elementary Forms of the Religious Life* (1912).[2] The only work that his German counterpart, Max Weber, published as a book during his lifetime was *The Protestant Ethic and the Spirit of Capitalism* (1904), but this was a precursor for his much

1. This paper is a translation and revised version of a Danish publication entitled "Moses og bodysnatchernes invasion (Durkheim, cognition og Deuteronomium)." *Collegium Biblicum Årsskrift* 2005: 33–45.

2. French original 1912 (*Formes élémentaires de la vie religieuse*); I quote from the 1995 translation by Karen E. Fields.

larger work on Chinese and Indian Religions and on ancient Judaism. Sociologists chose religion as a field suitable to demonstrate its abilities and thereby demand a place in the scientific world. Religion was well-suited as a subject for the new science because religion appeared both to be something eminently personal and individual, rooted as it seemed to be in private belief and conviction, something eminently irrational, and yet ubiquitous in human history. Sociologists wanted to demonstrate that under the apparent individuality there was a collective reality, and that seeming irrational convictions had their own kind of rationality—what may appear irrational to some seems rational to others. Durkheim's and Weber's sociologies of religion differed of course in important ways: Durkheim's focus was on archaic-primitive and elementary structures, while Weber's focus was on existing world religions. Durkheim found support in the early empirical anthropology, which was being developed at the time, while Weber leaned towards another young, but perhaps somewhat more solidly institutionalized, academic discipline: the history of religions. Both Durkheim and Weber made use of the Biblical exegesis of their time. Durkheim was deeply inspired by Robertson Smith's *The Religion of the Semites* (1889), and Weber's investigation into ancient Judaism was dependent upon the Biblical historical-critical exegesis of German Protestant scholarship—the most advanced at that time.

The fundamental assumption in Durkheim's understanding of religion has been and still is a source of inspiration for anthropological functionalism, structuralism and social or cultural anthropology. But, although Durkheim's contribution to a general understanding of religion may be well-known, it is often misrepresented on a crucial point. It is not unusual to see Durkheim paraphrased in such a way that what he assumed to be cultivated in religion "really" is "society" itself. As an example, Daniel L. Pals writes in his otherwise very useful textbook:

> Durkheim almost always claims that society determines, while religion is the thing that is determined. Society controls; religion reflects.... Durkheim insists that society powerfully shapes religious ritual and belief, while religious beliefs never seem able to do the reverse. (Pals 1996, 117)

This quasi-Marxist basis-superstructure-model rests on a fundamental misunderstanding.[3] Durkheim's idea was not that religion is an expression of society, but rather that religion and society are inseparable. Society for Durkheim was basically a "moral" fact, and morality consisted in a stable set of representations, that is, all kinds of elements of signification: norms, rules, ideas, and narratives shared by a given group of human beings. Durkheim was open to counting everything specifically human as "society" except economic activity. And religion, as far as it is accessible to a scientific understanding, is a set of representations: norms, rules, ideas, and narratives. One may better understand what Durkheim meant by "society" (French: *société*) if one translates it as *culture*.[4] "Society" is *not* a given

3. Durkheim rejected this understanding explicitly in the final chapter of his *Elementary Forms*, 426: "The last thing to do is to see this theory of religion as merely a refurbishment of historical materialism.... [C]ollective consciousness is something other than a mere epiphenomenon of its morphological base, just as individual consciousness is something other than a mere product of the nervous system. If collective consciousness is to appear, a *sui generis* synthesis of individual consciousnesses must occur. The product of this synthesis is a whole world of feelings, ideas, and images that follow their own laws once they are born."

4. Alternatively to "society," Durkheim used the concept of "morality" in this encompassing sense; in the following, however, I use the term "society." There are certainly connections here back to German preromantic ideas of a collective mentality, "Geist": cf., Burke (2004).

group of individuals and/or the statistical average of what people in a given group actually think and do. Society is rather the ideas and narratives which the members of a certain group have in common and which they regard as a relevant and legitimate possession, and the norms and rules to which they feel committed, at least in principle. Therefore, society is neither something which can be reflected by religion, nor can society be controlled by religion; religion *is,* or is *part of,* society.

To Durkheim, society is what human beings have created, that which does not exist in nature, in the world of biology, chemistry, and physics. These uniquely human traits include not only myths and ethical norms, but also regulations for actions, like language, food recipes, and rules of conduct. Human beings without society are but animal bodies: they may possess life in a biological and psychological sense, that is, they may be able to have sensations, move, seek food, mate, communicate by means of signals, perform parental care, but such bodies are not truly human. Through imitation, upbringing, and education, the biological-psychological mechanisms are supplemented with the uniquely human, which is society, i.e., norms, rules and concepts. These are exclusively human and so far invented not by any specific person, but transmitted anonymously as tradition. Representations are variable, but changes and new representations always transform and moderate previously existing representations. Durkheim's contrast is not society versus religion, but society, including or identical with religion, versus society-deprived bodies. In order to become a human being, a proto-human body must be introduced into a society, which becomes imprinted in the body's consciousness. To any concrete body, society always is something external, something strange, which through upbringing and education becomes something internal, something which becomes a part of the body itself. No singular individual being discovers or invents a language, logic, religion, rules of conduct, etc. One is always born into an already existing language, logic, etc. The paradox consists not only in the external part becoming a part of the individual, but also that it is regarded as the most valuable part. To be "human" is exactly to not be an animal; but it is as an animal, as a non-informed body that a human being is truly "in-dividual," that is, something that cannot be divided. To "come to oneself" means to think and act not according to one's own ideals, but to a shared set of ideals. A human being is at its best when it is acting, feeling and reasoning according to norms and rules common to a group.

But society as an external factor, which becomes internal, does not of course remove biology and psychology: the body continues to have sensations, be moving, etc. Durkheim never tired of stressing that a human being is a *homo duplex.* Norms and rules regulate the biological-psychological body, and concepts and narrations provide material for fantasies and explanations. The external elements can become a part of a person to such a degree that it will not be experienced as something strange. Normally, however, there will be a degree of external constraint that reminds the body that norms and rules are established by somebody or something else than itself. On the other hand, it will often, if not always, be in the body's own interest to let it be dominated by external criteria. They make normal decision-making fast and easy, and they further the body's chances for staying in the vicinity of other bodies. To the extent that society is a yoke, it is a yoke easy to carry because it is in the body's own interest to do so.[5]

5. Durkheim, 1995, 209: "society has its own specific nature that is different from our nature as individuals, it pursues ends that are also specifically its own; but because it can achieve this end only by working through us, it categorically demands our cooperation. Society requires us to make ourselves its servants, forgetful of our own interests. And it subjects us to all sorts of restraints, privations, and sacrifices with-

Society, according to Durkheim, is something which is at the same time concrete and non-material. It exists just like language, logic and politeness exist, as far as one can notice their effects and describe them as structures. And society is nothing else but the totality of these structures; accordingly, society has an independent existence without materiality and regardless of the human bodies. But on the other hand, it is an existence which presupposes the independent existence of human bodies and transmission from one body to another, or material fixation in texts, pictures, buildings, etc. For Durkheim it made sense to consider society as a "hyper-spiritual" form of existence (Durkheim 1996, 48). But society is not only external and independent with regard to a single human being, it is also bigger than any human being. In principle, language, mythology, etc. can and typically will contain much more than any individual can hold in its own consciousness. But initiation into a society will give the human being access to this surplus that it may be able to exploit, given the right circumstances. Therefore, Durkheim often describes the relationship between society and individual as a relationship of participation: the individual participates in a society, which is more than itself, which is non-material and therefore "spiritual," and which makes the individual into a human being to the extent that the individual participates and wishes to participate in it.

Society's "hyper-spiritual" weightlessness implies its evanescence. On the one hand, society or some of its elements may have enormous significance and in dramatic situations oppose the body's own animal interests, like when an individual chooses to renounce its own life in order to support or maintain a common value or a symbolical representation of it. On the other hand, imprinted rules, regulations, and narrations can also be weakened and forgotten. A society is always in competition with the body's needs and desires and its inclinations to choose faster and less costly solutions than those prescribed by its society. Also, single representations or a whole society can be out-competed by other representations or another society. If a body can participate in a society, and must do so, if it is to be regarded as "human," nothing prevents it from participating in several societies simultaneously. This may happen if there is a more relatively consistent system of norms, rules, concepts, and narrations, in which an individual can participate. These may not necessarily be in conflict, but most likely they often are. Alternative societies typically make alternative demands on the body, its behaviour, its conceptions, its attention, not least its time, for principal or practical reasons.

Thus, for a stable society to exist, it does not suffice that society's representations are installed into an individual, or that an individual is initiated into a society. After the initiation a process of weakening will set in, which will lead to erosion and ultimately disappearance of the representation unless some means of prevention is activated. Durkheim understood the maintenance of what an individual has learned as a sort of inverse echo, a resonance where an emitted representation is strengthened by being received back.[6] This

out which social life would be impossible. And so, at every instant, we must submit to rules of action and thought that we have neither made nor wanted, and that sometimes are contrary to our inclinations and to our most basic instinct." Cf., p. 351: "the individual gets the best part of himself from society—all that gives him a distinctive character and place him among other beings, his intellectual and moral culture. Let language, science, arts, and moral beliefs be taken away from man, and he falls to the rank of animality."

6. Émile Durkheim, 1995, 217f: "Once the individuals are gathered together, a sort of electricity is generated from their closeness and quickly launches them to an extraordinary height of exaltation. Every emotion expressed resonates without interference in consciousnesses that are wide open to external

happens constantly for the representations of daily life where all sorts of remarks are met with other peoples' affirmative utterances and gestures and thereby receive a confirmation which strengthens its place in the first individual's consciousness ("Nice weather today, isn't it?" "It certainly is!"). But representations less relevant to the micropraxis of daily life, such as religious, political or national representations, receive less often such "resonant" confirmations. They are much more exposed to erosion and must therefore be upheld with more effective means. Among these the oldest and most efficient is without doubt *ritual.* Through congregation, the membership of each individual is confirmed—the kind of knowledge which is highly exposed to erosion—and at the same time some or all of the representations that form this group's common reservoir of norms, rules, notions and narratives are confirmed. Durkheim assumed that groups unable to establish such periodically recurring reunions would not be able in the long run to resist erosion and effacement of what kept the group together; sooner or later such a society would cease to exist.[7] Rituals therefore are an elementary means to keep representations in the consciousness of a group of people. In pre-literate societies—Durkheim's primary interest in his *Elementary Forms*—this means maintaining their very existence.[8]

Especially in spontaneous reunions, but often also in more organised rituals, Durkheim surmised that a certain state of mind would arise among the participants, a state of excitement, perhaps cheerfulness, sometimes aggression, but in all circumstances a coordination of individual consciousnesses in adjustment more or less to each other. The mechanism here is the same as that which secures the daily confirmation of representations: the resonance produced by receiving a representation from another person. But the effect is much greater when a representation reverberates simultaneously from many individuals gathered at one spot. This state, Durkheim's famous "effervescence," can also be an occasion for new, common representations to appear and be transmitted to other individuals in the group; and the mere experience of effervescence would be the argument, at least for some time, for the relevance and legitimacy of these new representations.

Reunions, i.e. under normal circumstances rituals, counterbalance the dispersive tendency of daily life. In Durkheim's Australian ethnographic material, people maintain their life by hunting and gathering in small groups. This contrast between reunion and dispersion, which in the Australian case is equivalent to the "sacred" and "profane" contrast, is also a contrast between shorter and longer periods of time, between closeness (bodies in intensive exchange in a restricted area) and dissolution (bodies with little exchange in an extended area), and it is the contrast between excitement and boredom.

Durkheim's principal contrast was society vs. bodies equipped with biology and individual psychology alone. His concept for this clash of worlds, the already-mentioned *homo duplex*, may seduce the reader into assuming a kind of dualism. The relation is more complicated, however. First, it is perhaps more natural to understand *homo duplex* as *homo triplex*, where *homo* is a medium position between two equally broad areas: society and animality. These two sides are not always or necessarily in conflict. Often it is in the individual's own best interest to accept society's representations and their more or less forceless force. But they can also be in conflict under more or less dramatic circumstances. It

impressions, each one echoing the others. The initial impulse is thereby amplified each time it is echoed, like an avalanche that grows as it goes along."

7. As languages in today's rapid process of extinction, cf. Gibbs (2002).
8. "To remain in the existential game" (Rappaport 1999, 409; 423).

therefore seems plausible to posit a psychic unit, a subject, that decides whether to follow the demands of society or nature. Both of these forces are "transcendental," bigger than the subject itself, strange areas and powers in which the subject participates.

In this case Durkheim's constellation reminds one of a more famous, but later formulated constellation, Sigmund Freud's so-called second topographical model (super-ego/ego/id). The super-ego, which is both consciousness as a judging and punishing agency and as ego ideal, by which the ego may develop, has been installed. It is not innate: "Even if conscience is something 'within us,' yet it is not so from the first. In this it is a real contrast to sexual life, which is in fact there from the beginning of life and not a later addition"(Freud 1965, 77). Conscience comes about through identification not with the parents and other adults as such, but with their own super-egos, and the child becomes "the vehicle of tradition and of all the time-resisting judgements of value which have propagated themselves in this manner from generation to generation" (Freud 1965, 84).[9] The id on the other hand is "the dark, inaccessible part of our personality.... [M]ost of that is of a negative character and can be described only as a contrast to the ego. We approach the id with analogies: we call it a chaos, a cauldron full of seething excitations" (Freud 1965, 91).

This comparison may not be so surprising after all, since Freud and Durkheim were partly contemporary. However, both models may be local instances of a much older constellation. Animality as an absence of collective representation and therefore absence of society was certainly a negative factor according to Durkheim. Likewise, Freud could only describe the id as something in opposition to the ego. In other words, both look like special cases of a Platonic-Aristotelian, monistic metaphysics where the phenomena or the things exist because they participate in something transcendent, being or *bonum*, which is greater than and encompasses the things, and which endures where the things must pass away. The transcendent is what allots the things whatever being, goodness, or beauty they might possess as parts of the One, often compared to the sun, which inexhaustibly spreads out its light and warmth, thereby waking things to life—the things which must pass because they also participate in the opposite of *bonum*, which is nothing.

Durkheim's humanization-readied bodies are not really human beings until they are brought into participation with society's representations. On the other hand, the representations are accepted into and survive in these bodies. Inevitably, sooner or later the bodies fall out of society again. Durkheim's highly theoretical construction foresees the possibility of a total erosion of representations, where individuals are isolated from each other for a longer period. No matter what, the individual bodies will die their individual, biological deaths, and what collective representations they may accommodate will definitively be extinguished by the body's disintegration into a purely physical and fragmented state of being, unless they have been passed on to other bodies.

| | | | |
|---|---|---|---|
| *metaphysics* | being | world of phenomena | non-being |
| *Durkheim* | society | (ego) | animality; materiality |
| *Freud* | Superego | Ego | Id |

## Cognition

Here lies, however, the difference between Platonic-Aristotelian metaphysics and the Durkheimian scenario. While *bonum* exists eternally, Durkheim's society only exists

9. In his *Elementary Forms*, not least in the chapter on the Soul, Durkheim emphasized (e.g. 266) that society constitutes a real transcendence, which has a passing stay in the individual consciousnesses.

because bodies exist. On the one hand, representations form a sphere in which the bodies can only participate when met at the time of birth and which (often) survive death. But on the other hand, representations only exist inside consciousnesses, at least before the invention of writing. Without bodies that can store norms, rules, ideas and narratives, the representations would not exist.[10] Therefore, Durkheim can also describe the relation between society and body as an invasion: society finds a place in consciousness and takes, to a certain degree, control over the body.[11] I propose to call this vision a cultural "body-snatcher-theory": human thoughts, rules etc. come from the outside, from an already existing culture, transmitted directly through other human beings or indirectly through different media: writing, pictures, electronic devices, etc. But only through storing can culture attain an independent existence. The body is thus more or less abducted by culture, thriving as it were on the bodies and in the minds of its victims.

Durkheim's reasoning anticipated current cognitive theory. There are obvious similarities to meme theory. Like genes, memes only exist in order to maintain their existence, and they only maintain it through copying themselves and finding their way into human bodies. Memes and meme-plexes can also conflict with each other, since working memory and attention is not unlimited. Memes come and go; some become a success for a shorter or longer period, others get a short life or none at all.

Memetics is a theory which focuses on the transmission of representations and the conditions for their relative successes. Basically representations are regarded as analogous to viruses, not necessarily with respect to their harmful effect on the host organism, but with respect to their way of life and dissemination. A virus, which happens to reproduce itself, has been successful in invading an organism with the sort of cells in which it can install itself, take over, and use for its reproduction. Accordingly, only those representations will survive that find a consciousness where they can survive and from where they can disseminate to other consciousnesses. As with Durkheim, the stable structures in this scenario are outside vs. inside, guest (or intruder, invader) vs. host, immaterial vs. material.

The basic inspiration in Dan Sperber's epidemiology[12] is similar to meme theory, but here consciousness is not regarded as a clean blackboard, and signification is not transmitted as pure imitation. It is more plausible to assume that a consciousness assimilates representations through internal reconstructions of possible meanings and that the result is a transformation rather than an imitation of the representation. What are imitated are not specific acts or utterances, but their assumed meanings or functions. When individual rewritings nevertheless typically converge and a certain collective consensus is formed, the reason is the prevalence of so-called "attractors," which adjust the representations to those assumptions that will be the most plausible for those specific consciousnesses. There must be physiologically conditioned attractors because human brains draw comparable conclu-

10. "No doubt, the men could not live without the gods; but on the other hand, the gods would die if they were not worshipped" (Durkheim 1995, 350).

11. "Because society can exist only in and by means of individual minds, it must enter [*pénètre*] into us and become organized within us" (Durkheim, 1995, 211); cf., (1995, 351): "We now glimpse the profound reason why the gods can no more do without their faithful than the faithful can do without their gods. It is that society, of which the gods are only the symbolic expression, can no more do without individuals than individuals can do without society."

12. For cultural epidemiology, see e.g. "An epidemiology of representations [7.27.05]. A Talk with Dan Sperber": http://www.edge.org/3rd_culture/sperber05/sperber05_index.html (seen 25.11. 2005); cf., Sperber 1996.

sions in spite of cultural, ecological, economical, political or social differences. But there are also culturally determined attractors. Already existing representations will further the assimilation of those new representations that seem most plausible compared to already accepted ones, and they will tend to transform new representations into something similar to the already ingrained ones. Attractors counteract the tendency to intractable mutations that human understanding inevitably produces.

Merlin Donald's evolutionary theory of cognition places special emphasis on memory and culture. Through the evolutionary process that has lead to modern human beings, the specific hominid and human skills of mimetic, mythic (or oral-linguistic) and finally theoretic memory have settled on the top of the more basic procedural and episodic memories, which humans have in common with other species.

Donald's theory is based on physiology and archaeology and makes no reference to Durkheim's *Elementary Forms*.[13] Nevertheless, one can maintain that it rejoins Durkheim's theory in at least two ways. First, it does so by establishing mimesis as a fundamental pre-human/human mode of signification, communication and memory, temporarily and logically prior to language, that is, to symbolic thought; this resembles Durkheim's assumption of effervescent gatherings, gestures and the like as the precondition for specific human representations. If spontaneous gatherings are the precondition for the gradual development of representations, they must have been brought about by some other mental skill, and it seems plausible to assume that an enhanced mimetic capacity was the determining factor here in comparison to other species.[14] In other words, phylogenetically, ritual precedes myth and theology. This is of course what a long-established theory in studies of religion has assumed at least since Robertson Smith (cf., Strenski 2006, 129), one of Durkheim's most important predecessors, in the late nineteenth century. Donald's results provide this model with an evolutionary-cognitive foundation.

Second, to Donald the biological evolution of human consciousness was made possible thanks to the development of culture, that is, supra-individual mental spaces in which a number of individual consciousnesses have participated. The way Donald describes culture and its anthropological implications comes very close to Durkheim's concept of society, including his *homo duplex* anthropology:

> Unlike that of other species, the human mind has a collective counterpart: culture. We stubbornly adhere to the idea that we are distinct individuals, yet we are also highly cultural beings. Indeed, humanity might be defined as the only species on earth that combines individual with collective cognitive processes and in which the individual can identify with, and become part of, a group process.... The life of the human imagination oscillates between these two polar extremes, individual and corporate. (Donald 2001, xiii)

Individual consciousness and culture are mutually dependent: without the one, the other could not be. Also to Donald, culture can be regarded as an invasion (cf. Durkheim's concept of 'penetration') of the individual consciousness from an outside force:

> the symbolic tools that are installed in our minds, such as languages, which exist solely because of a virtual invasion of our brains by external symbolic systems. They are all exten-

13. At least not in his two monographs, *Origins of the Modern Mind: Three Stages in the Evolution of Culture and Cognition* (1991) and *A Mind So Rare: The Evolution of Human Consciousness* (2001).

14. In certain respects, Merlin Donald's concept of mimesis resembles very much René Girard's, whose own theory of the origin of ritual, religion and culture has obvious connections to Durkheim's, even if it was not directly inspired by it; see Girard (2004, 190). Critical to Donald's lack of attention to the Girardian concept of mimesis, Gans (2003).

> sions of culture.... Part of the equipment is outside, and part inside, our heads. Culture leads, and mind follows. In other ways, mind leads. Symbolic culture is nothing without the human mind, while the human mind, without culture, remains locked incoherently upon itself. (Donald 2001, 157)

An outstanding theme in Donald's thinking is the invention of writing. The possibility of storing information in an external medium enormously enhances the quantity of what can be remembered, and it liberates memory and symbolic messages from the organic substrate of the human brain. It tentatively abolishes history by fixing ideas that would otherwise be lost together with the person who had them in the first place, and it is also much more effective in sharing ideas and participating in others' ideas than oral communication can be. But what also emerges in this narrative of the emancipation of the symbolic is, not surprisingly, an almost Platonic metaphysics, quite close to Durkheim's. The anthropological duality leads to a cosmological duality, with human bodies as the meeting points and mediators of heavy matter and almost weightless spirit:

> The actual physical presence of mind is very small. The phenomena that we call mental are caused by a tiny fraction of supercomplex living matter called nervous systems. Evolutionary theorist E.O. Wilson has written that the total biomass of humans ... is only a fraction of a fraction of a percent of one ten-billionth of the mass of the planet. Moreover, our brains are only one-fiftieth of that.... This delicate reticule of nervous systems stretches across the biosphere like a spiderweb, apparently too fragile to support the weight of a universe. But it does, nevertheless. All things cognitive emanate from that infinitesimal fraction. (Donald 2001, 96–97)

## Deuteronomy[15]

Theorizing about universes of *sui generis*-existing representations penetrating into the consciousnesses of individual human beings seems obviously relevant for understanding religions. But being an instance of theorizing about religion, it is not very far from similar expressions in religious traditions. An example is Deuteronomic theology as it is formulated in the parenetic parts of the book of Deuteronomy of the Hebrew Bible.

The literary form in chapters like 4–11 is highly stylized and repetitive with its endless and quasi-identical expressions, and they can appear unfocused or boring. It is also these texts, however, that present some of the Hebrew Bible's most fundamental theological reflections. The literary setting is the last days of Israel's 40 years of wandering in the wilderness, and the book consists mainly of Moses' long farewell speech before his death and before Israel's entering the land and its accommodation to a new, agriculturally based life. The content in itself is not very broad. Its main theme is Israel's liberation from Egypt, the wanderings in the wilderness, the meeting with Israel's god Yahweh at the mountain of Sinai, and the future existence in the land predetermined by Yahweh for Israel's settlement. This line of themes can be embellished with many subplots and embedded episodes, as they are in the previous books of the Pentateuch, but the Deuteronomic reformulation mostly sticks to the main thematic sequence. Therefore, what makes up the bulk of the theological material here is not the theological core as such or its re-narration, but a practical concern for cognitive themes: teaching and learning, memory, mental awareness and

15. The approach to Deuteronomy in this chapter was developed independently of the brilliant work of Jan Assmann, e.g., 2006, 16–21.

transmitting information. It teems with passages which reflect on what must be known. An example is 4:1–10, where Moses is presented as the great teacher (Weinfeld 1991, 200), that is, the great meme-disseminator:

> So now, Israel, give heed to the statutes and ordinances that I am teaching you to observe.… You must neither add anything to what I command you nor take away anything from it, but keep the commandments of the Lord your God with which I am charging you.… See, just as the Lord my God has charged me, I now teach you statutes and ordinances or you to observe.… But take care and watch yourselves closely, so as neither to forget the things that your eyes have seen nor to let them slip from your mind all the days of your life; make them known to your children and your children's children, how you once stood before the Lord your God at Horeb, when the Lord said to me: "Assemble the people for me, and I will let them hear my words, so that they may learn to fear me…and may teach their children so."[16]

Far from being a matter of course, the Yahweh religion is very much something to be learned. Some of the story the Israelites (that is: the first generation of post-exodus Israelites) have seen themselves with their own eyes; but the Torah must be learned, and the real teacher is Yahweh himself. If the content of the Yahweh religion mainly consists in narratives and rules, all this is nevertheless something external in relationship to individual consciousnesses. The rules and norms, which come from the heavenly Yahweh and which no Israelite seems able to produce by him- or herself, and the stories, which the generation of the desert wanderings have experienced themselves, but which must be likewise external for all coming generations—all those representations must enter the bodies and remain there.

Deuteronomy contains different models for how to achieve learning in the land. Deut 31:10–12 lays down that every seventh year, in the year of remission of old debts, all Israelites must assemble at the central sanctuary. At this occasion the Torah will be recited for everyone, men, women, children and alien residents, "so that they may hear and learn to fear the Lord your God." The precondition for this is the external symbolic storage, which is the written Torah, an intact code (knowledge of the relevant writing and the language)[17] and a rudimentary, biologically based recollection of the duty to reopen the storehouse (a recollection that remembers that something is to be remembered, even if it does not remember exactly what). The recitation disseminates representations from the central storehouse out into individual consciousnesses where they, in a more or less identical shape, are supposed to remain stored until, after seven years, they have been so worn down that they are ripe for repetition.

This slow rhythm contrasts with the much more intensive rhythm with which the Torah must be brought to attention according to Deut 6. Here, the brief *shema* (the later Jewish "credo": Hear, O Israel: The Lord is our God, the Lord alone") is followed by a thorough instruction in daily praxis:

> Keep these words that I am commanding you today in your heart. Recite them to your children and talk about them when you are at home and when you are away, when you lie down and when you rise. Bind them as a sign on your hand, fix them as an emblem on your forehead, and write them on the doorposts of your house and on your gates.

While learning in Deut 31 is collective, ritual, and embedded in a slow pace, learning according to Deut 6 is individual, with a much quicker pace. The Torah must be learned

16. Translation according to the *New Revised Standard Version*.

17. The relevance of which, however, is rarely thematized in the Hebrew Bible.

by heart (literally: "be on your heart") and written on the body, and it must be present everywhere at all times. Perhaps, the prescription of this remarkable frequency with which the Torah must be maintained is connected to its specific textual context. Yahweh's land, of which the Israelites are ready to take possession, contains all the riches of civilization: cities, houses, hewn cisterns, vineyards and olive groves. But the riches imply also dangers. With satiety follows the risk of forgetting Yahweh, who entrusts the land to the Israelites, but who remain its true owner (v. 12). Satiety must be seen in close connection to the activities that lead to it: the economy—as it clearly appears from the passage in Deut 8:7–14, explicit with regard to the future agricultural basis for Israel's life: "a land of wheat and barley, of vines and fig trees and pomegranates, a land of olive trees and honey...." This opposition is in line with Durkheim, to whom economy *sensu stricto* (working the material world in order to obtain means for survival) was the single sort of activity, which did not belong to a collective set of binding representations (Cf., Durkheim 1997, 48–52; cf., *Elementary Form*, 352; 421).

To Durkheim originally, the economy was an independent sphere defined by the non-religious, non-collective "profane" sphere associated with the burdens and dullness of the individual's daily struggle for survival. Inevitably, society's representations would be diluted and eventually disappear in a life which was totally absorbed in economic activities. Rituals were the elementary mechanism that recalled and re-strengthened the collective representation, and Deuteronomy 31 seems to agree. The seven years mentioned in the text refers to the economy. For six years, people can lend out and borrow money; but in the seventh year debts must be released. It is in this exceptional year when the population is assembled in the sanctuary at the great, yearly feast (that is: cut loose from daily, economic praxis) that the Torah must be recited.

Deuteronomy 6 on the other hand demands constant remembrance. Obviously this would be impossible for a collective, but it may, at least in principle and in theory, be expected from each singular (adult, male) member of the people. The distance between the extremely slow and the extremely fast rhythm must be due to a difference in understanding of the opposite of the learned, namely oblivion. According to Deuteronomy 31, the Torah seems to be able to remain in memory through seven years; according to Deut 6, fatal erosion may set in within a few minutes. Perhaps Deuteronomy 31 deals with the erosion, which quotidian economic activities inevitably will cause. In this case, the structure is identical with Durkheim's basic constellation. Deuteronomy 6, on the other hand, may deal with an alternative, where a set of representations are not threatened with general oblivion, but rather with competition from other, alternative representations.

This seems to be the case in Deut 6 and many other instances in Deuteronomy, which connects Yahweh's oblivion with the worship of other gods. Deuteronomy 8 is especially explicit. It mentions two sorts of false gods: the Israelites themselves (v. 17: "Do not say to yourself: 'My power and the might of my hand have gotten me this wealth.'") and the more common "other gods" (v. 19: "If you do forget the Lord your God and follow other gods to serve and worship them, I solemnly warn you today that you shall surely perish.") Of course, Yahweh as the active partner will cause this destruction over Israel. But in another, Durkheimian understanding, the destruction will be but oblivion in itself. Yahweh's oblivion will not necessarily lead to the destruction of Israel's material life; but it will lead to the destruction of Israel's life, the destruction of this particular form of life, this specific religion and culture. "Israel" is a moral, spiritual quantity, like a language, a fashion, or a given knowledge of something. A language exists as far as it is used, and when no one uses it, it will disappear (unless, of course, it has been fixed in an external symbolic storage with the necessary information as how to decode it again). "Israel" is not a quantity of Israelite people in the meaning of bodies with consciousness, for those bodies can live on, even if "Israel" disappears, as something else, such as Babylonians, Philistines or Egyptians.

The situation in Deuteronomy connects neatly with Durkheim, meme theory and Merlin Donald. In the confrontation between the Yahweh religion and the worship of other gods, representations fight for space in the memory of every single Israelite. Yahweh religion offers strong representations (or memes): a people wandering through the wilderness, led by its god who reveals himself on the top of a mountain surrounded by fire; the people receive direction for daily life from this god, who ultimately leads them to a fertile land. Nevertheless, these representations were seemingly not considered to be strong enough to compete against alternative representations, other gods and other religions. Deuteronomy knows at least that no one is born a worshipper of Yahweh, and no Yahweh worshipping person, even so instructed, will remain so. Therefore, in the parenetic passages in Deuteronomy the important thing is not the doctrine itself. The fundamental "narrative of salvation," as it is sometimes called by modern interpreters, formulated in passages like Deuteronomy 6:21–23 and 26:5–9, contains merely three or four verses; and the doctrine can perhaps be compressed even more into the *shema* formula: "The Lord is our God, the Lord alone." In itself the doctrine is neither implausible nor difficult to understand. The problem for the Deuteronomic parenesis is not that it is difficult to understand, the way St. Paul's letters later were (2 Peter 3:16), but that it is difficult to remember, in spite of its brevity. In focussing on memory, Deuteronomy touches on something that is relevant to all religions.

A religion will always be a product of consciousness. What is special about Deuteronomy is not the learning (that is: to enter or re-enter the religion) in itself which is so important. The truly critical point is to protect it, or give it the necessary means for protecting itself, once it already has been read into a number of consciousnesses, against other, alternative sets of representations, be they religious or not. No religion exits by virtue of its plausibility alone. Even if the population whose consciousnesses have been colonized by a religion and willingly participate in it is shielded by other and competing representational universes, there will always be so much else that demands attention, except for rare moments where everyone is gathered in an effervescent, commonly shared attention.

The scenario imagined by Deuteronomy 6 to be necessary in order to maintain the Yahweh religion in the minds of its participants and thereby also keep an effective shield active against infection from other representations that try to find a new home, is impressive enough, but it does not sound realistic for others than very dedicated members of a Yahweh worshipping elite. Even Deuteronomy 31 with its much slower (and maybe more agrarian rhythm) agrees with the possibly more urban Deuteronomy 6 in not relying on biologically based memory alone. Both texts place what is to be remembered in an external memory field; no doubt both have the most advanced information technology at the time in mind, alphabetic writing, although they don't say so explicitly. Writing is primarily a strong means against oblivion, and alphabetic writing was a much stronger form of memory storing than the tested, but in the first millennium B.C. also outdated, memory technologies of semasiographic and syllabic writings systems, because the code is so much easier to learn and therefore potentially available to a much greater part of a population.

In the Roman Empire, a religious mass extinction took place. Mesopotamian, Egyptian, Anatolian, Syrian-Phoenician-Punic, Greek, Roman, Celtic and probably many others disappeared as official religions. No doubt, many features survived as marginal religiosity and "folk religion" or melted into one of the few surviving religions. The only religions to come out of the Roman Empire intact were Christianity and Judaism. The religious mass extinction was not caused by external events like meteor impacts or volcanic outbursts; but it was accompanied by Christianity's virtual monopoly. That Judaism survived as well may be due to a kind of co-evolution: unlike other religions, Judaism was recognized as the predecessor of Christianity. Both had realized the importance of alphabetic writing as an external memory storage. This is certainly not a sufficient explanation of Christianity's success in late Roman time in itself, but it is probably an important factor. Religions that will stay in the existential game cannot be content with just gaining access to individual consciousnesses in brains (or hearts). They must also establish an immune response against other religions or otherwise competitive ways of thinking and acting. And they must be able to survive. Deuteronomic theology may be one of the earliest instances of a systematic reflection on this problem.

## References

Assmann, Jan. 2006. *Religion and Cultural Memory: Ten Studies*. Standford, CA: Stanford University Press.

Burke, Peter. 2004. *What Is Cultural History?* Cambridge: Polity Press.

Donald, Merlin. 1991. *Origins of the Modern Mind: Three Stages in the Evolution of Culture and Cognition*. Cambridge, Mass.: Harvard University Press

Donald, Merlin. 2001. *A Mind So Rare: The Evolution of Human Consciousness*, New York: W.W. Norton.

Durkheim, Émile. 1995. *The Elementary Forms of Religious Life* (transl. Karen E. Fields), New York: The Free Press. (French original: *Formes élémentaires de la vie religieuse*, 1912).

———. 1996 [1898]. "Représentations individuelles et représentations collective." In *Sociologie et philosophie*, 1–48. Paris: Quadrige, PUF.

———. 1997. *Leçons de sociologie*. Paris: Quadrige, PUF.

Freud, Sigmund. 1965. *New Introductory Lectures on Psycho-Analysis*. Translated by James Strachey. New York: Norton. (German original: *Neue Folgen der Vorlesungen zur Einführung in die Psychoanalyse*, 1933).

Gans, Eric. "What's New and Old about the Origin of Language (I)", Chronicles of Love and Resentment, No. 283. 2003: http://www.anthropoetics.ucla.edu/views/vw283.htm (accessed 14/07/2006).

Gibbs, W. Wayt. "Saving Dying Languages." Scientific American Digital, august 2002. www.sciam.com (accessed 24.11.2005).

Girard, René. 2004. *Les origines de la culture*. Paris: Desclée de Brouwer.

Pals, Daniel L. 1996. *Seven Theories of Religion*, Oxford: Oxford University Press.

Rappaport, Roy A. 1999. *Rituals and Religions in the Making of Humanity*. Cambridge: Cambridge University Press.

Sperber, Dan. 1996. *La contagion des idées*. Paris: Odile Jacob.

———. "An epidemiology of representations [7.27.05]. A Talk with Dan Sperber." http://www.edge.org/3rd_culture/sperber05/sperber05_index.html (accessed 25.11. 2005).

Strenski, Ivan. 2006. *Thinking about Religion: An Historical Introduction to Theories of Religion*. Oxford: Blackwell.

Weinfeld, Moshe. 1991. *Deuteronomy 1-11*. The Anchor Bible 5. New York: Doubleday.

# — 10 —

## Fantastic Re-Collection: Cultural vs. Autobiographical Memory in the Exodus Narrative[1]

Laura Feldt

### Introduction

Memory is sometimes seen as a mainly inner phenomenon taking place in individual minds. But what a memory stores, how it organizes its content, and the period of time for which it stores a specific content, is not only a question of individual cognition, but also of external, social and cultural frames, according to recent insights in both cultural and cognitive theory. This article wishes to examine the interrelations between cultural and autobiographical memory and the role of religious narrative with respect to this problem, focusing on explicit means for transmission of cultural memory to individuals in the Exodus narrative of the Hebrew Bible. The thesis is that the Exodus narrative offers an example of how a religious medium may affect the mind. In the Exodus narrative, questions of memory, ritual, narrative and the fantastic are intricately interwoven. For that reason it constitutes an interesting area of study in relation to the current debate in the cognitive study of religion revolving around the questions of the transmission of counterintuitive representations and belief. This paper is a contribution to this important topic. In the following, I introduce the two perspectives employed in this article, current theories of cultural memory and theories on the fantastic from contemporary literary criticism, after which I will proceed to the case study.

### Cultural Memory

The French sociologist Maurice Halbwachs developed the concept of collective memory in the 1920s (Halbwachs 1925). Halbwachs disregarded the bodily, neuronal and brain physiological basis for memory. Instead, he investigated the social frames without which no individual memory could constitute itself. "Es gibt kein mögliches Gedächtnis ausserhalb derjenigen Bezugsrahmen, deres sich die in der Gesellschaft lebenden Menschen bedienen, um ihre Erinnerungen zu fixieren und wiederzufinden," Halbwachs holds (Halbwachs

1. I wish to thank Professor Armin W. Geertz for reading a previous version of this chapter as well as Professor Joseph Bulbulia for a fruitful discussion.

1985, 121). Memory is something that grows in humans in the socialization process. Only individuals "have" memory, but this memory is collectively shaped. Memory—also personal memory—comes into being only in communication and interaction within the frames of social groups. Halbwachs thus places the collective as the subject of memory, and he coins terms such as "collective memory" and "national memory." He sees memory as playing a decisive role in the formation of collective identity.

Basing their point of departure on Halbwachs, Jan Assmann (1992) and Aleida Assmann (1999) developed the concept of *cultural memory* as something different from the terms "memory," "social memory" and "collective memory" as used by Halbwachs. While the Assmanns are indebted to Halbwachs in important ways, they put his insights into a new framework in which collective memory is seen as a cultural matter that is played out within various social frameworks. The term cultural memory emphasizes the extent to which shared memories of the past are the products of mediation, textualization and multiple acts of communication. These are not just regrettable deviations from the spontaneously produced memory of the participants, but rather a precondition for the operation of memories across generations, for the production of collective memories in the long term. Jan Assmann distinguishes between two phases of collective memory: communicative memory, corresponding to the phase when multiple narratives by participants and eyewitnesses circulate and compete with each other, and cultural memory proper, corresponding to the much longer phase when all eyewitnesses and participants have died out and a society has only relics and stories left as a reminder of past experience (Assmann 1992, 48–66). To offer an example, we may say that our shared memories of World War I are predominantly the product of films, books, commemorative ceremonies, etc. As stressed recently by Rigney, cultural memory in this way is always external in the sense that it pertains by definition to other people's experiences as these have been relayed to us through various media and multiple acts of communication.[2] To the extent that cultural memory is not a product of direct experience, it is by definition a matter of vicarious recollection (Rigney 2005).

This type of research indicates that we also need to look at memorial practices and mnemonic techniques at a cultural level and see how these interact with individual cognition. I suggest that the Exodus narrative provides an interesting case in that regard.

## The Fantastic

"The fantastic" or "fantasy theory" refers to a literary critical perspective developed in the twentieth century which concerns itself with the cultural phenomenon of the fantastic, especially in the form of so-called fantastic literature (see Todorov 1973; Brooke-Rose 1981; Jackson 1981; Cornwell 1990; Lachmann 2002). It is a literature of monsters, metamorphoses, miracles, strange and unnatural occurrences, apparitions, supernatural beings, etc. Although definitions vary within the field of fantasy theory, recent conceptualizations of the fantastic see it as a mode of narration that cuts across genres, or as a set of strategies that narratives can use to variable degrees (Lachmann 2002; Ivanovic *et al.* 2003). It is a fuzzy-edged category within which we may identify and describe sub-types characterized by specific determinants (semantic, pragmatic, etc.) in a more detailed way, while simultaneously retaining their basic relatedness within the overall category of the fantastic. Fantastic

2. Rigney's Foucault-based position thus seems to be somewhat akin to K.J. Gergen's social constructionist position on self-memory; a position in which accounts of memory gain their meaning through their usage, not within the mind nor within the text but within social relationships (Gergen 1994, 81, 100).

elements may appear in all types of media to varying degrees, yet we may still distinguish works in which the fantastic mode is dominant (e.g. Tolkien's *The Lord of the Rings*, James' *The Turn of the Screw*, the Exodus narrative) and in which it is not (e.g. *Hamlet*, the Book of Ruth). Fantastic literature is thus seen as a type of literature which actively creates new and alternative worlds by transgressing or violating the boundaries of the real, the normal and the natural. The fantastic is seen as a function of language and cognition; as a mode of narration attempting to describe the other and the absent.

This theory formation may thus supply a literary perspective to the debate on counterintuitiveness in the study of religion. Many narratives may be fantastic narratives, but only a specific subgroup of these are then specifically religious fantastic narratives (similarly with counterintuitiveness: only some of it is religious; cf. Pyysiäinen 2001), and the important thing for scholars of religion is to sort out the specific religious determinants. The fantastic provides literary-critical concepts for the specific types of counterintuitive representations and embeds them in a narrative framework which offers an opportunity to analyze them, not in their singularity, but as part of a set of such representations put to specific use in a narrative. Yet the idea of counterintuitiveness and the underlying distinction of the intuitive vs. the counterintuitive still forms the basis of the category of the fantastic (see Feldt 2006a). But the fantastic is more than a question of representations; it is a question of the use of representations in narrativity. The advantage of speaking of the fantastic as a narrative mode is that we may then distinguish between narratives dominated by the fantastic and such that although they may implicitly rely on some violation of intuitive cognitive categorization, they are not dominated by it.[3] In other words, it enables us to construct a typology of the spectrum of fantasy dominance in narratives. Were we to do that, we would find that the Exodus narrative is a religious narrative highly dominated by the fantastic. Upon analysis we will also see that it explicitly deals with questions of memory and belief.

## The Exodus Case

Major strands in biblical and Jewish tradition attest that it is the narrative of the fantastic events taking place outside of the promised land and before the foundation of the monarchy, rather than the narrative of statehood, that is decisive for the identity and self-understanding of the collective of "Israel." The peculiar thing about the Exodus narrative[4] is thus not only its central importance as the narrative traditions about the origins of Israel and its founding as an ethnic unit and of its religion and institutions, but also this same narrative's fantastic character: we find all the major[5] miraculous, counterintuitive/supernatural occurrences of the Hebrew Bible in this text. However, none of the eyewitnesses to these formative events come to enter the promised land according to the narrative of the Exodus. Therefore the collective memory of the founding events must be transformed from a biographical to a cultural memory (see Assmann 1992, 48–66). The Exodus narrative deals with this problem by establishing specific mnemonic practices for the purpose of shaping the memories of individuals in later generations; a set of practices meant to (re-)produce cultural memory in

3. We may also encounter religious narratives in which the fantastic is not dominant, e.g. the biblical Book of Ruth.

4. In this context "the Exodus narrative" is understood as roughly corresponding to chapters 1–18 of *Shemot*, the second book of the Hebrew Bible, i.e. the pre-Sinai-pericope.

5. Of course we find fantastic events elsewhere in the Hebrew Bible. I only mean to say that the major ones occur here. Further analyses of the fantastic more broadly in the Hebrew Bible may be found in Feldt 2011.

individuals, belief and social cohesion. The majority of them are linked to the Passover ritual which becomes the central identity marker for the Israelites, cf. Ex. 12:14, where the ritual is called *zikkārôn*, memorial. Religious narrative and ritual are interwoven in this text and combined with explicit reflection on the instigation of memory and belief. For this reason the Exodus narrative offers interesting material with respect to the question of how religious narrative may affect memory. I will argue that the two basic questions that this narrative attempts to answer are, first, how is remembrance ensured, and second, how is belief ensured in relation to the fantastic events. Both of these questions are intimately linked to the question of social cohesion or the construction of a collective identity.

## The Fantastic in Exodus

The Exodus narrative is constructed as an image-intensive narrative full of multivocal and ambivalent imagery about a group of people's intense, shocking, violent, emotion-evoking experience at a specific point in time. Opposite is a survey of the types of fantastic strategies used in the narrative.

Let me now analyze these types of fantastic strategies in more detail (see Table 10.1).

### 1. Metamorphoses

From the starting position of everyday reality, a metamorphosis represents a violation of the normal order of things.[6] Metamorphoses are spectacular boundary-transgressions that the text itself presents as disturbances of order and inversions of existing assumptions about phenomena (Lachmann 2002, 7). The metamorphoses of Exodus come in two kinds: category metamorphoses and kind metamorphoses. The *category metamorphoses* consist of phenomena changing their basic ontological category, for instance from artefact to animal (staff to snake), natural object to body fluid (water to blood), natural object to animal (dust to gnats), and natural object to illness (soot to dust to rash to boils). The following are attested: tool → animal, natural object → body fluid, element → animal, natural object → natural object → illness. Second, *kind metamorphoses*, i.e. metamorphoses within the same basic ontological category,[7] changing its status or kind, for instance such as from one state of an element to another (bitter water to fresh water), or from one physical condition to another (healthy hand to sick hand), or from one kind of element to another (water to dry land). As for changes of state, we see moves from a positive state or condition (e.g. healthy) to a negative state (sick) and from a negative state (bitter water) to a positive state (fresh water).

Similar processes are seen in the mental metamorphoses, which, in a sense, represent kind metamorphoses in that a change of attitude or heart does not fundamentally violate

6. Metamorphosis is a powerful means for *phantasmagenesis*, says Lachmann (Lachmann 2002, 116).

7. In the distinctions between ontological category, kind concept and cultural concept, I rely on research in cognitive categorization, see Hirschfeld and Gelman 1994; see also Boyer 1994, 2000; and Sørensen 2000a and 2000b concerning religious representations. Research in cognitive categorization seems to indicate at least five basic intuitive ontological categories: PERSON, ANIMAL, PLANT, ARTEFACT, NATURAL OBJECT. This level is considered unconscious, intuitive, automatic and universal, and it concerns the experience of phenomena and forces in the world (up-down, the characteristics of physical objects, etc.) and encompasses both ontological concepts (person, animal, plant, etc.) and kind-concepts (cat, hammer, etc.) The second type of assumptions, the culturally specific conceptualizations, concern culturally specific structurations of semantic elements like "mother," "fuel" or "love," etc., which depend on the basic categorization but where the specific construction and use is culturally variable.

1. *Metamorphoses*

a) *Physical*

- A staff turns into a snake (Ex. 4:1–5)
- A healthy hand turns into a leprous hand (Ex. 4:6–8)
- Water from a river turns into blood (Ex. 4:9)
- Dust turns into mosquitoes (or gnats) (Ex. 8:12–15)
- Healthy cattle turn into sick cattle (except certain specific cattle owned by the Hebrew people) (Ex. 9:1–7)
- Soot from an oven turns into a dust cloud which produces a rash that produces boils in humans and animals (Ex. 9:8–12)
- An ocean turns into dry land (the parting of the sea) at a specific moment (Ex. 14:21–22)
- Bitter water turns into fresh water (Ex. 15:22–25)

b) *Mental*

- Changes of inimical attitude into a friendly attitude in an entire people (Ex.11:3; 12:36)
- The hardening of a specific person's heart (Pharaoh) (*passim*)
- The creation of panic in a specific army at a specific point in time (Egyptian army when attacking Hebrew people) (Ex. 14:24–25)

2. *Inexplicable events or apparitions*

- A bush burns without being consumed, and it speaks, too (Ex. 3:1–6)
- A strange type of bread falls like dew (rain) in the morning in specific amounts (Ex. 16:1–36)
- Water produced from a rock (Ex. 17:1–7)
- A pillar of cloud appears during the day to guide the Israelites (Ex. 13:21–22)
- A pillar of fire appears during the night to guide the Israelites (Ex. 13:21–22)
- The supernatural murder attempt on Moses (Ex. 4:24–26)
- The supernatural killing of all firstborn humans and animals among the Egyptian people (Ex. 11:4–8; 12:29–33)
- The sudden appearance of a flock of birds at the camp site (Ex. 16:2–13)
- Moses' raised arms cause a military victory (Ex. 17:8–16)

3. *Hyperbolizations*

- Extreme fertility in a specific group of people (Ex. 1:7, 10, 12, 20)
- Frogs appear in extreme amounts at a specific time and place (Ex. 7:26–8:11)
- Flies appear in extreme amounts ditto (Ex. 8:16–28)
- Grasshoppers devouring all produce appear in extreme amounts ditto (Ex. 10:1–20)
- Sudden hail storm (thunder, hail, fire, lightning) which kills humans, animals (but not the Israelites and their animals) and vegetation, at specific time/place (Ex. 9:13–35)
- A three-day tangible [Hebrew verbal root *mshsh*] darkness in all of Egypt except in a specific place (where the Israelites live) (Ex. 10:21–29)

4. *"Coincidences"*

- That Moses as a child is found by pharaoh's daughter and 'adopted' by her (Ex. 2:5–10)
- Moses' coming to sit down by a specific well in Midian (Ex. 2:15ff)
- That the location of the burning bush incident being the same as that of the revelation at Mt. Sinai (Ex. 3:12)
- That Aaron is on the way to meet Moses when the deity speaks to Moses in the burning bush (Ex. 4:14)

Table 10.1 Major types of fantastic strategies used in Exodus 1–18.

or transgress any basic ontological category. In the Exodus narrative, the narrator (and s/he only) claims that the mental metamorphoses of especially Pharaoh and to some extent the Egyptians have a supernatural cause and therefore they too entail a category transgression.

The metamorphoses in Exodus are staged as boundary-transgressions, disturbances, violations of the normal by means of various strategies: their spectacular character is signalled in the text by the "persuasion effect" that they are ascribed, i.e. upon seeing or experiencing these *signs*,[8] people will, according to the deity Yahweh's utterances in the narrative, come to believe that the phenomena in question have a supernatural cause.[9] The spectacular character is further signalled by means of the emotional reactions of the personae of the text, especially fear[10] and an oscillation or uncertain wavering state between belief and disbelief.[11]

## 2. Inexplicable Events or Apparitions

With regard to the inexplicable events or apparitions attested in Exodus, the question of what makes these events strange or inexplicable immediately arises. How does the text alert us to the fact that something out of the ordinary is going on? This is done by means of inversions or violations of standard assumptions about the phenomena in question either with regard to its basic ontological category, with respect to its kind (cf. above), or with respect to culturally specific assumptions.

If we look at the inexplicable events in Exodus from this angle, and go through the individual events, we see both violations of *ontological* categories, e.g. a voice (PERSON) coming from a bush (PLANT); *kind concept* violations, e.g. a fire (kind-concept) that does not consume, bread (kind-concept and cultural concept) falling from the sky like rain (bread is dry, human-made and related to the earth); and *cultural concept* violations like the sudden, voluntary appearance of a flock of birds that people normally hunt for (prey) at the campsite. Some of the events even violate more than one type, for instance a pillar (cultural concept) cannot be made of cloud (violation of kind and cultural concept) nor can it guide people on their way (violation of basic ontological category ARTEFACT).

If we examine the inexplicable events as a group, only three different semantic fields are used: the alimentary,[12] the atmospheric/natural phenomenon,[13] the anthropological/personal.[14] The *alimentary* type represents systematic inversions of what constitutes ordinary food according to the Hebrew Bible: manna is an imaginary, fantastic type of bread raining from the sky. It is a type of food which, contrary to ordinary types of food, does not require work:[15] it is free, it is collected and not harvested, it cannot be stored,[16] a type of food which is not in limited supply—everybody gets what s/he needs. In all of these respects, the phenomenon of manna is oppositional to ordinary bread, and thus it represents

8. Hebrew *'ōth.*

9. E.g. Ex. 4:1–9. However, that is not exactly how it goes. The personae of the narrative react to these signs with doubt, hesitation, suspicion and belief. We will return to this below in the section entitled "Purposes and Effects of the Fantastic in Exodus."

10. Ex. 4:3; 14:25.

11. Ex. 7; 8; 9; 14:31; 15. Cf. Feldt 2006b.

12. Manna, water, birds, Ex. 16–17.

13. Burning bush, pillar of cloud, pillar of fire, Ex. 3, Ex. 13.

14. Murder attempt on Moses, murder of firstborn, Ex. 4, Ex. 12.

15. Cf. Gen. 3:17.

16. It rots if stored, cf. Ex. 16:20.

a systematic inversion of the normal in the Hebrew Bible: farming and subsequent bread-making (cf. Jensen 2000, 239). The appearance of the flock of birds is equally oppositional to ordinary forms of food in that it is also a work-free type of food which inverts not only cattle farming but also hunting, in that the prey here approaches the hunter, not the other way around. The water produced from the rock by striking it with the staff also represents a magico-miraculous type of nourishment, which is work-free, instantly produced *ex nihilo* and oppositional to the ordinary. As such, all three types of alimentary apparitions thus represent inversions of the ordinary, everyday world of culture, which includes bread, farming, hunting and gathering. The *atmospheric/natural phenomenon* type of apparition also represents inversions of the known and natural: neither fire nor clouds normally behave in this manner. What we see is a transfer of expectations from the PERSON category to the NATURAL OBJECT category, enabling clouds and fire to guide. Further, there is a transfer of expectations from the ARTEFACT to the NATURAL OBJECT category, allowing us to see clouds and fire as pillars. The anthropological type of apparition, the supernatural persona Yahweh and the demon Mashchit[17] as killers of the Egyptian firstborn, draws upon standard inferences and expectations to the category of PERSON, yet violates the category at certain points: no two persons can kill that many children in such a short period of time, they are invisible, and yet they are assumed to have intentions, etc.

What is represented by means of these events is thus a world in which both ontological categories, kind concepts and cultural concepts are violated and transgressed, in which culture, the normal and the known, is fundamentally inverted.

### 3. Hyperbolizations

By means of the hyperboles, the fantastic here unfolds itself in the representation of the extreme and impossible, for the reiterated hypertrophy leads beyond the merely improbable towards the impossible: Characteristic of all of the hyperbolic processes presented in the narrative is that they bring along with them the possibility that they *could* be little else than exceptional natural phenomena on an unfamiliar scale. However, it is the unprecedented convergence of calamities, the reiterations of hypertrophy, that make them into *fantastic* events. The hyperbolizations occur in the natural phenomenon area (human fertility is increased to an extreme degree, frogs, flies and grasshoppers appear in extreme amounts at a specific time and place, etc.). But the closeness of the phenomena to natural events and processes is noteworthy. This closeness opens a possibility for doubt or hesitation as to the cause of the phenomena: is there *really* a supernatural cause for these events, in terms of an *explicit* intervention? The closeness to ordinary, natural phenomena can be said to create a space for a "double readability."[18]

### 4. "Coincidences"

In addition to the literary figures that provide the verbal and argumentative means to create the fantastic, there is also the "lebensweltliche Faktor" of *coincidence* or *chance*, says Lachmann. "Coincidence" marks the questionable moment in which the other, the unexpected and inexplicable, enters into an existing, normal process and disturbs

17. In Hebrew Bible scholarship there is an ongoing discussion as to whether a demon is present in the text, which is ambiguous (cf. Ex. 12:23 vs. 12:29). Further discussion in van Seters (1994,114).

18. This is also apparent in the scholarly commentaries on the text—some interpret the phenomena as natural, some as supernatural.

its coordinates. The opposition coincidence/non-coincidence is central; it is an opposition that the fantastic text reacts to by placing the question of whether it is one or the other in the foreground (Lachmann 2002, 117). The ambivalence of a coincidence may be used positively by the fantastic narrative in such a way that its (possible) contingency and its (possible) meaningfulness are brought into play together, so that the barrier between them is transgressed in both directions. This oscillation between the poles creates an effect of insecurity or uncertainty. In Exodus, the events are staged not as coincidences, but rather as meaningful events, and so the strange events and appearances are not *Trugbilder*, but rather explained by the text as signs of the presence of the deity. In this way, the meaning ascription that transforms unknown into known, and inexplicable into explicable, acquires a soteriological aspect (Lachmann 2002, 137). This is what happens in the present narrative when the supernatural persona or the narrator offers authoritative interpretations of the "coincidences" as caused[19] and willed by him, yet the ambivalence of the coincidences remains present as a destabilizing factor.

## Summary

In the Exodus narrative, a clear interpretation of the fantastic events is given in the narrator's voice, as we shall see in the next section. However, there are also discontinuities in the text which function as an underlying discourse that comments upon or even contradicts the authoritative, overt ideology of the text—and this discourse is connected to the types of fantastic strategies used and the way they are introduced in the narrative. As we saw above, some of the fantastic strategies used to generate the phantasms played with the possibility of other interpretations than the supernatural, namely the events being un-extraordinary and normal. However, their fantastic, unusual character or supernatural cause was still asserted in the text. Familiar objects, experiences and processes like childbirth or grasshoppers were staged as strange, unusual, unfamiliar, removing them from the world of everyday experience. The fantastic character of the natural processes provoked us to see them in new ways, while simultaneously retaining the possibility that they might be nothing more than natural events. In this way, they contributed to creating a discourse of ambivalence or doubt, a fair distance from the clear belief and a common, ethnic identity posited by narrator and deity that we will encounter below. This discourse of doubt, amplified in the reactions of the protagonists (see below), plays an important role in the text, for it is one of the ways in which the fantastic strategies are used to represent the supernatural persona; it is part of how the phantasms assist in the verbal and narrative representation of the transempirical. This will become clearer as we review the purposes and effects of the fantastic, below.

## The Purposes and Effects of the Fantastic in Exodus

The purposes of the fantastic events are staged as the transempirical persona, Yahweh's own words in direct speech, and they are uttered at various points throughout the narrative. However, they may be summed up in six points:

1. Rescue from suffering and slavery: 3:7–11; 6:6
2. Land donation/appropriation: 3:7–11; 6:8

19. This, however, is not done for all the "coincidences," for instance the finding of the child Moses by pharaoh's daughter.

3. Ethnogenesis: 3:7–11; 6:7; 7:5; 7:17; 8:19; 9:17; 12:13; 14:4; 14:18
4. Demonstration of divine power: 11:9; 14:4; 14:17
5. Inducement of belief: Israelites: 4:5. Egyptians and Pharaoh: 7:5; 7:17; 8:19; 9:14, 16–17; 12:13; 14:4; 14:18
6. Instigation of memory and memory practices: 10:2; 12:14

If redundancy is emphasis, then we may conclude that ethnogenesis (3) and inducement of belief (5) are the most important purposes according to the supernatural protagonist. If we have a closer look, we see that the remaining purposes are connected to these two as means. The first purpose (1) forms the enabling premise for the subsequent ones. The second purpose (2) couples the collective, ethnic identity to a land area which belongs to the deity according to the narrative.[20] This purpose is, then, intimately linked to the over-arching purpose of ethnogenesis, the creation of a common identity for the Israelites as a people. This collective identity is specified as the property of the transempirical persona (Ex 3:7, 10; 5:1; 6:7; 7:4, 16; 8:1, 8:20–23; 9:1, 13–17, 27; 10:3–4; 12:31; 16:4), which means that the counterintuitive/transempirical and the normal/empirical are connected from the outset according to the narrative.

The sixth purpose is crucial because this is what provides the causal link between the counterintuitive/transempirical realm (dominated by the transempirical persona Yahweh) and the normal, everyday realm of human life. Pascal Boyer (among others) has pointed out that mere counter-intuitiveness is not enough to instigate "religion," specification of the causal link between the counterintuitive and the normal is necessary in order to produce religion (Boyer 1994, 42–43, 125–154). Thus, the function of the Passover ritual may not be to convey any specific meaning but rather to establish a causal link between the supernatural and the ordinary, and that causal link is memory. The fourth purpose, demonstration of divine power, is connected to the inducement of belief: the fantastic events are meant to induce belief because according to the supernatural protagonist they demonstrate divine power. The second over-arching purpose—that of the inducement of belief—is crucial because one would think that belief would be the causal link between the counterintuitive and the normal. But according to the narrative, as I will demonstrate in the following, it does not work like that. The fantastic events are, according to the deity, meant to provide orientation, social cohesion (a strong collective self), and belief, but the fantastic produces neither. The fantastic events produce belief *and* doubt, orientation *and* disorientation, and the amount of disorientation and doubt produced seems even to exceed that of belief. This is where memorial narrative comes in. I will return to the question of memory and memorial practices shortly, but first I will demonstrate the effects of the fantastic events as divided onto the personae of the narrative (Table 10.2):[21]

20. The coupling of deities with specific land areas or cities was a common feature of religiosity in the ancient Near East.

21. While the distinctions between personae and narrator are common in narrative analyses, it seems proper to note that they are used here in accordance with Mieke Bal's theory of narrative (Bal 1997). She makes an analytical distinction in the analyses of narrative between *text*, *story* and *fabula*. The narrative analysis offered here focuses on the *text*-layer.

| Persona | Reaction to the fantastic |
|---|---|
| Moses | **Hesitation, puzzlement:** Categorizes his vision of the burning bush as "an apparition" [Hebrew mar'eh]: 3:3f<br>**Doubt:** doubts that the Israelites will listen to him and that they will believe in Yahweh: 4:1; 4:10; 4:13; 6:12; that Pharaoh will listen: 6:12; 6:30<br>**Fear:** 6:3<br>**Suspicion:** Interprets Yahweh's actions as evil, doubts his intentions: 5:22<br>**Exhortation to Pharaoh to believe:** 8:6; 9:29<br>**Belief:** 10:6; 12:23; 16:6<br>**Recognition of Yahweh's ethnogenetic purpose:** 12:23<br>**Exhortation to remember:** 12:27; 13:3–9<br>**Exhortation to people to believe:** 14:14; 15:26<br>**Praise:** Ch. 15 |
| The Israelite people | **Doubt:** 15:24; 16:2<br>**Suspicion:** Suspects the deity's intentions of being evil, murderous, interprets the fantastic as lethal threat: 14:11; 16:3; 17:2–3<br>**Rejection of ethnogenetic purpose:** Prefer slavery in Egypt to starvation in the wilderness, wish they would have died in Egypt with the Egyptians (i.e. that there would have been no ethnogenesis): 14:12; 16:3 |
| The king, Pharaoh | **Rejection, disbelief:** *passim*<br>**Acceptance, belief:** *passim* |
| The Egyptian people | **Pain/suffering:** 12:30<br>**Acceptance of ethnogenesis:** 12:31–32<br>**Fear:** 12:33; 14:25<br>**Belief:** 8:15; 10:7 |

Table 10.2 Effects of the fantastic.

As we see, the amount of doubt, suspicion, rejection and disorientation is surprisingly great in all of these speakers, in direct opposition to the supernatural persona's stated purpose. People are disoriented, confused, bewildered. The fantastic events are staged as emotionally and cognitively disorienting experiences; shocking, violent and emotion-evoking episodes.[22] And as staged by the Exodus narrative, such experiences do not lead to belief.

But naturally, the narrative does not leave us with doubt and ambivalence. When we look closer at the Exodus narrative, we discover a pattern in the reactions to the fantastic. It seems that being closer to the events results in less belief. Direct experience seems to lead to doubt and disorientation (e.g. Ex 14:10–12 and cf. above), whereas narration or inter-

22. Might they be likened to similar episodes in rituals? Sometimes, it is assumed that such extreme events provide the novice with "a new awareness of the structure of the cosmos and their place within it" (Whitehouse 2000, 20). But as Houseman points out, it is not clear what this new awareness consists of. There seems to be a lack of evidence that novices absorb a great amount of new information. Rather, it seems that the extreme experiences question the system's core ideas, rather than introducing new ones (Houseman 2002, 19). Clearly, this could resemble Victor Turner's point about ritual symbolism leading to awareness of and reflection upon axiomatic cultural values (Turner 1967). This point seems to be corroborated by the Exodus material, where the surprising revelations and fantastic events have the effect not only of orienting—their intended purpose according to the supernatural persona (which is often accepted by commentators, even though the connection between miracles and belief is rather tenuous)—but clearly also of disorienting. The presence of both makes possible a questioning of core cultural ideas, is reflection-enabling, and therefore makes possible both belief and rejection.

pretation by Moses or the narrator leads to belief and orientation. The narrator is a type of omniscient external narrator who may survey past, present and future effortlessly, and s/he is the only speaker who is never disoriented (Table 10.3).

| **3rd person external narrator explanations** | The fantastic signs are signs of Yahweh's concern, remembrance, and commitment to the people: 2:23–25; 4:30; 14:30<br>The signs produce belief: 4:31; 9:20; 12:50; 14:31<br>Hearing about the signs produces belief: 4:31<br>A promise of signs/purposes leads to rejection: 6:9<br>Yahweh's wonders are better than the wonders of the Egyptian magicians: 7:13<br>The fantastic events are related to origins: 9:24<br>The wonders are ethnogenetic: 9:26; 10:23<br>Israelites trust Yahweh: 14:8<br>Israelites don't trust Yahweh and Moses: 14:10; 16:20; 16:27 |
|---|---|

Table 10.3 The narrator.

The quandary between orientation and disorientation, belief and doubt, or the state of doubt and hesitation that the eyewitnesses to the events find themselves in, raises the central question of belief: if they are indeed bewildered and disoriented, how is belief, a strong collective self, and orientation ensured?

## Dis/orientations: Levels of Narration[23] in Exodus

As we have seen above, the fantastic events do not have the function of founding and orientating the Israelite collective self. Social cohesion is not established by the events at the Exodus, and they do not induce belief, and when they do, then only momentarily. The events are *also* disorienting, disconcerting, and provoke doubt and hesitation. As hinted above, the two types of responses to the fantastic events (belief/doubt, orientation/disorientation) correspond to levels of narration. There is a level of narration and a point of focalisation where Israel *is* formed as a strong collective self, and *is* oriented in time and space by the narrative of the fantastic events, and where people do come to believe. But this occurs only at a level of narration twice removed from the events, as it were: in the narration of the narration of the events, that is, when the narrative narrates that it narrates. When retold, re-narrated, the fantastic narrative gives a sense of direction, induces belief and orients people's lives in time and space. There is a distinction in the text between those to whom this is narrated, the narratees, and those who experience the events, the eyewitnesses. At the level of narration and point of focalization staged as the point of view of the contemporaries, the eyewitnesses, we find disorientation, confusion and disintegration as a result of the fantastic events, whereas the narratees are presented as believers. The construction of the levels of narration seems to tell us that with narration comes distance and thus orientation. Table 10.4 below illustrates this point.

23. The concept of levels of narration as well as other narratological concepts in this article are used in accordance with Bal 1997.

| Level of narration | Narration of narrated events and experience | Narration of events and experience |
|---|---|---|
| Teller's time | "Present," moment of recollecting | "Past," recollected moment |
| Position of teller in relation to told | Distance | Proximity |
| Point of focalization | Later generations | Contemporaries |
| Effect of the fantastic | Belief, Orientation Differentiation, Clear identity for Israel | Doubt, disorientation, tension or dialectic between disorientation, a weak collective self, and belief, orientation, and a strong collective self |

Table 10.4 Levels of narration.

On the one hand the text communicates the importance of one, common self for the collective of Israel, but on the other hand this necessity is argued by means of a memorial narrative of the past, which shows clearly that this same identity was *not* there.[24] This means that "the category 'Israel' is an ideal, normative category—more than a historical factual category" (Jensen 1998, 17). The consequence is that what the Exodus narrative portrays as "*past*" (the right column above) is really a description of the "present." And the "*present*," (the middle column in the diagram above) is, rather, the future, because it signals what it desires to achieve; it is a utopian vision. The collective self of this narrative is not there in the "present" either, but is, rather, a fantasy for the future.[25] The Exodus narrative is, as most commentators agree, a founding epic embodying Israel's memory of its fantastic origins, but it also discloses an interweaving of two conflicting perspectives in relation to that fantastic origin: one of unity, belief, orientation, and satisfaction/fulfilment, and one of disintegration, disbelief, disorientation, and disappointment.[26] What the diagram shows is that narration, according to this cultural model, is crucial for the formation of belief, orientation and stabilization of the collective self. The oscillation between belief and doubt present in the text is absolutely necessary, for without it the problem of the text would not be clearly stated: how do we secure belief and orientation of the collective self? The text implies that this can be achieved by means of memorial narrative. But that is jumping ahead of things. For how exactly is the link between fantastic events and memory established?

## Narrative Integration, Ritual and Mnemonic Practices

The interplay between narrative and embedded ritual here turns out to be important, for it is in ritual that the link between memorial narrative and the fantastic is established. The participants in the ritual are staged as eyewitnesses to the fantastic events by means of spe-

24. As H.J.L. Jensen explains, history begins and ends in catastrophes *because* Israel did not worship its god, because Israel was not aware of its true collective identity, worshipped other gods, mixed with other peoples in marriage and so lost its ethnic specificity; in short because Israel did not remember (Jensen 1998, 17–18).

25. The Exodus narrative explores a possible future by means of a narrative about the past, about origin; the fantastic strategies are used to create an origin, which may alter the future.

26. It is thus a religious narrative which has room for both a relativistic tendency (cultivating a dialogue between differing understandings of experience) and a fundamentalistic tendency (laying down one coherent, dominant narrative). For the distinction relativistic vs. fundamentalistic, see Ochs and Capps (1996, 32).

cific bodily practices that replicate the original Passover meal: food prepared in the same way, eaten at the same time of day, wearing the same type of clothing, and, importantly, in the same atmosphere of fear (Ex. 12:11).[27] All of these are likely to have emotional effect. The ritual is performed vicariously for the eyewitness generation, and the participants are meant to mimic the original situation in detail as staged in the narrative. But *narrative* is also used as a means to integrate past, present and future, eyewitnesses and narratees, initiates and initiators, fathers and children. The fathers (eyewitnesses) in the Exodus narrative did not come to enter the land (they died in the wilderness) and so were never able to participate in the re-tellings in the land that the narrative with embedded ritual prescribes (Ex. 12:14, 24–27). This point stresses that the positions of eyewitness vs. narratee, and father vs. son, are ideal and transferable: the present sons will in due course occupy the father-position, the eyewitness position. In and through the ritual, *narrator* position is then merged with eyewitness position, so that the *vicarious* "eyewitnesses" (the narratees in the text-external world) come to possess and transmit the authoritative interpretation that the narrator of the Exodus possesses and transmits in the Hebrew Bible narrative (but which the *eyewitnesses* in the story did not possess).

An important religious trait of this narrative comes out in the direct reader appeals forming part of the embedded ritual. This is where the text attempts to transgress the boundary between narrated world and reader's world by means of *deixis*: explicit reference to the reader/listener by means of the second person pronoun, see Exodus chapters 12 and 13 (*passim*). Yahweh explains how to perform the ritual and how to explain it to the later generations, the children of the eyewitnesses of the Exodus. But the eyewitnesses of the Exodus narrative (including Moses) cannot have told it to their children themselves because they died in the wilderness, and therefore the use of deixis in the ritual prescriptions functions as a means for establishing integration between cultural narrative and autobiographical narrative. But in and through the embedded ritual, further means for establishing the narrative integration that blends or merges narrator (possessing the authoritative interpretation of the events) and eyewitness (characterized by confusion and disorientation) positions are specified. Let us have a look at the mnemonic practices established in Exodus (the majority of which are linked to the Passover ritual) to further pursue the question of memory.

The narrative thematizes the question of memory explicitly: the Israelites are presented as forgetful and in constant need of new fantastic events, ever more miracles. No matter how often they are presented with miracles, wonders, violent killings and other extreme events, no matter how awesome and fantastic the things that take place, they do not have a lasting effect, because the Israelites forget them almost immediately and doubt the supernatural agent that performed them.[28] This leads us to look for explicit strategies for the preservation of memory in the narrative. The Exodus narrative instigates a set of mnemonic practices, no less than nine different ones, used for the transmission to and education of later generations.

1. Familial, (vicarious) autobiographical narration (father to son): 10:2; 12:26
2. Oral repetition:[29] 13:9

27. *Ḥippazon* connotes a combination of fear and haste. Eating with *ḥippazon* is unusual, a violation of table manners. See Propp (1998, 397).

28. For examples, see the shifts from Ch. 12 (Yahweh's killing of all firstborn in Egypt) to Ch. 14:10–12; from Ch.14–15 (the separating of the sea) to Ch.15:24; from Ch. 15:25 (transformation of bitter water to fresh water) to 16:3, from Ch.16:4ff. (manna) to Ch.16:20, 27; from 16:4ff. (manna) to 17:2, etc.

29. The Hebrew text says that *tôrat Yahweh* is to be in the mouth of the individual, which signals orality.

3. Marking/offering all firstborn of humans and cattle as Yahweh's (the Consecration of the Firstborn):[30] 13:13

And explicitly within the framework of the Seder-meal/Passover ritual:

4. Bodily markers: 13:9, 16
5. Temporal markers: 12:14, 17, 42
6. Alimentary markers: 12:4–10, 15, 20
7. Vestimentary markers: 12:11
8. Emotional markers: 12:11
9. Boundary markers (liminal symbolism) in relation to familial houses: 12:7, 13

These are the mnemonic sites that reflect and manage the threat of oblivion in the narrative; a set of practices meant to produce memory and social cohesion. As we see, the majority of them are linked to the Passover ritual, which becomes the central identity marker for the Israelites. Memory means to confess or affirm one's adherence to the group by re-telling the narrative and performing the ritual as a remembrance of the events of the Exodus (cf. Ex 12:14).[31] These mnemonic techniques show how to maintain memory, how to avoid oblivion; they are the *how* of memory in Exodus.

Let us now consider the relations between cultural memory and individual memory.

## Cultural Memory—Individual Memory?

Cultural memory in the form of religious narrative is always external in the sense that it pertains to other people's narratively mediated experience as relayed to us through various media and multiple acts of communication. Even if people undergo comparable experiences, the cultural memory of those experiences is the ongoing result of public communication and the circulation of memories in mediated form. We have here seen an example of how memory circulation and transmission may be culturally formed. How does this relate to research in autobiographical memory?

In his book from 2002, Scott Atran argues the need for narrative consolidation when it comes to individual memory of intense religious rituals. Individual memories need social support and narrative consolidation in order to be recalled readily (Atran 2002, 161–163). And if this is true for intense initiation rituals based in stressful and striking memories,[32] then it is probably even more so in the case of less intense rituals. And why not also in the case of image-intensive, fantastic religious narrative? Research in autobiographical[33]

---

*Torah* here connotes "teaching" rather than any particular law, see Propp (1998). *Tôrat Yahweh* may mean both "teaching from" and "teaching about" Yahweh. The latter is more likely than the former, in view of the preceding narrative taking place before the Sinai event, and probably refers to Yahweh's extraordinary actions at the Exodus.

30. Which could not have been celebrated with the Egyptians in pursuit, since it is a seven day festival (Propp 1998, 429).

31. According to this cultural model, oral narratives and autobiographical narration and ritual in a family setting have the power to establish enduring bonds between individuals and deity.

32. Atran argues that we should not confound memories of stressful events with memories of striking events. In the case of striking events, the need for narrative consolidation is even greater (Atran 2002, 162–165).

33. For a definition see Neisser and Fivush (1994, 1).

memory and narrative and consciousness suggest that narrative plays a crucial role in the formation of mind (Fireman *et al.* 2003; Kihlstrom 2003; Ochs and Capps 1996). Professor of psychology Mark Freeman has shown how autobiographical memory is necessarily narrative in character, that imagination is inevitably involved and combined with the inputs of others, literary conventions and cultural scripts (see also Bruner 1991). But the conflation of autobiographical memory, narrativity and imagination does not necessarily imply untruthfulness. Life narration and self memory are (also) *about* imagining and meaning making (*poiesis*) (Freeman 2003, 125).[34]

The Exodus narrative combined with the Passover ritual is an annually repeated performance, but there is no logical doctrine connected to it, and what is to be remembered—the fantastic experience—is very imaginative. The Exodus narrative tells of emotionally arousing experiences and autobiographical memory of a spatiotemporally specific kind; of sacred and revelatory events with a once-in-a-lifetime quality. The fantastic events do not ensure social solidarity among co-participants contrary to what the narrator and the deity tell us. Narrative consolidation and memorial practices connected especially to the body are stipulated as necessary in order for the fantastic events to ensure social solidarity and group cohesion, belief and orientation. The narrative seems to show that autobiographical memory can be staged, that it can be vicarious,[35] and that rare but exceptionally salient experience may be culturally mediated. Importantly, it also shows that memory may concern the future and how to live a good life more than the past and what happened then.

This religious narrative forms a memorial narrative of a specific type meant to be supplanted to individuals in later generations. This happens by using specific textual and practical strategies that place later generations in a vicarious eyewitness position that blends with the narrator's position in the Hebrew Bible and so comes to possess the authoritative interpretation of the fantastic events. Those who experienced the events (according to the narrative) are doubtful, weak in memory, confused and disoriented, while those who "experience" the events through narrative mediation are believers, strong in memory, and oriented. Individual memory thus seems to need cultural aid in the form of narrative and bodily practices according to this religious narrative, in order to secure belief and transmission of the fantastic (Table 10.5).

| **Type of memory** | *Autobiographical memory* | transforms into → | *cultural memory* | transforms into → | *autobiographical memory* |
|---|---|---|---|---|---|
| **Means of transformation** | | narrative | | narrative and bodily practices | |

Table 10.5 Memory transformations as staged by the Exodus narrative.

34. Remembering the past means redescribing the past. Such redescriptions may be true in the sense of truths that we *now* assert of the past. Yet they need not necessarily have happened (see also Hacking 1995, 249, reference in Freeman 2003); an important point in the present context. Freeman concludes that autobiographical narratives are not only about what happened when, and how the events can be emplotted, but also about how to live a good life.

35. Neisser and Fivush (1994, 2) stated that self-narratives may concern things that we do not remember (although he generally argues that it does matter what "really" happened). Within memory studies it is generally acknowledged that not only in relation to cultural memory but also in relation to individual, episodic memory, vivid recollections can be completely fabricated; we may have convincing memories of things that never happened (references in Neisser 1994, 5).

## Concluding Remarks

This reading of the Exodus narrative offers a case of autobiographical memory that is narratively and ritually mediated, or, in other words, culturally staged. It seems that fantastic religious narratives may produce specific effects in individual memory when coupled with ritual action that enhances transmission and belief. The Exodus narrative presents a set of cultural techniques for anchoring cultural memory to autobiographical memory. The narrative first narratively stages the move from autobiographical memorial narrative to cultural memorial narrative, but subsequently, it provokes the move from cultural memorial narrative to *vicarious* autobiographical narrative by means of specific techniques—textual (deixis), narrative (familial narration), and ritual (primarily bodily) practices. Narrative is thus used as one of the primary means for creating continuity between past, present and imagined worlds. Narrative is also the fundamental mediating link between individual and collective selves. This reading thus confirms what has been suggested by others, that narrative plays a crucial role in the formation of mind between the individual and the social. The individual self here becomes integrated into the collective self, using narrative as a primary vehicle for self-definition.[36] The narrative with its embedded ritual constitutes membership in the Israelite community and is used as a resource for socialization. It offers a case of individual memory being formed by religious narrative, the memory thus produced being vicarious but nevertheless efficient.

This leads to the question of whether it is only collective ritual performances (as suggested elsewhere, see Whitehouse 2000 and the Review Forum in *Journal of Ritual Studies* 16, 2002) that may provide participants with surprising, emotion-arousing experience—the building-blocks of episodic memory? Is there really such a great difference between "narrating" and "pretending"? May not skilfully imagined and narrated actions be just as intense as actions actually performed? Is not the violence in the intense initiation rites, so central in the recent memory debates in the study of religion, while severe and distressing for participants, still staged, make-believe?[37] Scott Atran says:

> Extreme religious rituals of initiation arouse existential anxieties by culturally *mimicking* and *manipulating* the seemingly capricious and uncontrollable situations that naturally provoke them: terror and risk of death from unidentifiable sources, the menace of infirmity and starvation through physical ordeal and deprivation, the injustice of whimsical oppression, sudden isolation and loneliness. (Atran 2002, 165, my emphasis)

What I wish to emphasize here are the words "mimicking" and "manipulating," for in initiation rituals even of the most extreme type, initiates do not usually die, starve to death, or loose health permanently. It is make-believe, the violence is staged. Is the difference to narrated, fantasized action really that great? Such actions, whether narrated or ritually performed, are not "real life," they are one step away from actuality (see Schechner 1993, 313–314). And this leads us back to the question of the entanglement of memory and imagination that we have discussed here.[38] Philosopher Paul Ricoeur suggests that

> ...imagination and memory have as a common trait the presence of the absent and as a dif-

36. Interestingly, Louis J. Cozolino suggests that the self as a matrix of learning and memory is organized into narratives. Underlying these narratives are hidden layers of neural networks. He sees the brain as sculpted in the context of social relationships, and the self as reflective of and shaped by social interactions, even if functioning on the basis of neural processing (Cozolino 2002, 154, 167, 170-171).

37. Ochs and Capps have also suggested similarities between narrative and rituals in relation to socialization (Ochs and Capps 1996, 31).

38. K.J. Gergen says "...unless it is theatre, it will not be recognizable as self-remembrance at all" (1994, 95).

ferential trait, on the one hand, the bracketing of any positing of reality and the vision of something unreal, and on the other, the positing of an earlier reality. (Ricoeur 2004, 44)

If the difference between memory and imagination lies in the bracketing of the positing of reality and the vision of something fantastic on the one hand, and the positing of an earlier reality on the other, then the interesting thing about memory in the Exodus narrative is that this difference seems to be collapsed. This leaves us with a fantastic type of recollection. Exodus stages its narrative and its fantastic events as something in between memory and imagination, and thus it "shares the authority of recollection with the play of imagination" (Schechner 1993, 318). This is a blend which shows itself to be very powerful. The narrative combined with the embedded Passover ritual stabilizes the encounter between imagination and memory by anchoring it in familial narration and bodily practices.

## References

Assmann, J. 2000. *Religion und kulturelles Gedächtnis*. München: Verlag C.H. Beck.

———. 1992. *Das kulturelle Gedächtnis. Schrift, Erinnerung und politische Identität in frühen Hochkulturen*. München: Verlag C.H. Beck.

Atran, S. 2002. *In Gods We Trust: The Evolutionary Landscape of Religion*. Oxford: Oxford University Press.

Bal, M. 1997. *Narratology*. Toronto: University of Toronto Press.

Bal, M. L. Spitzer and J. Crewe, eds. 1999. *Acts of Memory: Cultural Recall in the Present*. Lebanon, NH: University Press of New England.

Biblia Hebraica Stuttgartensia *Torah, Nevi'im uKetuvim*. Abbr. BHS, text critical edition of the original Hebrew Masoretic mss. for the books of the Hebrew Bible.

Boyer, P. 1994. *The Naturalness of Religious Ideas: A Cognitive Theory of Religion*. Los Angeles: University of California Press.

———. 2000. "Functional Origins of Religious Concepts: Ontological and Strategic Selection in Evolved Minds." *Journal of the Royal Anthropological Institute* 6(2): 95–115.

Brooke-Rose, C. 1981. *A Rhetoric of the Unreal: Studies in Narrative and Structure, Especially of the Fantastic*. Cambridge: Cambridge University Press.

Bruner, J. 1991. "Self-making and World-making." *Journal of Aesthetic Education* 25(1): 67–78.

Cornwell, N. 1990. *The Literary Fantastic. From Gothic to Postmodernism*. London: Harvester/Wheatsheaf.

Feldt, L. 2006a. "Signs of Wonder—Traces of Doubt: The Fantastic in the Exodus Narrative." In *Fremde Wirklichkeiten: Literarische Phantastik und Antike Literatur*, edited by Nicola Hömke and Manuel Baumbach, 311–338. Heidelberg: Universitätsverlag Winter.

____. 2006b. "Dis/orientations: Fantastic Memory in the Exodus Narrative." In *Literature and Memory: Theoretical Paradigms, Genres, Functions*, edited by A. Nünning *et al.*, 95–111. Tübingen: Francke Verlag.

____. 2011. *Signs of Wonder—Traces of Doubt. The Fantastic in Religious Narrative from Exodus to Elisha*. London: Equinox.

Fireman, G.D., T.E. McVay and O.J. Flanagan. 2003. *Narrative and Consciousness: Literature, Psychology, and the Brain*. Oxford: Oxford University Press.

Freeman, M. 2003. "Rethinking the Fictive, Reclaiming the Real: Autobiography, Narrative Time, and the Burden of Truth." In *Narrative and Consciousness: Literature, Psychology, and the Brain,* edited by G.D. Fireman, T.E. McVay and O.J. Flanagan, 115–128 Oxford: Oxford University Press.

Gergen, K.J. 1994. "Mind, Text, and Society: Self-memory in Social Context." In *The Remembering Self: Construction and Accuracy in the Self-Narrative, edited by* U. Neisser and R. Fivush, 78–104. Cambridge: Cambridge University Press.

Hacking, I. 1995. *Rewriting the Soul: Multiple Personality and the New Sciences of Memory*. Princeton, NJ: Princeton University Press.

Halbwachs, M. 1925 [1975]. *Les cadres sociaux de la mémoire*. Here cited from German translation: *Das Gedächtnis und seine sozialen Bedingungen*, 1985, Frankfurt am Main.

Hirschfeld, L.A. and Gelman, S. (eds.) 1994. *Mapping the Mind: Domain Specificity in Cognition and Culture*. Cambridge: Cambridge University Press.

Houseman, M. 2002. Review of H. Whitehouse, *Arguments and Icons: Divergent Modes of Religiosity* (Oxford: Oxford University Press). *Journal of Ritual Studies* 16(2): 18–22.

Ivanovic, C., Lehmann, J., May, M. (eds.) 2003. *Phantastik – Kult oder Kultur? Aspekte eines Phänomens in Kunst, Literatur und Film*. Stuttgart: J.B. Metzler.

*Journal of Ritual Studies* 2002. Review Forum on *Arguments and Icons: Divergent Modes of Religiosity*, by Harvey Whitehouse (Oxford University Press 2000), 16(2): 4–40.

Jackson, R. 1981. *Fantasy: The Literature of Subversion*. London: Methuen.

Jensen, H.J.L. 1998. *Gammeltestamentlig religion*. Copenhagen: ANIS.

Kihlstrom, J.F. 2003. "Memory, Autobiography, History." On-line paper, Berkeley, at <http://ist-socrates.berkeley.edu/kihlstrm/rmpaa00.htm>

Lachmann, R. 2002. *Erzählte Phantastik. Zu Phantasiegeschichte und Semantik phantastischer Texte*. Frankfurt am Main: Suhrkamp.

Lemche, N.P. 1996. *Die Vorgeschichte Israels*. Biblische Enzyklopädie 1. Stuttgart: Kohlhammer Verlag.

Neisser, U. and R. Fivush. 1994. *The Remembering Self: Construction and Accuracy in the Self-Narrative*. Cambridge: Cambridge University Press.

Ochs, E. and L. Capps. 1996. "Narrating the Self." *Annual Review of Anthropology* 25: 19–43.

Propp, W.H.C. 1999. *Exodus 1–18: A New Translation with Introduction and Commentary*. The Anchor Bible. New York: Doubleday.

Pyysiäinen, I. 2001. "Cognition, Emotion and Religious Experience." In *Religion in Mind: Cognitive Perspectives on Religious Belief, Ritual, and Experience,* edited by Jensine Andresen, 70–93. Cambridge: Cambridge University Press.

———. 2002. "Mind and Miracles." *Zygon* 37(3): 729–762.

Ricoeur, P. 2004. *Memory, History, Forgetting*. Translated from the French by K. Blamey and D. Pellauer. Chicago, IL: University of Chicago Press.

Rigney, A. 2005. "Plenitude, Scarcity, and the Production of Cultural Memory." *Journal of European Studies* 35(1/2): 209–226.

Schechner, R. 1993. "Ritual, Violence and Creativity." In *Creativity/Anthropology*, edited by S. Lavie, K. Narayan and R. Rosaldo, 296–320. Ithaca, NY: Cornell University Press.

Sørensen, J. 2000a. "Den magiske illusion? En kognitiv tilgang til magiske ritualer." *Tidsskriftet Antropologi* 41: 65–77.

———. 2000b. *Essence, Schema and Ritual Action: Towards a Cognitive Theory of Magic*, Ph.D.-dissertation, Aarhus University.

Todorov, Tz. 1975 [1973]. *The Fantastic: A Structural Approach to a Literary Genre*. Ithaca, NY: Cornell University Press.

Turner, V. 1967. *The Forest of Symbols: Aspects of Ndembu Ritual*. Ithaca, NY: Cornell University Press.

van Seters, J., 1992. *Prologue to History: The Yahwist as Historian in Genesis*. Louisville: Westminster John Knox Press.

———. 1994. *The Life of Moses: The Yahwist as Historian in Exodus-Numbers*. Kampen : Kok-Pharos Publishing House.

Whitehouse, H. 2000. *Arguments and Icons: Divergent Modes of Religiosity*. Oxford: Oxford University Press.

# — 11 —

# Parables, Cognitive Shock, and Spontaneous Exegetical Reflection

Douglas L. Gragg

In the chapter on "Cognitively Costly Religion" in his book on the theory of modes of religiosity, Harvey Whitehouse summarizes male initiation ritual among the Mali Baining of Papua New Guinea. These initiations, he writes, "involved a period of isolation of the novices at a remote location in the forest, the endurance of painful ordeals, and the imposition of secrecy (enforceable on pain of death)." During the initiations, he continues,

> a series of strange objects (mostly masks and instruments) were revealed to novices. They were taught to make such artifacts themselves and were required to paint some of them using blood agonizingly extracted from their tongues. The culminating rites involved an elaborate series of dances in which novices wore the masks they had constructed. The heavy frames of their costumes were supported by a pin (made of sharpened cassowary bone) that was forced through the skin at the base of each dancer's spine.

Most important for Whitehouse's theory is his further observation that

> throughout the process of initiation, novices were given to believe that every detail of the masks, instruments, rituals, tortures, and dances had deep inner meanings that only the elders were capable of grasping fully. The unraveling of these meanings was largely a matter of personal revelation rather than direct instruction, and this was expected to take place very gradually (over a lifespan). (2004, 52–53)

Such initiation rites, found in many human cultures past and present, along with similar types of low-frequency, high-arousal ritual, are characteristic of what Whitehouse calls the "imagistic" mode of religiosity. Because of its "shock-effect," he observes, ritual experience in this mode etches itself into the episodic memory of participants and creates a lasting bond among fellow "survivors" of the ordeal. The shock created by the ordeal is primarily cognitive. Participants struggle over time to make sense of the trauma that was imposed upon them by members of their own community, even of their own clan or family. Whitehouse refers to this process of highly motivated sense-making as "spontaneous exegetical reflection." The natural result is a diversity of insights, or "revelations," among initiates over time about the meaning of the ritual.

In this paper, I first want to call attention to an analogy between the effect Whitehouse attributes to rituals in the imagistic mode of religiosity and the effect many contemporary scholars of early Christian literature attribute to parables that are traceable to the historical Jesus. I will then offer two proposals:

1. The reason these parables—and Whitehouse's imagistic rituals—have such a strong effect is that the cultural assumptions they violate or challenge reflect expectations that are rooted in standard human cognitive architecture.
2. These deeply rooted expectations remain resilient in the face of challenge and sometimes lead to (mostly unconscious) resistance. Results of this kind of resistance are discernible in the (mostly unconscious) "corrections" that some parables of Jesus underwent as they were transmitted (mostly orally) in the earliest Christian communities.

## A Striking Analogy

The idea that parables of Jesus had an effect on hearers analogous to that of imagistic rituals is not immediately apparent from the way parables are represented in the early Christian texts, traditionally called "gospels," in which they were first preserved in written form. There one can discern already the results of a process of transformation, in which parables were turned into allegories or example stories. Allegorization and moralization, in fact, quickly became established as standard approaches to the interpretation of parables among Christian expositors, and these approaches were not seriously and effectively challenged as historically dubious until the late nineteenth century.

A.B. Bruce (1882) and Adolf Jülicher (1899) were among the first to argue successfully that the parables of Jesus were not intended as allegories, in which each element of a story corresponded to some feature of a conceptual schema external to the story. They insisted, instead, that each parable had a point to make on its own (and a single point at that), which could be restated as a general moral or religious truth. Discovering the original point of a parable, they argued, required first recovering its original wording by stripping away interpretive devices contributed by the gospel writers.

C.H. Dodd (1936) and Joachim Jeremias (1947) agreed that a parable made a single point that became clear once the parable had been critically extricated from the secondary narrative context of one or more gospels, but they were not satisfied to express that point as a general truth. They sought, rather, to reconstruct the kind of situation in the public ministry of Jesus that the point of a parable would have addressed. For Dodd and Jeremias that point was generally polemical, growing out of conflict between Jesus and his opponents.

By the midpoint of the twentieth century, critical scholars were in agreement (1) that allegorization was not the original intent of the parables of Jesus and (2) that one must analyze the parables apart from their literary contexts in the gospel narratives in order to understand their original rhetorical function. A new interest in philosophical hermeneutics among New Testament critics and the application of new literary-critical methods to the analysis of parables led to serious doubts, however, about the contention that a parable could be reduced to a simple moral or religious "point" that could be restated in discursive form as a general or situational truth.

The new literary critics took their cue from a remark by C.H. Dodd that he had not fully developed. Dodd had defined the parable as a metaphor or simile drawn from nature or common life, arresting the hearer by its vividness or strangeness, and leaving the mind in sufficient doubt about its precise application to tease it into active thought (1936, 5).

These critics found in this strangeness and ambiguity of Jesus' parables (and of his metaphorical speech in general) clues to their original rhetorical function. Robert Funk, one of the first to develop these clues, suggested that parable as metaphor does not illustrate meaning but "raises the potential for new meaning." It "redirects attention...by means of imaginative shock" and "shatters the conventions of predication in the interests of a new vision" (1966, 138–139).[1]

The first point I want to make, then, is simply that the similarity of this description of the effect of Jesus' parabolic discourse to Whitehouse's analysis of the effect of imagistic ritual is striking and worthy of investigation. I am not interested here in offering a careful defense of the proposal of New Testament critics that parables of Jesus had this kind of effect on hearers historically (though I do find the proposal persuasive). I only want to suggest that if one works on the assumption that parables did have such an effect, comparison with the effect of imagistic rituals is instructive. Examining a few sample parables at this point will make the analogy more vivid and provide material for further analysis.

Most of the parables depict scenes from the everyday life of Jewish peasants in an agrarian society. For example, one of the longer parables, preserved only in the Gospel of Matthew, concerns a vineyard owner who hired laborers at various points in the course of a day. He agreed with those hired at the beginning of the day to pay them "one silver coin" (a denarius, the standard day's wage) and agreed to pay those hired at later points in the day "whatever is just." At the end of the day he paid those hired last first and gave them each a silver coin. Thinking, then, that they would receive more, those hired at the beginning of the day became indignant when each of them, too, received one silver coin:

> "These guys hired last worked only an hour but you have made them equal to us who did most of the work during the heat of the day." In response [the owner] said to one of them, "Look, pal, did I wrong you? You did agree with me for a silver coin, didn't you? Take your wage and get out!" (Matthew 20:12–14)[2]

What were listeners to make of this? Technically, the vineyard owner is right, but do the disgruntled workers not also have a point?

One of the best known parables of Jesus is preserved only in the Gospel of Luke. Removed from its interpretive context in Luke's narrative, it reads like this:

> This fellow was on his way from Jerusalem to Jericho when he fell into the hands of robbers. They stripped him, beat him up, and went off, leaving him half dead. Now by coincidence a priest was going down that road; when he caught sight of him, he went out of his way to avoid him. In the same way, when a Levite came to the place, he took one look at him and crossed the road to avoid him. But this Samaritan who was traveling that way came to where he was and was moved to pity at the sight of him. He went up to him and bandaged his wounds, pouring olive oil and wine on them. He hoisted him onto his own animal, brought him to an inn, and looked after him. The next day he took out two silver coins, which he gave to the innkeeper, and said, "Look after him, and on my way back I'll reimburse you for any extra expense you have had." (Luke 10:30–35)

The shock of this parable for its original hearers would have been caused by the identity

1. This understanding of parables is worked out in detail especially by John Dominic Crossan (1973; 1975) and Bernard Brandon Scott (1989; 2001).
2. For arguments that the parable ended here originally, see Scott 1989, 281–298. Translations of the parables are taken from Miller 1994 unless otherwise attributed.

of the "hero" of the story. We may assume that Jesus' audience of Jewish peasants identified sympathetically with the victim of robbers and either smiled knowingly or sighed disapprovingly at the story's depiction of the behavior of priest and Levite. We may also assume that the listeners anticipated a conclusion to the story in which the unresponsive clerics were either punished somehow or perhaps shamed by a layperson (someone like themselves) who helped the injured man. Instead, Jesus surprised his listeners by introducing a despicable Samaritan as the hero of the story. What did he mean by this?

In several of his parables, Jesus used shocking images to depict the rule of God. In one of the shortest, quoted here as it appears in the Gospel of Matthew, he said:

> Heaven's imperial rule is like leaven that a woman took and concealed in fifty pounds of flour until it was all leavened. (Matthew 13:33)

Leaven was generally a negative image in Jewish tradition, representing corruption (Scott 1989, 324–325). What could Jesus have meant by comparing God's rule to the spread of corruption or, as we might put it today, to the spread of a dreaded infectious disease? Another of his parables appears to recommend shrewd dishonesty:

> There was this rich man whose manager had been accused of squandering his master's property. He called him in and said, "What's this I hear about you? Let's have an audit of your management, because your job is being terminated." Then the manager said to himself, "What am I going to do? My master is firing me. I'm not able to dig ditches and I'm ashamed to beg. I've got it! I know what I'll do so doors will open for me when I'm removed from management." So he called in each of his master's debtors. He said to the first, "How much do you owe my master?" He said, "Five hundred gallons of olive oil." And he said to him, "Here is your invoice; sit down right now and make it two hundred and fifty." Then he said to another, "And how much do you owe?" He said, "A thousand bushels of wheat." He says to him, "Here is your invoice; make it eight hundred." The master praised the dishonest manager because he had acted shrewdly (Luke 16:1–8a).

Would it be possible to say definitively what any of these parables "means"? With all deference to generations of interpreters who have tackled that assignment with great confidence, I think we would have to answer negatively. Like rituals in Whitehouse's imagistic mode of religiosity, parables of Jesus eluded straightforward interpretation. They administered cognitive shocks by violating convention, provoking extended reflection on their meanings. Borrowing the language, quoted earlier, that Whitehouse used to describe the exegetical reflection stimulated by imagistic rituals, we could say of the response to parables as well: "The unraveling of these meanings was largely a matter of personal revelation rather than direct instruction, and this was expected to take place very gradually (over a lifespan)" (2004, 53). With this analogy between parables of Jesus and imagistic rituals in focus, I will now develop two proposals.

## First Proposal

Parables like those of the historical Jesus and imagistic rituals have their intended effect only when two conditions are met. First, the ritual or parable must convey convincingly to targeted individuals that significant meaning is in play. Various devices are used in both ritual and parable to achieve this. Second, the ritual or parable must be set over against some strongly held assumption or expectation. The male initiation rites of the Mali Baining work because of the boys' expectation that community, clan, and family can be counted on to provide nurturing support and protection from harm. A ritual in which those very caregivers inflict pain and injury violates that expectation and demands explanation. Similarly, the parables I cited earlier work because they violate expectations that wages should be commensurate with labor, that shrewd dishonesty is bad behavior, and so on.

What is the source of such expectations? Are they socially constructed, as many have argued, imposed on individuals by culture? Although I believe—with Dan Sperber (1996, 9-31), among others—that it is a mistake of category to speak of abstractions, such as "culture" or "society," as subjects of causation, I do not doubt that ideas or beliefs held by most individuals in a population are generally hard for other individuals in that population to resist. I would attribute this, however, to a propensity for conformity that emerged as adaptive in early stages of human evolution (see Ridley 1996, 171–193). I propose, in fact, that *all of the assumptions which parables and imagistic rituals violate or challenge derive their tenacity from the fact that they reflect, in one way or another, expectations rooted in the standard human cognitive architecture that was cobbled together on the savannas of Africa.* This is easy to argue in the case of expectations about care and protection among kin (relevant to Mali Baining initiation rites), fair distribution of resources (relevant to the parable of the vineyard laborers), or aversion to cheaters (relevant to the parable of the dishonest manager).[3] Assumptions that Samaritans are bad characters or that leaven is a negative image, on the other hand, cannot be regarded as universal. Do the parables of the Samaritan and the leaven, then, count against my proposal that the expectations violated in parables of Jesus are rooted in standard human cognition?

Here it is important for us to be good semioticians and distinguish clearly between deep and surface semantic structures. What was shocking about the parable of the Samaritan was that a bad character was made the hero of the story. It is only incidental—a feature of the surface structure of the story—that Jesus and his audience regarded Samaritans as such characters. This is clear from the fact that, in later (gentile) Christian tradition, in which this prejudice against Samaritans was not shared or even understood, the parable had no shock value and so functioned as *a different kind of story*. Already in the Gospel of Luke, for example, there is no sign that listeners were disturbed by the fact that the hero of the story is a Samaritan. The story is used simply to illustrate the theme of neighborly compassion. Luke has Jesus ask his discussion partner, after the parable has been related, "Which of these three [priest, Levite, or Samaritan], in your opinion, acted like a neighbor to the man who fell into the hands of the robbers?" When the man says it was the one who showed him compassion, Luke has Jesus reply, "Then go and do the same yourself" (Luke 10:36–37). The parable is now an example story.

The image of leaven should be seen, similarly, as a surface-level expression for a deep-structural value—in this case, corruption. It was comparing the rule of God to corruption that made the parable of the leaven shocking. In later Christian tradition, in which leaven did not automatically suggest the idea of corruption, the parable lost its shock value and became a positive image for the advance of God's rule. This is especially obvious in the apocryphal Gospel of Thomas, where the parable reads,

> The Father's imperial rule is like a woman who took a little leaven, hid it in dough, *and made it into large loaves of bread* (Thomas 96:1–2; my emphasis).

The parable has become an allegory.

The parables of the Samaritan and of the leaven, therefore, do not contradict but rather

3. See, for example, Cosmides and Tooby (1992).

support my proposal. Once levels of semantic structure are properly distinguished, it is only necessary to point out that suspicion of bad characters and avoidance of corruption or contamination are demonstrably standard human cognitive predilections, rooted in our shared evolutionary history.[4] Consequently, the shock-effect of these two parables could have been preserved by substituting images that did represent to a new audience bad character and corruption, respectively, but that is not what happened. Perhaps later gentile interpreters just didn't "get the joke" (so to speak). Even if they had, they would not likely have been motivated to preserve the original shock of the parable if my second proposal is correct.

## Second Proposal

The first proposal was that the reason parables and imagistic rituals have such a strong effect is that the cultural assumptions they violate or challenge reflect expectations that are rooted in standard human cognition. I have illustrated this by reference to four sample parables, two of which (the ones concerning the Samaritan and the leaven) seemed at first to be exceptions. My second proposal is that *these expectations—precisely because they are rooted in standard human cognition—remain resilient in the face of challenge and sometimes lead to resistance.* I will illustrate this by noting the fate of the parables concerning the vineyard laborers and the dishonest manager in Christian tradition.

The parable of the vineyard laborers, which had set before its original hearers a puzzling dilemma regarding fairness, became for the author of the Gospel of Matthew a story warning disciples against presumption before God. There the parable is used as part of a response of Jesus to Peter who has just asked what he and other disciples will get for all of their sacrifices. Matthew has Jesus begin by assuring the disciples of their reward:

> Let me assure you that, in the new world, when the Son of Adam sits on his glorious throne, you who have followed me will also sit on twelve thrones, judging the twelve tribes of Israel. Anyone who has left home, brother, sister, father, mother, child, or land because of me will receive a hundred times as much and will inherit eternal life....

Matthew then has him add a word of warning:

> But many who are first will be last, and many who are last will be first.
>
> (Matthew 19:28–30; my translation).[5]

The parable follows immediately so that readers are encouraged to interpret the vineyard owner's rebuke of the disgruntled laborers' presumption as a figure for God's rebuke of presumption on the part of disciples. To make sure readers get the point, Matthew repeats the warning of 19:30 as a kind of conclusion to the parable: "So the last will be first, and the first will be last" (20:16; my translation). If scholars who argue that the parable originally ended with verse 14a are correct (see note 2), Matthew also added remarks of the vineyard owner that emphasize the contrast between his own *generosity* and the grumblers' *envy* (20:14b–15).

Focusing on the owner's generosity as a figure for divine grace, later Christian writers, such as Irenaeus, Origen, and Augustine, moved further in the direction established by Matthew, interpreting the parable as an elaborate allegory. On their reading, God (the

4. On the instinct to avoid corruption, see Rozin, Haidt and McCauley (1993).

5. For an extended defense of the interpretation of Matthew 19:30 as a warning *to disciples*, see Gragg (1990, 139–151).

owner) calls believers (the laborers) into his kingdom (the vineyard) to receive eternal life (the denarius wage). They interpreted the different hours of hire variously as successive epochs in human history or stages in an individual human life. They understood the parable to affirm that such differences in time of call are relativized by the reality of an equal "wage" for all—eternal life, freely given by God.

Let me make the point as clear as possible. Each of the disputants in the story (the owner and the disgruntled workers) has a claim that hearers could be expected to regard as valid because of hearers' expectations, both about fair distribution of resources and about social hierarchies, that are deeply rooted in human cognition. The juxtaposition of competing valid claims is what would have caused hearers to experience cognitive disorientation. In the Gospel of Matthew and in the allegorical interpretation of later Christian writers, this tension in the story is softened, if not completely removed, by the fact that the dispute is "resolved" in favor of the owner. This is achieved by neutralizing the laborers' claim, either by interpreting it in negative terms or by ignoring it altogether. The former approach is evident in the Gospel of Matthew, where the parable became a (negative) example story, warning disciples against presumption before God. The patristic commentators went the other way, disregarding the dispute as incidental to the story, whose purpose (for them) was a straightforward celebration of divine grace. What I am proposing is that the drive to resolve the dispute is best explained by the tenacity of the expectations that made the story shocking in the first place. Since neither claim could be dismissed without some qualification that would satisfy these expectations, the story was "corrected" in the course of its transmission (perhaps even unconsciously) through qualification of the laborers' claim.[6] As a result, the parable lost its ability to create cognitive disorientation and so to provoke and sustain extended exegetical reflection.

The parable of the dishonest manager, in which his shrewd dishonesty is commended, proved to be much more difficult—though not impossible—to "correct." It is hard to read without a chuckle the tortured attempts to do this that follow the parable in the Gospel of Luke. No fewer than three different interpretive conclusions, designed to soften the offense of the story, are offered. It is as if readers are being invited to "take their pick":

> For the children of this world exhibit better sense in dealing with their own kind than do the children of light.
>
> I tell you, make use of your ill-gotten gain to make friends for yourselves, so that when the bottom falls out they are there to welcome you into eternal dwelling places.
>
> The one who can be trusted in trivial matters can also be trusted with large amounts; and the one who cheats in trivial matters will also cheat where large amounts are concerned. So if you couldn't be trusted with ill-gotten gain, who will trust you with real wealth? And if you can't be trusted with something that belongs to another, who will let you have property of your own? No servant can be a slave to two masters. No doubt that slave will either hate one and love the other, or be devoted to one and disdain the other. You can't be enslaved to both God and a bank account. (Luke 16:8b–13)[7]

Can there be any doubt that here the very expectations the parable was designed to subvert have mounted a vigorous counterattack? Here the effort to domesticate the parable is more

6. It is not important for the argument whether the author of the Gospel of Matthew was the first to make this move or it had been made already in the course of oral transmission of the parable (I favor the latter view).

7. For a slightly different way of breaking this down, see Scott (2001, 86–88).

obviously conscious than it was in the case of the parable of the vineyard laborers. My proposal, though, is that the effort is *driven by* largely *unconscious* expectations rooted in standard human cognition.[8]

This "domestication" of the parables of Jesus in early Christian tradition can be interpreted as both a loss and a gain to that tradition. Parables with the power to provoke reflection and generate new meaning lost much, if not all, of that power. At the same time, the transformation of such parables into example stories or allegories in the course of transmission rendered them more serviceable in the effort to inculcate orthodox Christian teaching. Early Christianity—as opposed to the movement that had gathered around the historical Jesus—operated, after all, not in Whitehouse's imagistic mode of religiosity but in the mode he has identified as "doctrinal." Deployed in that mode, parables served best when they were found (or construed) to *illustrate* or *clarify* teachings supplied by authoritative leaders (the fact that the parables were represented as stories told *by Jesus* gave them particular authority in that context). Parables that provoked free and diverse exegetical reflection, on the other hand, could only have been regarded as obstacles to effective catechesis. Let me again be as clear as possible. I am *not* suggesting with these last observations that transformation of parables into example stories or allegories was a *result* of doctrinalization. I am suggesting, rather, that parables so transformed under the pressure of expectations rooted in standard human cognition became usable in the inculcation of doctrine in a way they would not have been otherwise.

## Conclusion

I have called attention in this paper to an analogy between the effect of rituals in Whitehouse's imagistic mode of religiosity and the effect of parables as many scholars believe they were employed by the historical Jesus. That effect is, among other things,[9] to create cognitive shock or disorientation that makes a lasting impression, motivating extended exegetical reflection. I have proposed that parables (at least as employed by Jesus) and imagistic rituals have such a strong effect because the cultural assumptions they violate or challenge reflect expectations that are rooted in standard human cognition. I have also proposed that these deeply rooted expectations remain resilient in the face of challenge and sometimes lead to effective resistance such as is discernible in some of the transformations parables of Jesus underwent in the course of their transmission. It is important to note that each of the two proposals is testable. I have offered some limited empirical support for them here. It would be possible also to test them *experimentally*, but that is a job for another day and, likely, another investigator!

8. As in the case of the parable of the vineyard laborers in the Gospel of Matthew, it is not important for the argument whether the author of the Gospel of Luke was the first to provide these interpretive conclusions (I favor the view that he was not).

9. One of those "other things," as Whitehouse has demonstrated in the case of imagistic rituals, is the creation of a lasting bond among participants. Although I have not been able to develop the point in this paper, it could be fruitful to consider whether the parables and other metaphorical discourse of Jesus could have been sufficient *in themselves* to create a communal bond, at least among his closest disciples.

## References


Bruce, A.B. 1882. *The Parabolic Teaching of Christ*. London: Hodder and Stoughton.

Cosmides, Leda and John Tooby. 1992. "Cognitive Adaptations for Social Exchange." In *The Adapted Mind: Evolutionary Psychology and the Generation of Culture*, edited by Jerome H. Barkow, Leda Cosmides and John Tooby, 163–228. Oxford: Oxford University Press.

Crossan, John Dominic. 1973. *In Parables: The Challenge of the Historical Jesus*. New York: Harper and Row.

———. 1975. *The Dark Interval: Towards a Theology of Story*. Allen, TX: Argus Communications.

Dodd, C.H. 1936. *The Parables of the Kingdom*. New York: Charles Scribner's Sons.

Funk, Robert W. 1966. "The Parable as Metaphor." In *Language, Hermeneutic, and Word of God*, 133–162. New York: Harper and Row.

Gragg, Douglas L. 1990. *The Parable of the Workers in the Vineyard and its Interpreters: A Text-Linguistic Analysis*. Ph.D. Dissertation. Atlanta: Emory University.

Jeremias, Joachim. 1947. *Die Gleichnisse Jesu*. Zürich: Zwingli-Verlag.

Jülicher, Adolf. 1899. *Die Gleichnisreden Jesu*. 2 vols. Freiburg: J.C.B. Mohr.

Miller, Robert J., ed. 1994. *The Complete Gospels: Annotated Scholars Version*. Revised and expanded edition. Santa Rosa, CA: Polebridge Press.

Ridley, Matt. 1996. *The Origins of Virtue: Human Instincts and the Evolution of Cooperation*. New York: Penguin Books.

Rozin, Paul, J. Haidt and C.R. McCauley. 1993. "Disgust." In *Handbook of Emotions*, edited by Michael Lewis and Jeannette M. Haviland, 69–73. New York: Guildford Press.

Scott, Bernard Brandon. 1989. *Hear Then the Parable: A Commentary on the Parables of Jesus*. Minneapolis, MN: Fortress Press.

———. 2001. *Re-Imagine the World: An Introduction to the Parables of Jesus*. Santa Rosa, CA: Polebridge Press.

Sperber, Dan. 1996. *Explaining Culture: A Naturalistic Approach*. Oxford: Blackwell Publishers.

Whitehouse, Harvey. 2004. *Modes of Religiosity: A Cognitive Theory of Religious Transmission*. Walnut Creek, CA: AltaMira Press.

# — 12 —

## Icons and Agency in the Georgian Orthodox Church[1]

Anders Nielsen

As you enter an Orthodox church, your vision is often overwhelmed by the extraordinary amount of icons decorating the walls of the church.[2] In front of the icons you see churchgoers making more or less the same gestures of crossing and prostrating themselves, kissing the icons and mumbling low prayers. The objects of these acts are "dead" artefacts that by their very nature are unable to heed or reciprocate the actions performed by the person before them. If you are invited to an Orthodox Christian's home, you are very likely to be shown his "icon corner," the domestic altar or "little church" in front of which the same actions one sees in the "real" church building can be performed.

Thus this use of icons can be understood as the decisive factor giving these religious images their impact as being more than just decorative pictures on the wall of the church. I suggest that this practice is exactly what gives the icons their saliency as constituting, or, one could say, epitomizing Orthodox Christianity. The massive presence of these religious images, and the way they are used, forces questions such as: "Why these icons, and why these acts seemingly directed at mere pictures that cannot respond?"

### The Orthodox Theology of the Icon

Hitherto most of the literature dealing with Orthodox icons has had an explicit theological interest in its subject, i.e. the icon is seen from within the Christian Orthodox tradition. Leo-

1. An earlier draft of this article was presented as a paper at a Ph.D. course on "Religion, Cognition and Culture," April 18–20, 2006 at the Sandbjerg estate, Aarhus University, organized by Armin W. Geertz and Jeppe Sinding Jensen. I would like to thank Armin, Jeppe and the other participants for a very stimulating discussion. I would also like to thank Annika Hvithamar and Anders Lisdorf for invaluable, constructive criticism since the very start of my project. Finally, I would like to thank Brian Russell Graham, Lund University, for helping me make the English of this article more readable.

2. Fieldwork was carried out during three months from July to October 2003 and again in November 2004 and October 2006 at the Svetitskhoveli Church in Mtskheta. Empirical material was gleaned through participant observation together with qualitative interviews of fourteen ordinary churchgoers. Focus throughout the fieldwork was on the practice of embedding icons in the Orthodox Church, and on how the Orthodox Georgians themselves conceptualize this practice. When referring to "the Georgians" in the rest of this chapter, this means primarily my fourteen interviewees, but also a larger number of Georgian churchgoers whom I observed during my fieldwork. See Nielsen 2004 and 2006.

nid Ouspensky, a Russian Orthodox icon painter and theologian, stresses and proclaims the icon's function as manifesting Orthodox Christianity (Ouspensky 1978). Ouspensky builds upon John of Damascus' theology from the seventh and eighth centuries, which assigns the icon a central position in Christianity. According to this theology, the icon epitomizes in itself Christianity's theological message. Orthodox icons are teaching aids which can depict central aspects of Orthodox Christianity through easily understood imagery. At the same time it is very important for Orthodox theology to see icons not as mere symbols, but as sacred objects in themselves. Thus the icon is understood as an expression of the Christian incarnation, "the divine" taking physical form and manifesting itself in matter (St. John of Damascus 1980, 23). Icons have great impact in the Orthodox liturgy, where they are understood as a quintessential part of the cult and seen as intermediaries that help bring forth a union between the divine and the human, a function which leads to their being accorded the same standing as the Bible (Ouspensky 1978, 139–141). Of utmost importance to this theology is the establishment of a distinction between the material and what is depicted on the material. The icon being a sacred object, it is for Orthodox theologians still a necessary point that it is not the icon's *material* in itself (i.e. paper, wood or paint) that is being honoured, but the *person* or *event* depicted on the icon (St. John of Damascus 1980, 82–90; Ouspensky 1978, 138). This is a central point to which the Orthodox Christian theologians draw attention when they want to defend the cult of the icons against outside accusations of "idolatry."[3]

## Non-Theological Approaches to the Study of Orthodox Icons

What has been lacking up until now in the literature on Orthodox icons are approaches that take the popular use of icons seriously; approaches from outside the Orthodox tradition focusing on religious practice, and not preoccupied with judging whether this practice can be said to be theologically correct or not. Kenna (1985) has a small section describing the practice in relation to icons on a small Greek island, but the overall essay is still focused on Orthodox theology and iconographic analysis. Belonging to a small group of more historically-oriented scholars is Moore (1977), who has a short chapter describing how the popular use of icons saw these holy images as possessing apotropaic powers that could protect the religious practitioners.

For Belting (1994), popular use gave the icons their great impact in early Christianity, in spite of the official church's initial condemnation (Belting 1994, 144–146). Along with relics, the icon was seen as a kind of *pars pro toto*, a living image and representation of

3. The theology of the icon emerged as a direct result of the Iconoclastic Controversy that raged in the Byzantine Empire in the eighth and ninth centuries. The iconoclasts attacked the cult of the icons, claiming that it was idolatry to worship material objects as divine. Defending the cult of the icons, the iconophile theologians maintained that the proper way of addressing icons is through "relative worship," which is "veneration" or "proskynesis," whereas "absolute worship," "latreia" is to be conferred to God alone (see St. John of Damascus 1980; Sahas 1986; Strizhev 1998). Despite her declaration that hers is an art historian's approach, Isar's (2000) attitude is that of the theologian, and it is as a theologian that she tries to distinguish the cult of the icons from "idolatry," i.e. "false worship." Claiming to be speaking about "the Byzantine beholder," Isar sustains her argumentation by quoting Byzantine *theologians* and the Bible as the authoritative references throughout the article. Thus it is clear that this is the perspective of the Byzantine *theological beholder*, i.e. the religious elite, and Isar distinctly writes from a position within this tradition, thereby herself contributing to the Christian Orthodox theological tradition. For more theology of the icon see Galavaris (1981); Florenskij (1995); Ouspensky and Lossky (1999); Sendler (1988).

the deceased that perpetuated the saint's extraordinary powers after death (Belting 1994, 59–61, 195, 299). The use of icons was legitimized through the miracles that were ascribed to them (Belting 1994, 49–57). Only in the sixth century, Belting claims, was the cult of the icons finally fully integrated and admitted to the official church by the majority of the theologians, who then endorsed a long established practice (Belting 1994, 1, 144ff). It was this part of the church that came out victorious in the subsequent Iconoclastic Controversy in the eighth and ninth centuries (Belting 1994, 144–172). Tarasov (2002) gives a historical account of the popular use of icons in Russia from the sixteenth century to the revolution in 1917. Tarasov's study shows that the icons' relevance in the everyday life of the Russians was guaranteed by their allegedly protective and curing powers that helped to keep misfortunes and evil powers at bay (Tarasov 2002, esp. 37–118). Tarasov's study is supported by Ryan (1999), who in his work on magic in pre-Revolutionary Russia portrays both official church and the popular use of icons in divinatory and apotropaic practices (Ryan 1999, esp. 228–254).

The theologian Ouspensky sustains these findings when he gives the theological unease about the excessive use of apotropaic icons decorating the everyday clothing of Byzantine citizens as one of a number of reasons for the Iconoclastic Controversy (Ouspensky 1978, 103).

## Introducing the Cult of Icons in the Georgian Orthodox Church of Today

Various practices put the icons to use as an integrated part of the religious life in the Georgian Orthodox Church today. First and foremost, icons make up an important part of the interior of the church. A typical Georgian church will have icons both as frescoes on the walls and as portable images that can be moved around inside and out of the church. Orthodox churches have a clear division between the altar, only accessible to the clergy, and the nave. The altar screen or iconostasis makes up the partition wall between the altar and the nave. In contrast to the classical Russian wooden iconostasis that will have up to 5 rows of icons, a typical Georgian Orthodox church will have only one row of icons placed inside the stone iconostasis. The "icon of the day," the icon depicting the saint(s) or event whose feast day is commemorated according to the church calendar, will be displayed in front of the iconostasis, on an icon stand called a *proskynetarion*. In front of all the other icons inside the church are candle stands, where wax candles are lit to accompany the prayer.

When an Orthodox Georgian enters the church, he will usually make a round to greet the most important icons by offering candles, kissing the icons and making the sign of the cross and prostrating himself in front of them. This will take place in front of the icons to which he has a special relationship, e.g. a miracle-working icon, or the icon depicting his namesake patron saint. When a person is baptized in the Orthodox Church, he is given a new name, always the name of an Orthodox saint. Most of the Georgians I talked to believed that this saint was one's own personal patron: by carrying the saint's name you stand in a special relationship to your patron; the saint's feast day is celebrated as a kind of extra "birthday party"; and, importantly, you can "see" and "interact with" this saint through the icon.

And it is usually in front of the icons that the Orthodox Georgian will perform his prayers. These prayers can consist of words formulated by the person himself, or he can read ready-made prayers addressing the saint or even the icon itself from prayer leaflets sold inside the church.

Icons are also an integrated part of the Orthodox Church's liturgical practice. During the services, the priest censes first the icons and then the entire congregation, thus making the icons participate in the rituals on almost equal terms as the human participants. Icons are also consecrated inside the church. This can be done in various ways. A typical consecration ritual has the priest reading over the icon and sprinkling consecrated water on it. An icon can be consecrated by placing it on the altar during the liturgy. When the liturgy is finished, the icon is consecrated and sanctified.

## The Cult of Icons in the Georgian Orthodox Home

Every Orthodox Georgian will have icons in his own house as well. The icons can hang on all the walls of the house, but the most important icons will always be those that are placed in the distinctive *icon corner*. The icon corner must face in an eastern direction, as this is where Paradise is located according to the Orthodox Christian tradition. A priest will often have consecrated the individual icons or the whole icon corner. The icon corner is often spoken of as "the sanctuary of the home" or "the little church," for it is here the Orthodox Georgian performs his daily prayers and the same prescribed actions in relation to the icons as in the church. In addition to icons, the icon corner can be equipped with crucifixes, wax candles, oil lamps, censers, prayer books and other objects that can be seen in the church as well (and bought there). Thus the icon corner is meant to expand and distribute the distinctive sphere of holiness characteristic of the church building, channelling that special power into the Georgian Orthodox home.

Therefore an icon corner will have the same icon reproductions as in the church. An asset of the icon corner in relation to the church, however, is that the former can be personalized by the Orthodox individual choosing which icons should make up the sacred corner. The family's patron saints, for instance, will often be present here through their depictions on the icons, a feature that cannot be taken for granted in the local church. The icon corner gives the Orthodox Georgian easy access to practice his religious life outside the church building. The Orthodox Georgian can follow his Church's prescriptions and say his morning and night prayers in front of his icon corner. Furthermore, he can turn to his own icons and pray to the depicted saints whenever he wishes.

## Alfred Gell's Anthropology of Art

Alfred Gell's anthropological theory of art (Gell 1998) constitutes the theoretical foundation of my study. Applying Gell's theory to the empirical material provided my analysis with a new terminology that served as a valuable tool with which to approach the material. Gell's approach to the anthropology of art focuses on the practical and mediatory role of art objects in a social process (Gell 1998, 6). In Gell's terminology, art objects can be seen as "person-like agents" (Gell 1998, 9). Like other "persons," these art objects can exercise "agency" and also be the objects of "agency." For Gell, a "social other" doesn't necessarily have to be another human being, but can be a "thing." Gell uses notions like "person" and "agency" in a purely relational and context-dependent way; a "person" is an agent that interacts with another "person" in a social process (Gell 1998, 17–19, 22). In this domain "'objects' merge with 'people' by virtue of the existence of social relations between persons and things, and persons and persons *via* things" (Gell 1998, 12). According to Gell, this is in line with Maussian exchange theory, where the gift is understood as an object with personhood, an artefactual extension of the giver's person (Gell 1998, 9).

From Peirce, Gell borrows the technical term "index" to substitute inflated words like "art

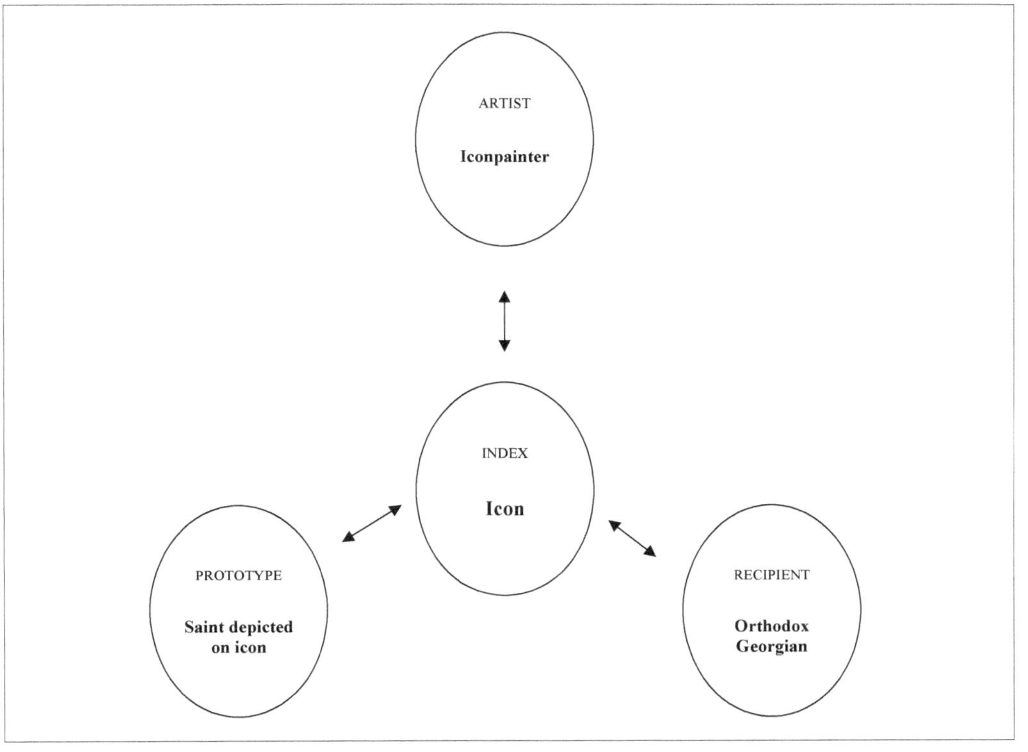


Figure 12.1 The centrality of the index (here the Orthodox icon) where and through which the agents meet.

objects," "artworks" and the like. Often we will have difficulties categorizing the "things" we study as "art." Furthermore, Gell argues that the whole concept of art with its focus on "aesthetics" instead of social qualities blurs the whole anthropological enterprise. An "index" can be virtually anything material (or non-material), a sign that motivates the observer's causal inference, or abduction of agency (Gell 1998, 13). "The *index is itself seen as the outcome, and/or the instrument of, social agency*" (Gell 1998, 15). By our normal inferences, we will see the index as a sign of the dispositions and intentions of a social other, in other words an index is always an index of agency in Gell's understanding (Gell 1998, 15).[4]

Because indexes are artefacts, they will often be understood as "secondary agents," physical expressions or "objective embodiments" of the "primary agents" that have intended the indexes. "Secondary" is not to be understood as "second rate," for it is through these secondary agents that the primary agents exist as social agents (Gell 1998, 20–21). Indexes are secondary agents in so far as they borrow their agency or are understood as mediators of some external source, i.e. the primary agent (Gell 1998, 36).

4. In his theory of signs, C.S. Peirce distinguished between "index," "symbol" and "icon," three signifiers that refer to the signified, each in its own way. An "index" is a physical part of what it signifies, or at least causally contingent on it, whereas a "symbol" is related to the thing signified through convention. An "icon" (which here has nothing to do with the Christian Orthodox religious image) is related to the signified through likeness (Peirce 1960, 156–173). Gell puts emphasis on Peirce's notion of the index as the most conclusive signifier, as it is of importance for Gell's own anthropological theory to stress that the material artefact as an index is understood as a "real" physical part of what it signifies. Gell's choice of the "index" over "symbol" and "icon" can be seen in line with Peirce's own writing, which maintains that all signs will possess indexical qualities (Peirce 1960, 172).

In Gell's anthropological theory of art, the relations between "social agents" and "patients" will form part of a nexus between four terms, each of which can be considered as a social patient (Gell 1998, 28).

The model on Figure 12.1 emphasizes the centrality of the index. Here Gell's four terms have been translated into my empirical material:

1. Index = Icon
2. Artist = Icon painter
3. Recipient = Orthodox Georgian
4. Prototype = Depicted saint

Any of these four terms can be seen as either a social agent or a patient in relation to one another (and to themselves)—what is important is that they meet in or through the index.

In conclusion, it is through an index like the icon that the other three agents in the Orthodox Church can interact.

Figure 12.2 reproduces the table Gell worked out for the Art Nexus that could help order and classify empirical material (Gell 1998, 28–29). The table rests on the premise that all four of the terms can be considered as social agents of different kinds and as such can be agents or patients in relation to one another.

Figure 12.3 shows the Art Nexus table applied to my material. The boxes are to be understood as social spaces of action where the icon as index will always be present, sometimes

AGENT

| PATIENT | Artist | Index | Prototype | Recipient |
|---|---|---|---|---|
| Artist | Artist as a source of creative act / Artist as witness to act of creation | Material inherently dictates to artist the form it assumes | Prototype controls artist's action. appearance of prototype imitated by artist. Realistic art. | Recipient cause of artist's action (as patron) |
| Index | Material stuff shaped by artist's agency and intention | Index as cause of itself: 'self-made' / Index as a 'made thing' | Prototype dictates the form taken by index | Recipient the cause of the origination and form taken by the index |
| Prototype | Appearance of prototype dictated by artist. Imaginative art | Image or actions of prototype controlled by means of index, a locus of power over prototype | Prototype as cause of index / Prototype affected by index | Recipient has power over the prototype. Volt sorcery. |
| Recipient | Recipient's response dictated by artist's skill, wit, magical powers, etc. Recipient captivated. | Index source of power over recipient. Recipient as 'spectator' submits to index. | Prototype has power over the recipient. Image of prototype used to control actions of recipient. Idolatry. | Recipient as patron / Recipient as spectator |

Figure 12.2 The Art Nexus. Reproduced from Alfred Gell, *Art and Agency*, Oxford, 1998, 29.

AGENT →, PATIENT ↓

| PATIENT \ AGENT | Icon painter | Icon | Saint | Orthodox Georgian |
|---|---|---|---|---|
| Icon painter | Icon painters 'generate & test' / Icon painters 'self-reciprocity' | Icon reveals itself to icon painter during the process of painting. "The 'inherent' image reveals itself in the hand of the icon painter" | Saint dictates how icon painter should depict him/her. 'Realistic' depiction. | Orthodox Georgian commisions icon painter to paint an icon |
| Icon | Icon created by the icon painter's agency and intention, talent. "Saint Luke as the first icon painter" | "Icon has appeared", revealed itself / "weeping icons" | Saint creates icon by himself, dictates which form icon should have. E.g. "sudarium"-"Christ's icon-not-made-by-human-hands" | Orthodox Georgian bows and crosses himself in front of icon and kisses it |
| Saint | Icon painter dictates saint's appearance. Icon painter's gift of expression is appreciated by depicted saint | | | "Prayer" and "praise" as ways of address. Orthodox Georgian asks Saint for help (or thanks for the help offered) |
| Orthodox Georgian | Orthodox Georgian's response is dictated by icon painter's skill | Icon has an effect on orthodox Georgian. Icon 'protects' and 'looks at" orthodox Georgian, makes him perform/ abstain from an action | Saint helps orthodox Georgian | |

Figure 12.3 The Art Nexus, as applied to Georgian Orthodox empirical material.

explicitly and sometimes more implicitly. The agents of each social space are presented horizontally, reading downwards, and the patients are presented vertically, reading across. This Art Nexus table is a useful model for categorizing one's material, but it's also important to notice that it can be difficult to distinguish between when it is the icon and when it is the depicted saint that is seen as the initiator or object of agency. Gell was aware of that fact, which is why he stressed the distinction between "primary" and "secondary agent," a notion that was explained earlier. If seen in terms of my fieldwork observations, the distinction between "primary" or "secondary" can seem a bit futile. Looking at our table, we can easily see that it can be difficult to establish if it is the icon "in itself" which is being addressed or the depicted saint. But Gell made this distinction in order to avoid "primitivism," that is, the recipients (here the Orthodox Georgians) should be fully aware that the religious icon they are addressing as they would a normal person, is not a "biological person," but an extraordinary artefact, in fact an artefact that crosses over to a personal domain. We shall return to this shortly.

## Icons as Part of a Georgian Orthodox Social Exchange

As well as observing Georgian Orthodox churchgoers' practice in relation to their icons, I conducted qualitative interviews with Georgian churchgoers, where they were asked to describe their practice in relation to their icons (and other religious artefacts as well), and also to try to describe how they conceptualize this practice. Based on my analysis of these interviews, I argue that icons are not to be understood as mere decorative objects of "art" whose function is restricted to serve as illustrations of Georgian belief. Nor should the actions religious Georgians perform in relation to their icons be seen as "symbolic communication," for the Georgians themselves perceive their icons as objectified representations with the "real presence" of the saints they depict. Being fully aware that the icon is in itself only an artefact, my Georgian informants nevertheless conceive this religious image as a fully operational stand-in for the saint depicted. Thus icons are understood as "person-like" indexes of agency that have the power to mediate and extend the personhood of the saints they represent.

Therefore it can be argued that Georgian Orthodox icons can be said to gain their impact from their *use*, which Georgians conceptualize as a kind of social exchange: the icon is revered through prayers and bodily actions that serve to stress the acting person's disparity of power and status in relation to the icon. In return, the icon as an artefactual embodiment of the depicted saint is understood to participate in a social interaction with its user, the Orthodox Georgian. The Georgians substantiate their conception of a social interaction with the icons by referring to a large number of miraculous events where the icons have been ostensibly instrumental.

The Georgians mention these miracles as bearing witness to the fact that the depicted saints are socially interested agents who care for the Georgians, through the icons. This care is shown in a number of more or less concrete ways, e.g. icons weeping holy myrrh over the present economical and social hardship in Georgia; icons working as protective, apotropaic devices to get cars unharmed through the Georgian traffic; an icon curing a mother's paralyzed hands; an icon procuring a new husband for an unmarried Georgian woman; or icons preventing domestic disputes.

## The Icon as Index of an Extended Mind / Distributed Personhood

To return to Gell, we can only meet a social other as a "person" through the indexes that bear witness to his agency. Thus every index can be understood as a distributed object of personhood, and we have access to this person only through the whole corpus of indexes that distribute his personhood. In the same way, a person's personhood is not limited to the person's biological life, but goes beyond that: the indexes are indicative of this as they distribute the person's biographical existence. Gell argues that every conception of "agency" coexists with an attribution of intention and "mind" to the detected "social other," e.g. the "theory of mind" (Gell 1998, 125–133). It is the indexes as material instantiations of personhood that extend the spatio-temporal existence of the given person. The indexes are intended by a mind, the person's mind, and thus the indexes externalize the person's mind. That is, the indexes are objectifications of an "extended mind."

Material instantiations, indexes, are where the religious practitioners have access to the superhuman agents, e.g. a very popular superhuman agent like the Virgin Mary, Mother of God. Figure 12.4 is a model providing just a few examples from many of where my Georgian informants purport to have met the Mother of God. The model is an analytical construction that serves to translate the informants' narratives to the overall analysis. Using Gell's terminology, it is argued that the Georgian informants infer these meetings as being indexical to what the Virgin Mary is usually known for in the Orthodox Church, namely

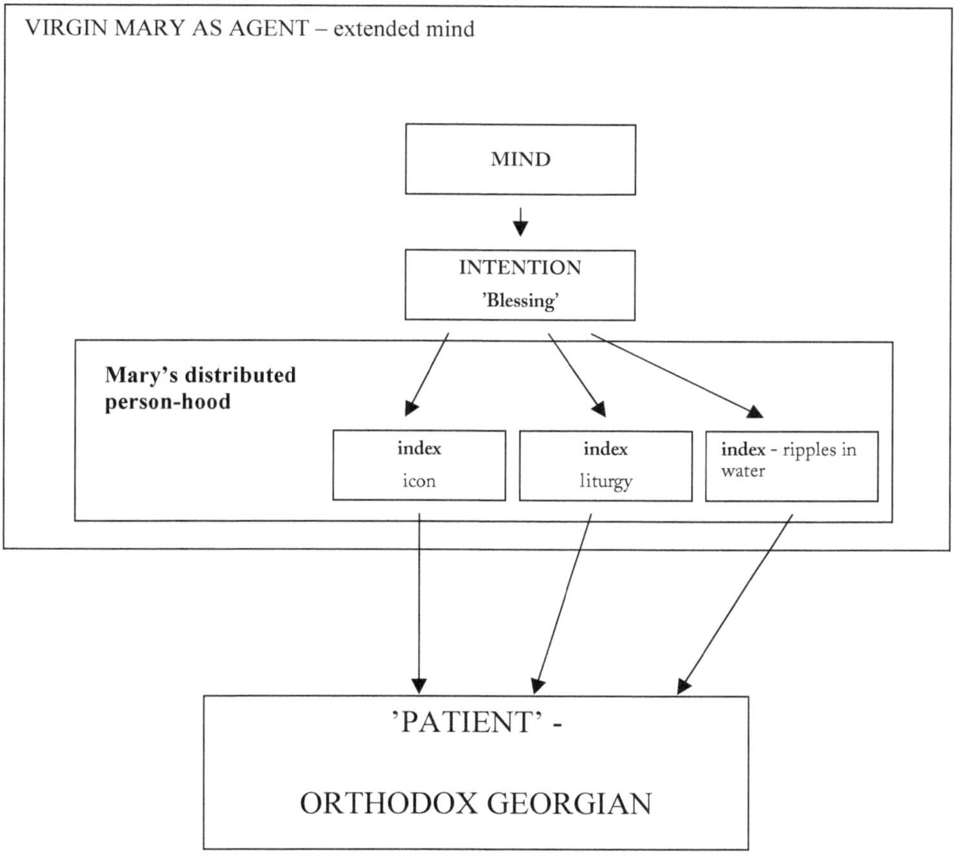


Figure 12.4 Virgin Mary's extended mind.

compassion and blessing. Many of my informants talked about the icon as Saint Mary's "face," and about having "eye contact" with the icon as with a real face. In order to show that the notion of index is not confined to images or other artefacts alone, my model includes other examples which show that an index can, in principle, be anything that documents the this-worldly presence of the primary agent.

"Liturgy" is included as a second example of an index. The Christian Orthodox liturgy is often understood as the meeting place between superhuman saints and human practitioners, and can thus be understood as indexical of a saint such as the Virgin Mary. Figure 12.4 provides a third example of an index that distributes Virgin Mary's personhood as a this-worldly agent. The "ripples in water" box refers to an episode during a christening ceremony where the Virgin Mary allegedly manifested herself in the baptismal font in the form of a well-known Georgian icon "The Iveria Icon of The Mother of God" and thus blesses the baby.

Figure 12.5 shows a photograph of the event that I obtained from one of my informants (the father of the baby). The photograph has appeared in a Georgian Orthodox church magazine and is now understood as an icon in itself. The religious interpretation of the ripples in the water as a miraculous index of the Virgin Mary's agency shows the broad scope

Figure 12.5 Ripples in the baptismal water or The Iveria Icon of The Mother of God manifests itself in the baptismal font during Mariam Urushadze's christening and thus confers blessings to the namesake child born on the very day the church calendar celebrates the Iveria Icon (February 25).

of the concept of index and widens our understanding of the Orthodox way of conceptualizing the notion "icon." Of note is the fact that the Georgian Orthodox Church interprets the ripples in the water as manifesting not only the Virgin Mary, but the Virgin Mary as they know her, that is, through another index, i.e. an icon depiction.

Ideally, the baptized baby will stand in an indexical relationship with her namesake saint, the Virgin Mary. For the Orthodox Georgians, this was secured by the baby being born on the feast day where a specific icon of the Virgin Mary is commemorated. The splashes of water in the baptismal font are perceived as an intentional miracle that adds to and strengthens this indexical relationship. So, by being given the saint's name and receiving her miraculous blessing at the same instant, the baby could grow up and ideally turn out to be not a material, but a human, index that would further distribute the Virgin Mary's personhood.

I hope to have demonstrated the applicability of Gell's anthropological theory of art and its relational and context-dependent concept of agency.[5] Some reviewers have criticized Gell's theory for being too structuralistic and only workable in closed contexts, thus leading to empirical reductionism (Arnaut 2001). Gell has been criticized for his total dismissal of aesthetics (Arnaut 2001; Layton 2003, 450; Bowden 2004; Rampley 2005)

5. See also Pinney (2001) and Harris (2001), who both show the usefulness of Gell's art nexus in their work with the agency of religious images in India and Tibet respectively.

and semiotics (Layton 2003), and for being too eager to demarcate his field of social anthropology (Rampley 2005) by some rather overstated arguments (Thomas 2001, 11). Further, some reviewers argue that the level of abstraction making Gell's a "high theory" renders it problematic as a tool of analysis (Arnaut 2001; Rampley, 2005, 548). Bowden (2004) concludes that Gell's anthropological theory is not about art at all, but about social relations and agency (Bowden 2004, 323 ff). This is an interesting point, for it is true that it will probably be difficult to cover all aspects of the "art world" with this theory. The question of whether or not a European (or indeed a Japanese) notion of an "aesthetic perception of art" could be studied using Gell's anthropological theory need not concern us here. But I hope to have shown that this art nexus can be very useful as a tool of analysis for the study of religious material culture. Whether the practitioners conceptualize their material artefacts as "art" or not (which my Georgian informants did not), their practice in relation to their material artefacts can still be analyzed using Gell's terminology. It helps when the practitioners themselves speak of the indexes as persons (like my religious informants did), but this is not a necessary precondition.

This anthropological theory's preoccupation with detecting social relations and agency is in line with some of the more recent theoretical contributions to the cognitive study of religion. What I would like to do now is to place the findings of the analysis into a wider frame of reference, and try to outline some of the directions this material could take if we were to refine Gell's theory to make it more useful for the comparative study of religion.

## The Icons' Counterintuitive Personhood

As already stressed, my Georgian informants are fully aware that their icons are not biological beings, persons of flesh and blood, but in a physical sense only dead artefacts, yet the icons are understood as "person-like agents" with whom the human agents can engage in a social exchange. These findings are substantiated by some of Pascal Boyer's publications in his attempt to develop a cognitive theory of religion.[6] Boyer argues that the superhuman agents in a given religion will always be characterized by their being "counterintuitive," that is, they violate our immediate ontological expectations of where and how agents can act (Boyer 1994, 29–60; 2001, 58–105).[7] These ontological violations can be observed in my Georgian material when for example an icon is said to weep, is given credit for the fact that the Georgians after seventy years of communism are going to church again, or can cure a mother's paralyzed hands.

It is exactly such ontological violations, Boyer argues, that help give religious ideas their impact, for such ontological violations are attention-demanding, that is, for the religious

6. Boyer even claims to have given a full account of the phenomenon "religion," as the title of his book seems to indicate (Boyer 2001). I am here applying Boyer's terminology as an analytical tool without subscribing to some of his programmatic claims. I follow Jensen (2003), who in an otherwise critical review of *Religion Explained* nevertheless acknowledges that Boyer's notions of the counterintuitive could serve as valuable analytical models in the study of religion.

7. Taking as his point of departure findings in cognitive psychology, Boyer claims that human thinking is domain specific. Human minds across the world, it is argued, share universal domains, where ontological categories like "person," "animal," "plant," "artefact" and "natural object" are separated, and therefore will trigger different inferences. Only the category "person" will activate intuitive inferences about intentionality and mind. The counterintuitive violations are produced by either breach or transfer of categories (Boyer 2000, 198–200).

practitioners as well. This is an important point, for the superhuman agents may be counterintuitive, but not "counter-cultural." The ontological violations will thus seem "natural" to the religious practitioners themselves (Boyer 1994, 34–36).[8] I had the feeling that it was a "natural thing" for my Georgian informants that their icons can weep, look at you, or help your friends get married. But even for the Georgians themselves, these acts are understood as violations of the intuitive expectations they would have about any artefact's possibility of performing such an act. The Georgians themselves mention their icons' actions as extraordinary actions that violate the icons' status as material artefacts. These acts are understood by the Georgians as "miracles" and, what was very important for most of them to stress to me, proof that icons are extraordinary entities and not just mundane, common works of art, and proof of the truth that the Georgian Orthodox Church embodies. My informants are, in other words, very aware that these ontological violations of icons as material artefacts are, yes, "attention-demanding," even though they in a Georgian context seem to be a natural phenomenon that takes place quite often.

In their everyday practice, the religious Georgians do themselves violate the icons' status as material artefacts. By crossing and prostrating themselves, kissing the icons and mumbling low prayers—all this in addressing the icon—the Orthodox Georgians recognize agency in a fashion which would be unacceptable in relation to any other material artefact.

When asked about the difference between icons and other works of art, a great many of my informants mentioned exactly this fact, namely that they would never perform these bodily actions or pray to other images, as these are "only art." The actions performed in front of the icons belong to a different domain, the person domain, that has to do with intersubjective relations.

## Strategic Information and Social Relevance in Orthodox Georgian Life Management

The ontological violations must be "minimal" and primarily have social relevance to the human agents in order to gain saliency and cultural success.[9] Superhuman agents will be seen as socially interested persons with access to "strategic information,"[10] otherwise they would not matter (Boyer 2002, 77–81). My material shows that icons are understood as "persons," i.e. persons with access to strategic information. When my informants mention that the icons see and hear, the important thing is that the icons see the present need in post-Soviet Georgia, see when you argue with your wife at home, listen to your daily

8. Some critics are ready to discard Boyer's notion of the counterintuitive on the grounds that religious concepts for the religious practitioners themselves are not at all "counterintuitive," but, on the contrary, "natural" and "intuitive" (see e.g. Bloch 2002). It seems, though, that Bloch has misunderstood the fact that Boyer's "counterintuitive" is a technical term, and thus misses the point. It is central to Boyer that these violations are an integrated, natural part of the religious culture. Boyer has tried to accommodate similar misunderstandings by introducing the notion "counterontological" to substitute the term "counterintuitive" (Boyer 2001, 74–75). This could, however, just perpetuate the discussion; for one could just as well claim that ontological categories would be different from culture to culture. Again, the point for Boyer is to underline that the religious practitioners are themselves aware that their ontological domains are being violated.

9. Boyer mentions two tables that both violate our categories and are thus counterintuitive: the table made of sausages and the table that understands our conversations. The former will not have the same social relevance as the latter (Boyer 2002, 72).

10. Information is "strategic" when it becomes socially relevant, i.e. important for social interaction (Boyer 2001, 171–177).

prayers and then take action to do something about it. The icons' social relevance derives from the fact that they, like socially concerned "persons," react to the present economical hardships in post-soviet Georgia by weeping in sympathy for others, making spouses stop arguing, and curing paralyzed hands.

Ilkka Pyysiäinen, whose thinking is very much in line with Boyer, has introduced the concept "life management" as a notion meant to refine Boyer's notion of the counterintuitive and underline the relevance for religious practitioners on a this-worldly, everyday basis (Pyysiäinen 2001, 37; 2002, 118).

Boyer also mentions that superhuman agents that are important in a given religion are those that are "personal," for it is with them that one can engage in a social exchange (Boyer 2002, 73). As with all other "normal" persons, religious practitioners will engage in social exchange also with superhuman agents (Boyer 2003, 121–122).

An example from my interviews illustrates this point. Towards the end of an interview, I explicitly asked an informant if his icons had any impact on his everyday life, and he responded in the following manner:

> No, not on everyday life. I… when I turn to the icons, I turn to them with my worries and problems. These are not everyday questions. These are, how should I say?… strategic questions. Questions of that sort are not "everyday" for me […] When I have turned to the icons with such questions, and, you could say, dreams… or very… spiritual questions, they have been fulfilled. (30-year old man, interviewed September 7, 2003.)

Clearly, this informant has not read Boyer or Pyysiäinen, but he turns to his icons when he has personal problems. And by addressing the icons, he makes them socially relevant in his personal "life management," a relevance that seems to be justified by the icons allegedly helping him out. The icons' actions may be counterintuitive and miraculous, but they are nevertheless "person-like," for their actions belong to the person domain. As such, the seemingly ubiquitous miracle in Georgia of the icon giving off holy myrrh is explained "personally" as the icon "weeping," that is, weeping like a real person. Through this action, the icon as a material artefact transcends the person domain. Icons possess strategic information which they can use to make an impact on Orthodox Georgian lives. Like other "persons," icons need to get to know the things first, that is, through personal communication and interaction. Thus my informants say that the icons in a sense can see anything we do, that they "are watching us." But the same informants say that the icons "need to be told first," for example, through prayer. Some of the informants argue that it is through their prayers the icons listen to them. Here the personal aspect is underlined by the informants themselves conceptualizing their practice in front of the icons as a social exchange.

It can be argued that it is these counterintuitive actions violating the icons' artefact domain that transfer the icon to the person domain, which is where the icons become socially relevant in Orthodox Georgian everyday "life management."

## Counterintuitive Actions Embedded in Icons

That religious actions can be seen as no different from other actions that are part of a social exchange with other agents is also what lies at the core of Lawson and McCauley's theory of ritual competence (Lawson and McCauley 1990, 84–87; McCauley and Lawson 2002, 8–13). Culturally postulated superhuman agents (or CPS-agents) act counterintuitively, but still as participants in a cultural system of actions embedded in a religious conceptual system. Thus religious actions are always embedded actions (Lawson and McCauley 1990, 122–123; McCauley and Lawson 2002, 18–19).

The religious conceptual system is understood as a "filter of action" which makes it possible for the rest of us to study this community of social exchange as a "religion," and not just as a disparate group of schizophrenic people whose ontological categories have collapsed. Hence, a religious practice of social exchange can be seen as a range of "embedded counterintuitive actions," that is, the religious practitioners perform a range of acts, each of which can be seen as ontological violations of the domain where these acts would normally belong. But the ontological violations are very natural within the frame of reference, that is, the religious frame of action where they are performed.[11]

Thus we can see the religious Georgian practitioners performing a series of acts in front of their icons which violate the icons' status as artefacts. But these same acts are also embedded actions, prescribed by the Christian Orthodox conceptual system. As Lawson and McCauley's theory only deals with the human practitioners' embedded actions, they do not focus on the acts allegedly performed by the CPS-agents. Significantly, my Georgian interviews seem to suggest that the acts performed by the CPS-agents are also "embedded actions" in the Georgian Orthodox conceptual system. It can be suggested that even miracles are embedded, that is, embedded in the icons. For these miracles are not arbitrary actions, but actions that for Georgians show the icons participating in social space as interested "persons." Thus, "weeping icons," which all of my informants claim to have seen, are a "natural" and ubiquitous expression of the icons' "embedded actions." Furthermore, the many stories of miraculous agency by icons in Georgians' personal lives bear witness to the fact that it is an embedded practice that icons as socially interested agents can intervene and help the other participants in the religious community of social exchange—the "normal," human agents.

Thus it can be concluded that Georgian icons are understood as counterintuitive agents with whom religious Georgians exchange actions embedded in the overall religious conceptual system. The icons are seen as counterintuitive agents that have access to socially relevant strategic information. The strategic information is socially relevant to Georgian life management because it determines the icons' embedded actions, this being the reason why the icons matter to the Georgians.

## Icons as Material Anchors in the Georgian Orthodox Church

Another way of approaching this empirical material could be to introduce Edwin Hutchins, who argues that cultural models are not just ideas residing inside our heads but are always embodied outside ourselves, "in the wild," and in physical artefacts (Hutchins 1995; 2005,

11. Even though they don't elaborate on it, there seems to be an important divergence between Boyer and Lawson and McCauley as to how much importance the overall religious conceptual system will have for the individual practitioner. According to Lawson and McCauley, the religious conceptual system will have to be represented within the individual practitioner when he performs his religious action (Lawson and McCauley 1990, 88–89; McCauley and Lawson 2002, 12–13). Boyer, however, sees religion not as "systems of action competence," but as more loosely organized practical communities of performance and interaction (Boyer 1994, 197; 2001, 156). Holding on to the aspect of action, I find that Lawson and McCauley's notion of "embedded actions" is a very useful tool, as it emphasizes the role of religions as communities of shared action and social exchange with a shared frame of reference. This shared frame of reference, the ideas about the counterintuitive superhuman agents participating in the actions, is what distinguishes the phenomenon of religion from other human phenomena.

1558ff). These physical artefacts are "methods of loci" or "material anchors" that function as landmarks in the social and physical environments surrounding us. These material anchors give stability to cultural models as memory cues that help individuals navigate the cultural systems of which they are a part (Hutchins 2005, 1563ff).

Orthodox icons can be understood as such "material anchors" or "methods of loci" that stabilize the cultural model "Orthodox Christianity" and make it cohesive. An (ideal) Orthodox Christian will make his prescribed prayers in front of his icon corner at home every morning and at night, that is, when he gets up and goes to bed, and will turn to his icons when he feels that he could use some superhuman help. When he enters the church, the Orthodox practitioner will often as the very first thing make a round to greet the icons —usually only those to which he has a special relationship—and light candles in front of them. The icons are thus landmarks in the Orthodox Christian's everyday trajectory, landmarks by which the Orthodox person can navigate in and recall the cultural model "Orthodox Christianity." Thus, icons are material instantiations that anchor Orthodox Christianity in the physical and social world—this world. Icons are relatively simple memory cues situating and embodying what might perhaps be somewhat abstract religious representations. They give "a natural habitation to an otherwise airy nothing" (paraphrasing Pascal Boyer 2001, 3).[12] Being "memory cues," the icons also help trigger certain responses in the religious practitioner. The icons trigger a whole sequence of more or less prescribed actions on the part of the human "navigator," and thus help embody the practitioner in his religious cultural system.

## Icons as Part of Georgian Orthodox Social Embodiment

This naturally brings me to my last group of theorists, Barsalou *et al.*, who introduced the concept of "social embodiment." This theory gives importance to bodily actions as the primary means by which all kinds of knowledge, including religious beliefs, are entrenched (Barsalou *et al.* 2003; 2005). The embodied theories of knowledge are understood as the entire physical context of cognition, not just bodily states (Barsalou *et al.* 2005, 24). Performing actions enhances memory, and embodiment can occur as both stimulus and response, understood as "simulation" (Barsalou *et al.* 2003, 55, 72ff; 2005, 31ff). Thus, bodily states widely interact with cognitive ones and enhance each other (Barsalou *et al.* 2003, 62).

Applying these findings to my material, I can claim that some of the religious beliefs in Georgian Orthodox Christianity must be situated in the icons. Following Barsalou *et al.* (2005, 45ff), it is a material religious object like an icon that gives a "concreteness benefit" and location to Georgian Orthodox Christianity. The icon is in itself situated religion, for as a religious locus it insulates the necessary religious knowledge and provides a stimulus for doing religion. Icons are religious loci, a material part of Orthodox Christian embodiment, for icons activate religious actions and their associated mental states. Thus the icon

12. Day (2004) takes the same line in calling for a reevaluation of the significance of material culture in religious cognition. More than being just media of transmission, religious artefacts are also media of thought. Religious material artefacts, he argues, are not to be understood as "thin cultural wrap arounds," rather they serve as a kind of non-neural cognitive scaffolding central to religious thought. Religious cognition is employed *through* material culture, as it helps the religious practitioners easily recall and give their otherwise invisible counterintuitive superhuman agents a visible location from where these agents can be interacted with (Day 2004, 116ff).

is a very concrete memory enhancer for religious action, that is, seeing an icon makes a Georgian Orthodox Christian "automatically" perform his actions every time.[13] Performing the prescribed prostrations in front of the icons helps drive the practitioner's submissive stance towards the icon (or rather the depicted saint). The social interaction with the superhuman saints in the Georgian Orthodox Church is, as my material suggests, to be understood as a social exchange where the human agents' ritual actions are meant to stress a certain notion of disparity of power in relation to the superhuman agents. The religious practitioners show this disparity of power in a very concrete way by making prostrations in front of the saints' icons. In conclusion, it is the presence of the icons and the bodily actions they stimulate that help entrench the Georgian Orthodox practitioners' religious belief and give stability to Georgian Orthodox Christianity.[14]

## References

Arnaut, Karel. 2001. "A Pragmatic Impulse in the Anthropology of Art? Alfred Gell and the Semiotics of Social Objects." *Journal des Africanistes* 71(2): 191–208.

Barsalou, Lawrence W., Paula M. Niedenthal, Aron K. Barbey and Jennifer A. Ruppert. 2003. "Social Embodiment." *The Psychology of Learning and Motivation* 43: 43–92.

Barsalou, Lawrence W., Aron K. Barbey, W. Kyle Simmons and Ava Santos. 2005. "Embodiment in Religious Knowledge." *Journal of Cognition and Culture* 5: 14–57.

Belting, Hans. 1994. *Likeness and Presence: A History of the Image before the Era of Art.* Chicago, IL: The University of Chicago Press.

Bloch, Maurice. 2002. "Are Religious Beliefs Counter-intuitive?" In *Radical Interpretation in Religion*, edited by Nancy K. Frankenberry, 129–146. Cambridge: Cambridge University Press.

Bowden, Ross. 2004. "A Critique of Alfred Gell on Art and Agency." *Oceania* 74(4): 309–324.

Boyer, Pascal. 1994. *The Naturalness of Religious Ideas: A Cognitive Theory of Religion.* Berkeley: University of California Press.

13. Another good example of a very visible material memory enhancer for religious action is the church building. When an Orthodox Georgian passes a church building on foot or by car, he often "automatically" makes the sign of the cross or even prostrations

14. During the sometimes rather lengthy liturgies in the Georgian Orthodox Church, I found it interesting to watch the children who are brought in by their Orthodox parents as part of their religious upbringing. The parents often taught the children how to act properly in the church, instructing them on which ritualized bodily actions should be performed at various times during the service. But often the children were given leeway to run about inside the church for themselves. It was inspiring to observe how some of the kids would watch what the adults do, e.g. which bodily actions the grown-ups perform in front of the icons. The children would then imitate the same rituals after the adults, sometimes deliberately mocking the former performers by overdoing their prostrations and making large signs of crosses in front of themselves. It was obvious that the children were playing. But in the process of simulating and reenacting the adult Orthodox practitioners, the children are nevertheless being socialized into the church's practice, and, before they have any understanding of what is going on, are already becoming part of the Georgian Orthodox social embodiment. Malley (2004, 45–48) has an interesting parallel in his fieldwork conducted amongst American evangelical Christians, where two-year-old children in the Sunday school sing the *Pat the Bible* song, while each patting their imitation-leather Bible. This teaching is meant to make the children 'respect the book' long before they can read what is inside it. The Bible being arguably the most important religious artefact for evangelical Christians, this analogy from a different Christian context is yet another indication of the importance of simulating and stimulating bodily actions as a means of embodying religious knowledge.

———. 2000. "Functional Origins of Religious Concepts: Ontological and Strategic Selection in Evolved Minds." *Journal of The Royal Anthropological Institute* (N.S.) 6: 195–214.

———. 2001. *Religion Explained: The Human Instincts that Fashion Gods, Spirits and Ancestors*, London: Vintage.

———. 2002. "Why Do Gods and Spirits Matter at All?" In *Current Approaches in the Cognitive Science of Religion*, edited by Ilkka Pyysiäinen and Veikko Anttonen, 68–92. London: Continuum.

———. 2003. "Religious Thought and Behaviour as By-Products of Brain Function." *TRENDS in Cognitive Sciences* 7 (3): 119–124.

Day, Matthew. 2004. "Religion, Off-Line Cognition and the Extended Mind." *Journal of Cognition and Culture* 4(1): 101–121.

Florenskij, P.A. 1995. *Ikonostas*. Moskva: Iskusstvo.

Galavaris, George. 1981. *The Icon in the Life of the Church*. Leiden: EJ. Brill.

Gell, Alfred. 1998. *Art and Agency: An Anthropological Theory*. Oxford: Clarendon Press.

Harris, Clare. 2001. "The Politics and Personhood of Tibetan Buddhist Icons." In *Beyond Aesthetics—Art and the Technologies of Enchantment*, edited by Christopher Pinney and Nicholas Thomas, 181–199. Oxford: Berg.

Hutchins, Edwin. 1995. *Cognition in the Wild*. Cambridge, MA: MIT Press.

———. 2005. "Material Anchors for Conceptual Blends." *Journal of Pragmatics* 37: 1555–1577.

Isar, Nicoletta. 2000. "The Vision and Its 'Exceedingly Blessed Beholder.'" *RES: Anthropology and Aesthetics* 38: 56–72.

Jensen, Jeppe Sinding 2003. "Pascal Boyer: Den ganske historie om al religion (nogensinde)." *Religionsvidenskabeligt Tidsskrift* 43: 53–69.

John of Damascus, St. 1980. *On The Divine Images: Three Apologies Against Those Who Attack the Divine Images*. Translated by David Anderson. New York: St. Vladimir's Seminary Press.

Kenna, Margaret E. 1985. "Icons in Theory and Practice: An Orthodox Christian Example." *History of Religions* 24(4): 345–368.

Lawson, E. Thomas and Robert N. McCauley. 1990. *Rethinking Religion: Connecting Cognition and Culture*. Cambridge: Cambridge University Press.

Layton, Robert. 2003. "Art and Agency: A Reassessment." *The Journal of the Royal Anthropological Institute* (N.S.) 9: 447–464.

Malley, Brian. 2004. *How the Bible works: An Anthropological Study of Evangelical Biblicism*. Walnut Creek, CA: AltaMira.

McCauley, Robert N. and E. Thomas Lawson. 2002. *Bringing Ritual to Mind: Psychological Foundations of Cultural Forms*. Cambridge: Cambridge University Press.

Moore, Albert C. 1977. *Iconography of Religions: An Introduction*. London: SCM Press.

Nielsen, Anders. 2004. *Brugen af ikoner i den georgisk-ortodokse kirke—En religionsvidenskabelig analyse med udgangspunkt i et feltarbejde i Georgien*. M.A. Thesis, University of Copenhagen.

———. 2006. "Ikonerne bekymrer sig også for Georgien," *Kontur—Tidsskrift for Kulturstudier* 14: 29–40.

Ouspensky, Leonid. 1978. *Theology of the Icon*. New York: St. Vladimir's Seminary Press.

Ouspensky, Leonid and Vladimir Lossky. 1999. *The Meaning of Icons*. New York: St. Vladimir's Seminary Press.

Peirce, Charles Sanders. 1960. *Collected Papers, Vol. II: Elements Of Logic*. Edited by C. Hartshorne and P. Weiss. Cambridge, MA: The Belknap Press of Harvard University Press.

Pinney, Christoper. 2001. "Piercing the Skin of the Idol." In *Beyond Aesthetics—Art and the Technologies of Enchantment*, edited by Christopher Pinney and Nicholas Thomas, 157–179.

Oxford: Berg.

Pyysiäinen, Ilkka. 2000. "No Limits? Defining the Field of Comparative Religion." In *Ethnography is a Heavy Rite: Studies of Comparative Religion in Honor of Juha Pentikäinen*, edited by Nils G. Holm *et al.*, 31–40. Åbo: Åbo Akademis tryckeri.

———. 2002. "Religion and the Counter-intuitive." In *Current Approaches in the Cognitive Science of Religion*, edited by Ilkka Pyysiäinen and Veikko Anttonen, 110–132. London: Continuum.

Rampley, Matthew. 2005. "Art History and Cultural Difference: Alfred Gell's Anthropology of Art." *Art History* 28(4): 524–551.

Ryan, W.F. 1999. *The Bathhouse at Midnight: An Historical Survey of Magic and Divination in Russia*. Stroud: Sutton Publishing.

Sahas, Daniel J., ed. 1986. *Icon and Logos: Sources in Eighth-Century Iconoclasm*. Toronto: University of Toronto Press.

Sendler, Egon. 1988. *The Icon: Image of the Invisible. Elements of Theology, Aesthetics and Technique*. Hong Kong: Oakwood Publications.

Strizjev, Aleksandr, ed. 1998. *Pravoslavnaja Ikona. Kanon i Stil': K Bogoslovskomu rassmotreniju obraza*. Moscow: Pravoslavnyj palomnik.

Tarasov, Oleg. 2002. *Icon and Devotion: Sacred Spaces in Imperial Russia*. London: Reaktion Books.

Thomas, Nicholas. 2001. "Introduction." In *Beyond Aesthetics—Art and the Technologies of Enchantment*, edited by Christopher Pinney and Nicholas Thomas, 1–12. Oxford: Berg.

# — 13 —

# Towards a Socio-Cognitive Approach to Religious Texts: A Case Study in Indian Epic Literature

James M. Hegarty

> What matters is that the human mind, regardless of the identity of those who happen to be giving it expression, should display an increasingly intelligible structure as a result of a doubly reflexive forward movement of two thought processes acting one upon the other, either of which can provide the spark or tinder whose conjunction will shed light on both.
>
> (Lévi-Strauss 1994, 13)

Lévi-Strauss here advocates the necessity of the bringing together of two streams of thought in order that there might be progression in the understanding of either, both, or something more than both, of them. This comment resonates with recent developments in the cognitive sciences which have placed far greater emphasis on the role of social interaction in the development of "systematic cognitive structures in dialogical formats" (Tomasello 2003, 214). The aim of this paper is somewhat less grandiose than these broad characterizations of cognitive process and concerns instead the possibilities for, and ramifications of, a closer engagement between the humanities and the cognitive sciences.[1] Specifically, it engages with the potential for a creative encounter between approaches to religious texts and certain cognitivist orientations to human society and its technological and expressive output. It is my hope that current thinking in both textual studies and the cognitive sciences might play "tinder" and "spark" to one another in a way that might "shed light on both."

Such broad goals require careful delimitation. In this paper, I will explore how Tomasello's characterization of the cultural origins of human cognition can be fruitfully combined with the work of scholars, such as R.B. Nair and Mikhail Bakhtin, who have focussed more particularly on the roles of narrative in human social and cognitive life. These theoretical considerations will be developed through an analysis of aspects of the structure and function of an extraordinary South Asian epic tradition, that of the *Mahābhārata*.

Michael Tomasello, in his *The Cultural Origins of Human Cognition,* offers the following

1. There is of course a burgeoning literature in this general area. My paper is doggedly textualist in its focus, however, and it is still largely the case that textualist approaches to religious literature remain largely un-influenced by recent developments in the cognitive sciences.

description of the historical and developmental dimensions of human cognition:

> human children grow up in the midst of these socially and historically constituted artefacts and traditions, which enable them to (a) benefit from the accumulated knowledge and skills of their social groups; (b) acquire and use perspectivally based cognitive representations in the form of linguistic symbols (and analogies and metaphors constructed from these symbols); and (c) internalize certain types of discourse interactions into skills of metacognition, representational re-description, and dialogic thinking.
>
> (Tomasello 2003, 10)

Tomasello bases this complex series of developments on a single species-unique adaptation "the ability and tendency of individuals to identify with conspecifics in ways that enable them to understand those conspecifics as intentional agents like the self" (Tomasello 2003, 202). This is what Tomasello refers to as the "analogy of the self" (Tomasello 2003, 212), and it is evidenced from the very earliest stages of human cognitive ontogeny in, for example, neonatal mimicking and proto-conversations. For Tomasello, it is this adaptation which is the fountainhead of the rich and complex forms of human cognitive and cultural activity. This adaptation is dependant on the successful social transmission of those technologies and expressive forms which developed from this basic human capacity to "pool…cognitive resources." Tomasello calls this process of faithful transmission the "ratchet effect" (Tomasello 2003, 5), in which innovation is integrated into, and preserved within, the body of socially significant knowledge.

This paper will explore how, and in what ways, the Sanskrit *Mahābhārata* is structurally particularly well suited to the achievement of Tomasello's "a-b-c" of human socio-cognitive development. I will show how the text is capable of supporting both the conservation and adaptation of socially significant knowledge. It will not, however, directly engage with Tomasello's core evolutionary hypothesis.

More generally, I will ask the question of whether or not Tomasello's characterizations of the forms and functions of human history and ontogenesis can form the basis of a more general orientation to religious narrative. I will further ask the related question of whether or not the analysis of certain forms of religious narrative can provide a foundation for critical engagement, and perhaps even modification, of Tomasello's position, and in so doing establish a "doubly reflexive forward movement of two thought processes acting one upon the other" (Lévi-Strauss 1994, 13).

Having introduced the core questions and aims of this paper, I will now outline its structure. Firstly, I will contextualize my research, and provide a brief introduction to, the Sanskrit *Mahābhārata*, a source which will be largely unknown to many readers. I will then provide an overview of some of the key questions I ask of the Sanskrit *Mahābhārata* and my proposed answers to these questions. I will further situate these questions in relation to an orientation to the study of narrative which emphasizes its cognitive and social functions. I will then proceed to a brief substantiation of my main hypotheses through a close reading of sections of the Sanskrit text of the *Mahābhārata*. I will conclude with a critical consideration of the ramifications of a socio-cognitive orientation to the study of religious texts.

The Sanskrit *Mahābhārata* is a vast epic poem that tells of the enmity of two groups of cousins, the Kauravas and the Pāṇḍavas, locked in a conflict over succession that culminates in a horrific war that results in the slaughter of almost the entire *kṣatriya*, or warrior, caste. The dates of the composition and development of the Sanskrit text are vague. The scholarly consensus is that the *Mahābhārata* dates from the centuries immediately before and after the commencement of the Common Era and that it shows clear signs of having being transmitted orally prior to, and perhaps while, it achieved written forms. The text is also a repository of diverse tales, asides and political, philosophical and theological doctrine.

The *Mahābhārata* has been consistently approached as a unity by generations of South Asians in both Sanskrit and the many vernaculars. It achieved a maximally inclusive Sanskrit form in the edition and commentary of Nīlakaṇṭha in the second half of the seventeenth century.[2] It is this enduring cultural commitment to the transmission of a consistently expanded, and expansive, text which guides my own engagement with the *Mahābhārata* and which makes of it such a rich resource for the study of the role of narrative in human social and cognitive life.

I shall argue that "what the *Mahābhārata* does" in terms of its narrative structure has a profound impact not just on what it "means" to successive interpreters but also on "how it means" to those interpreters and how it has continued to "mean" successfully and compellingly for something in the region of two millennia. I will argue for a reading of the Sanskrit *Mahābhārata* as a "socially and historically constituted" process which enables both the transmission and adaptation of religious knowledge. I will further suggest that the text presents an elaborate form of narrative activity that allows for potential participants to "benefit from the accumulated knowledge and skills of their social groups" (Tomasello's "a") as well as model forms of complex, perspectivally based, cognitive representation (Tomasello's "b") and meta-cognitive and dialogic thinking (Tomasello's "c"). Taken together, these arguments will contribute to the substantiation of my own core hypothesis concerning the Sanskrit *Mahābhārata*: that the text is a primary means for *both* the transmission and adaptation of religious knowledge in early South Asia. That it is, in short, both a narrative "storehouse" and a form of "travelling narrative theory"[3] of early South Asian religious knowledge. It presents its narrative content in such a way that core concerns and concepts are successfully transmitted whilst still being left open to challenge and creative development. This is primarily achieved in the Sanskrit text by means of a complex narrative "frame" structure.

The structure of the *Mahābhārata* is, then, based on the conspicuous use of narrative frames. "Narrative framing" is the placing of stories within stories. This form of narrative can be seen in, for example, the *Canterbury Tales.* Here the pilgrimage context of the travelling companions provides a frame for the subsequent narrations of individual tales such as that of the Knight or the Franklin (it can also be found in a vast range of other texts, from the *Thousand and One Nights* to the *Decameron*).

I am not the first person to observe that narrative framing is a key feature of the structure of the Sanskrit *Mahābhārata.*[4] Previous research in this area has connected the use of encompassing frame stories in the *Mahābhārata* to earlier Vedic ritual forms which placed rites within rites in such a way that certain rituals (known as *sattras)* became infinitely extendable (and indeed tales of sacrifices extending over hundreds and thousands of years are common in Sanskrit literature). The main reason for the connection was not just structural

2. For details see Gode (1942, 146–161).

3. This concept was developed by R. B. Nair. She defines the theoretical function of narrative in the following way: "they present particular hypotheses about phenomena in the world and present a paradigm to explain them" (see Nair 2002, 343).

4. See Witzel (1987) and Minkowski (1989).

but also thematic. This is because the *Mahābhārata* has not one but two major encompassing narrative frames and because both of these frames have as their setting *sattra* rituals.

The first frame is of a famous bard telling the *Mahābhārata* to a group of *Brāhmaṇas* engaged in an extended *sattra* in a forest. The bard then states that he heard the tale at another extended *sattra* ritual performed by a descendent of the main characters of the *Mahābhārata* and dedicated to the extermination of snakes. He proceeds to narrate this telling which then forms the second of the two major narrative frames. This is, however, literally and figuratively, only the beginning of the story. To proceed further with my analyses I must introduce some basic distinctions between narrative and conversational discourse.

In socio-linguistics there is a simple notation which neatly expresses the difference in the expressive structure of narrative and conversation. Conversation is based on a pattern of exchange which may be expressed simply as ABABAB, and is, in this way, fundamentally dialogic in structure. Narration may be represented, in the same way, as a linear sequence of episodes ABCDEF. Conversation is interventionist and open to challenge whilst narrative is sequential and obeys (or self-consciously disrupts) an internal logic of progression, of development. R.B. Nair in her study of conversation and narrative, *Narrative Gravity*, poses a fundamental question that pertains to the interrelationship of these two forms of discourse:

> How to fit the long, monologic, past-ist, narrator-centred and closed ABCDEF sequence of narrative into the short, dialogic, present-ist, deictically changing and open ended ABABAB alternation of conversation. Solve this puzzle and some of the elusive constraints on communication might begin to reveal themselves. (Nair 2002, 6)

The *Mahābhārata* engages in an elaborate modelling of conversational exchange. It is a modelling activity that extends across the entire text. This modelling activity characterizes both the encompassing macro-narrative, sacrificial frames of the text and the multiple occasions when subordinate narratives are related. These features of the text, I will argue, are precisely directed toward the development of model interpretative competencies in relation to the text. That is to say, through conversational expansion, elaborate meta-messages are introduced regarding the ongoing reception of the text. The inevitable linear ABCDEF structure of the bounded Sanskrit work is dialogized through massing of intra-textual materials that reflect the conversational ABABAB pattern of conversational communicative exchange.

The function of narrative expansion in the Sanskrit *Mahābhārata* will, then, be a central concern in this paper. It is here, I would argue, that we find the clearest connections to Tomasello's characterization of the social transmission of significant knowledge as well as the foundation for a constructive critique of his point of view.

In order, first of all, to demonstrate some of the basic functions of narrative framing within the *Mahābhārata*, I have selected as an example one of the most famous tales in all Sanskrit literature: the story of Śakuntalā.[5] The *Śakuntalopākhyāna* (lit: The subordinate narrative of Śakuntalā) is among the more prominent of the tales embedded within the Sanskrit *Mahābhārata*. It tells of the foundation of the Bhārata dynasty of which both the Kauravas and Pāṇḍavas are a part. The *Śakuntalopākhyāna* foregrounds certain structural and linguistic elements of the *Mahābhārata* that are of tremendous importance to the potential social and cognitive roles of the text.

The tale itself tells of the love affair of a certain King Duḥṣyanta and Śakuntalā as well as of the subsequent trials and tribulations of Śakuntalā in obtaining recognition for herself

5. Mbh.1.62–69.

and her child (the young Bhārata) from him. It is told by Vaiśaṃpāyana to king Janamejaya. My analysis will focus on the interrelationship of narrative materials by means of the frame structure of the *Mahābhārata*. In this way, I will demonstrate the dialogizing function of narrative frames. I will consider, first of all, the interlocutory structure of the text.

In the *Śakuntalopākhyāna*, the characters who speak directly, that is to say, those who have their speech introduced with their name and the perfect of the root √*vac* (*uvāca*) are relatively few. In a cast of eight (in order of appearance, Janamejaya, Vaiśaṃpāyana, Kaṇva, Śakuntalā, Duḥṣyanta, Indra, Menakā, Viśvāmitra), only six do so, but only four for anything more than a single interjection (Menakā and Kaṇva speak directly only once). Of the characters within the narrated tale, (ie. not Janamejaya or Vaiśaṃpāyana) only Śakuntalā and Duḥṣyanta speak directly for any period of time.[6]

The complexity of the interlocutory structure is still, however, formidable. For example, when the central character Śakuntalā narrates the circumstances of her birth she reports the speech of her step-father Kaṇva who in turn reports the speech of the god Indra to the *apsara* (nymph) Menakā. This "narration within a narration" (Kaṇva's report of the speech of Indra and Menakā) occurs within the narrative of Śakuntalā's response to (her suitor) Duḥṣyanta's request. This itself occurs within the narrative of King Janamejaya's request to hear the origins of his line, which occurs within yet another narrative, that of the request of the forest *Brāhmaṇas* to hear the Great Bhārata in its entirety! It is, then, clear that frame structures are prominent throughout the text. The Russian literary theorist Mikhail Bakhtin, in a quite different context, emphasizes that the following must be kept in mind:

> the speech of another, once enclosed in a context, is – no matter how accurately transmitted—always subject to certain semantic changes. The context embracing another's word is responsible for its dialogising background, whose influence can be very great. Given the appropriate methods for framing, one may bring about fundamental changes even in another's utterance accurately quoted. (Bakhtin, 1981, 340)

These "nested themes" and their "dialogizing background," these diverse implications, are developed through the structure and self-presentation of the entire *Mahābhārata* and not just in its patterns of report of direct speech. Thus the frame structure of the text has a direct impact on the potential implications of the Śakuntalā narrative. This can be seen through a consideration of the distribution of interjected vocatives, or terms of address, in the text.

Pavel Grintster[7] has provided us with the most detailed account of what he considers to be oral-formulaic features in the *Mahābhārata*. Grintster emphasizes the importance of the formulaic endings of verses and the function of supporting words in these endings, among

6. It is also of interest to note that in the critical edition of the *Mahābhārata*, the introduction of the speech of a character (name + *uvāca*) is outside the metre while in the other great South Asian epic, the *Rāmāyaṇa*, it is incorporated. While this may point to a more rudimentary mode of production, or context of origin, it has the important function of offering clearer "anchorage" to the potential participant during the often complex "frame shifting" that occurs in the text. This is further complicated by the specifics of the various manuscript editions of the text. Southern manuscripts, according to Sukthankar, the editor of the critical edition, generally omit the *uvāca* and preserve only the name of the interlocutor. However this does not undermine our basic point (and indeed in manuscripts where the practice varies, such as in Ś1 in the critical edition of the Ādi, Sukthankar argues that the use of the name only is simply an abbreviation of the fuller form with the verb). See *The Mahābhārata*, (vol.1, 1933, x).

7. I rely on the summary of de Jong (1984) for access to the work, in Russian, of Grintser.

these he includes "semantically neutral" vocatives (de Jong 1984, 17). These vocatives are used, by the narrator, to address the person who is currently hearing the narrative, e.g. "O Janamejaya!" I will demonstrate below, however, that a vocative, which is of itself semantically neutral (simply a proper noun or an honorific), may, in its positioning, invoke a narrative frame which exercises a profound force on the interpretation of the events of the embedded narrative and vice versa. Interjected vocatives are crucial to the implications of a given narration offered in a given narrative context.

Interjected vocatives occur often in the *Śakuntalopākhyāna* and include: *bharatasattama* ("O Best of the Bhāratas," Mbh.1.62.3), *naravyāghra* ("O Tiger Amongst Men," Mbh.1.62.7), *Janamejaya* ("O Janamejaya," Mbh.1.68.2) and *Bhārata* ("O Bhārata," Mbh.1.37,43,51). It is my contention that such interjected vocatives are capable of fulfilling their role, as metrical "padding," in a far from semantically neutral fashion in the *Śakuntalopākhyāna*. The first two ("O Best of the Bhāratas," "O Tiger Amongst Men") occur very early on in the narrative and thus emphasize the hierarchical narrative "descent" that has occurred by referring back to the narrative context within which the *Śakuntalopākhyāna* occurs. That is to say, they emphasize the shift that has occurred from the context of narration to the narration itself, and it is this sense, in terms of a model of multiple narrative frames, that we have a metaphorical "descent." The rest occur at the close of the tale and thus prepare the reader/participant for the narrative shift "back" (or perhaps "up") to the enframing narrative. They also reinforce the consciousness amongst participants of the capability of the narrative to make such "shifts." The frame is thus completed, and the tale of *Śakuntalopākhyāna* elegantly "bracketed."

It is the triple reference in the last few lines of the narrative to Janamejaya as "Bhārata"[8] that I wish to emphasize in this respect. This triple reference comes in the context of the acceptance of Śakuntalā's son by King Duḥṣyanta as heir, and the revelation, by Kaṇva of his great destiny as a "*cakravartin*," a great king, or "turner of the wheel": "When he marches forward, the sovereign wheel of that great-souled turner of the wheel shall roll on forever unimpeded."[9]

In this way, the shift from the primarily honorific interjected vocatives of the opening to the dynastic form of address at the close of the narrative emphasizes Janamejaya's status as a descendant of this great line. The vocatives help to assert an overarching narrative and dynastic continuity. They indicate a narrative cohesion that exercises a transformative force on the way in which the narrative operates. This is achieved by uniting interlocutors in the aftermath of the main *Mahābhārata* war, which is a context that encompasses not just intra-textual participants but also all subsequent audience members. From a hierarchically ordered series of narrations, we are led to a consciousness of the connections between characters and narratives beyond simply who is telling what to whom. This is a primary means of discerning the text's capacity to "prompt" potential participants, by this I refer to a means of encouraging a fuller engagement with the text's complex narrative patterning.[10]

8. Mbh.I.37/43/50.

9. Mbh.I.67.29:
paraṁ cābhiprayātasya cakraṁ tasya mahātmanaḥ
bhaviṣyaty apratihataṁ satataṁ cakravartinaḥ

10. For a treatment of the progressive delineation of the idealized participant in the Sanskrit *Mahābhārata*, see my 2002 article on this topic. For more detail on frame structures, see my "Extracting the Kathā-Amṛta (Elixir of Story): Creation, Ritual, Sovereignty and Textual Structure in the Sanskrit *Mahābhārata*." *Journal of Vaishnava Studies* 14, 2006, 39–61.

This patterning can, however, have an ironic force.

The above emphasis on the glory of the Bhārata dynasty is, I would argue, just such a context of maximized ironic content. When Kaṇva states that "the sovereign wheel of the great spirited Bhārata shall roll forever unimpeded," his words are framed by the post-war narrative context of Janamejaya and Vaiśaṃpāyana. These two know all too well that the line of the Bhāratas was to proceed far from unimpeded. The very positioning of Kaṇva's reported speech within the narrative frame of Vaiśaṃpāyana's narration to Janamejaya allows for an ironic conflict of themes. The theme of the unimpeded glory of the Bhāratas conflicts with the thematic content of the frame narrative positioned as it is *after* the fratricidal war between the Pāṇḍavas and Kauravas (both of whom are of the Bhārata dynasty). The shift from generalized vocatives of address, from Vaiśaṃpāyana to Janamejaya, to an exclusive use of the dynastic "O Bhārata" form heightens the impact of the embedded narration in its double emphasis on both continuity and discontinuity. Thus we can begin to discern a far-reaching ambiguity in the text, an ambiguity that is both the product of, and yet can only be resolved by, dialogic interaction at multiple narrative and meta-narrative levels.

The *Mahābhārata* demands a *deliberative response* to its complex implications from potential participants. The ABCDEF of linear narrative is made maximally conversational in force and structure by means of an ABABAB alternation across and through multiple narrative and meta-narrative levels. This "conversational force" aids in the construction of an idealized participant by means of the massing of materials which demonstrate explicitly and implicitly the responses of diverse intra-textual participants.[11]

The "conversation structure" of the narrative frames of the text is very clearly related to the development of further narrative materials within the *Mahābhārata*. This is clear in the way in which both the complex structure, and potentially ironic implications, of the *Śakuntalopākhyāna* provide all sorts of prompts and clues as to how to engage with the complex narrative structure of the text as a whole. In this way, I argue for a dynamic and open-ended *Mahābhārata* as a text with a complex orientation to its intra-textual and extra-textual participants.

From this brief excursus on the internal structure of the Sanskrit *Mahābhārata*, we can see how closely the text resonates with Tomasello's vision of sociogenetic process. This can be summarily demonstrated by recalling the "a-b-c" of Tomasello's orientation to human history and ontogeny: the *Mahābhārata* is a repository of diverse forms of religious knowledge (of cosmogony, cosmology, statecraft, philosophy and ethics), in this way it minimally stands as a means whereby social groups might "benefit from the accumulated knowledge and skills of their social groups" (Tomasello's "a"). The elaborate frame structure of the text allows for the development and contrast of multiple viewpoints and communicative contexts, it makes the text both ineluctably reflexive and doubly communicative. By this I mean that the *Mahābhārata* can, by means of its elaborate frame structure, incorporate new narrative materials as well as simultaneously model forms of complex "perspectivally based cognitive representation" (Tomasello's "b") which enable the development of skills in the analysis and synthesis of textual content (and, by extension, socially significant knowledge). The need for

11. Edward Said captures much of what I am trying to argue for in relation to his analysis of the work of Joseph Conrad. He says of Conrad, "I have described the extraordinary *presentational* mode of his narratives, how each of them, almost without exception, dramatises, motivates, and circumstances the occasion of its telling, how all of Conrad's work is really made out of secondary, reported speech, and how the interplay between appeals to the eye and the ear and is highly organised and subtle and constitutes that work's meaning" (Said 1979, 176).

potential participants to both participate in, and resolve, the complex narrative structure of the text is the key stimulus that the text offers for "dialogic thinking" (Tomasello's "c").

There is, in fact, a narrative in the *Mahābhārata* that makes the cognitive richness of the text explicit. The tale is one in which the snakes, gods, sages and demons all go to the great god Prajāpati for instruction. Prajāpati offers only one brief lesson, the syllable OM. Each class of being interprets the syllable differently hearing it as an exhortation, respectively, to biting, giving, self-restraint and pride. The text concludes:

> Having approached one teacher and been refined (*saṃskṛtāḥ*) by means of a single sound the snakes, gods, sages and demons were all disposed differently.[12]

This potential polysemy is however carefully constrained and contained within the *Mahābhārata* precisely because of the conspicuous frame structure, and the multiple contexts of interpretation it offers to extra-textual participants. These participants may, in turn, record their responses in the very fabric of the text, a process of narrative "inflation" which is reflected in the manuscript data for the Sanskrit *Mahābhārata*. The complex narrative structure of the text, when interpreted in relation to recent developments in the cognitive sciences, might make more than an idle boast, at least *in potentia*, of the *Mahābhārata*'s repeated claim about itself that "Whatever is here is elsewhere, but what is not here is nowhere else."[13] If Tomasello's vision of the evolutionary development of *Homo sapiens* is dependent on the "analogy of the self," then my understanding of the *Mahābhārata* is based on its function as the "analogy of a world." I see the *Mahābhārata* as a form of ongoing public memory practice which allows for both the preservation and adaptation of socially significant knowledge by means of its complex narrative structure.

It is my hope that we have come sufficiently far to see the potential contributions of Tomasello's view of human cognitive and social development to the study of the *Mahābhārata* and by extension to religious texts in general. Another cognitivist, Merlin Donald, has provided a compelling description of the human species which stands as something of a call to action for the religionist:

> The human brain is the only brain in the biosphere whose potential cannot be realised on its own. It needs to become part of a network before its design features can be expressed. The cognitive infrastructure of human culture includes many things.... By eliminating the need for the explicit encoding of many important parameters of collective cognition, and by embedding layer upon layer of tacit or implicit knowledge in a cultural network, they make it feasible to evolve in a distributed fashion the kinds of highly abstract symbolic systems that we now take for granted. The result is that we are plugged-in, as no other species before us. (Donald 2001, 324)

The study of religious texts, and in particular the study of widely distributed narrative traditions, can and must be approached in relation to these recent developments in the cognitive sciences. These developments are placing greater and greater emphasis on the very "cognitive infrastructure of human culture" which has been the focus of the humanities, and in particular the study of the religious dimensions of human social life, since their inception.

12. 14.26.6-11, line 11 is translated above:
ekaṁ śāstāram āsādya śabdenaikena saṁskṛtāḥ
nānā vyavasitāḥ sarve sarpadevarṣidānavāḥ

13. Mbh.1.56.33 and 18.5.38:
yad ihāsti tad anyatra yan nehāsti na tat kva cit

What, however, can the study of religious texts offer in the way of reciprocal critical engagement with some of the core postulates of Tomasello's approach? An initial area might be in the potential corrective a nuanced account of religious texts might have on some of the more crassly linear and instrumentalist assumptions of Tomasello's developmental orientation to human cognitive and social development. It is important, for example, to highlight the danger of the over-determination of the notion of the so-called "ratchet effect." This is Tomasello's assertion that "the process of cumulative cultural evolution requires not only creative intervention but also, and just as importantly, faithful social transmission that can work as a ratchet to prevent *slippage backward*" (Tomasello 2003, 5). Such a notion seems to rest too heavily on a somewhat simplistic view of the complex processes involved in the transmission and adaptation of significant social knowledge that are so apparent in the Sanskrit *Mahābhārata* and elsewhere. It is also true to say that such an orientation to human cultural development carries the political danger of, among other things, a form of resurgent neo-primitivism in the human sciences. It is important to note that Tomasello himself advocates a fuller encounter between what he calls "intellectual historians" and cognitivists (Tomasello 2003, 210). Such an encounter must be based on an acknowledgement of, and engagement with, the extraordinary work on inter-subjectivity and inter-textuality that already exists in the humanities.[14] There is extraordinary potential here for mutually enriching dialogue. It was my intention that this paper would represent a small contribution to such a dialogue.

## References


### Sanskrit Texts

The *Mahābhārata*, 19 vols. Plus 6. vols. of indexes, edited by V.S. Sukthankar, S.K. Belvalkar, P.L. Vaidya, *et al.* Poona: Bhandarkar Oriental Research Institute, 1933–1972.

### Secondary Sources

Bakhtin, M. 1981. "Discourse in the Novel." In *The Dialogic Imagination*. Austin: University of Texas Press.

de Condillac, E. 2001. *Essay on the Origin of Human Knowledge*. Edited and translated by H. Aarslef. Cambridge: Cambridge University Press.

de Jong, J.W. 1984. "Recent Russian Publications on the Indian Epic." *Adyar Library Bulletin* 34: 1–42.

Donald, M. 2001. *A Mind so Rare: The Evolution of Human Consciousness*. London: Norton Press.

Foucault, M. 1978. *The History of Sexuality*, vol. 1. London: Penguin.

Gode, P.K. 1942. "Nīlakaṇṭha Caturdhara, the Commentator of the *Mahābhārata*—his Genealogy and Descendants," *Annals of the Bhandarkar Oriental Research Institute* 23: 146–161.


14. One might point to a current in social theory that moves across, at least, three centuries and might include, for example, the subtle investigations of the interrelationships of perception, memory and consciousness in Condillac, the social psychology of Mead, the work of Lévi-Strauss, the "dialogics" of Bakhtin, or even the notion of distributed "power-knowledge networks" in the later Foucault. See Condillac (2001, esp. 11–31), Mead (1977, esp. 202) and Foucault (1978, esp. 98–99). Such a bringing together of social and scientific theory is already apparent in the work of both Donald and Nair. Nair's work has been a particular inspiration in my research into the narrative structure of the Sanskrit *Mahābhārata*.

Hegarty, J.M. 2002. "An Apprenticeship in Attentiveness: Narrative Patterning in the *Dyūtaparva* and *Nalopākhyāna* of the *Mahābhārata*," *Rocznik Orientalistyczny* 54 (1): 33–62.

Hegarty, J.M. 2006. "Extracting the Kathā-Amṛta (Elixir of Story): Creation, Ritual, Sovereignty and Textual Structure in the Sanskrit *Mahābhārata*." *Journal of Vaishnava Studies* 14: 39–61.

Lévi-Strauss, C. 1994 [1969]. *The Raw and the Cooked: Introduction to a Science of Mythology*. London: Pimlico Press.

Mead, G.H. 1956. *On Social Psychology*. Chicago, IL: University of Chicago Press.

Minkowski, C.Z. 1989. "Janamejaya's Sattra and Ritual Structure." *Journal of the American Oriental Society* 109: 401–420.

Nair, R.B. 2002. *Narrative Gravity*. Delhi: Oxford University Press.

Said, E. 1979. "The Text, the World, the Critic." In *Textual Strategies: Perspectives in Post-Structuralist Criticism*, edited by J.V. Harari, 161–188. London: Methuen and Co.

Tomasello, M. 2003. *The Cultural Origins of Human Cognition*. Cambridge, MA: Harvard University Press.

Witzel, M. 1987. "On the Origin of the 'Frame Story' in Old Indian Literature." In *Hinduismus und Buddhismus Festschrift fur Ulrich Schneider*, edited by Harry Falk, 380–414. Freiburg: Hedwig Falk.

# — 14 —

# Pole Position: Space, Narrative and Religion

Anders Lisdorf

In this essay, I will consider religious narratives, or myths, and their relation to space. The Australian Arrernte myth about Numbakulla, originally recorded by Spencer and Gillen around 1900, will serve as a reference point because it clearly involves a relation to space and because it has been richly commented, most notably by two of the greatest historians of religion, Mircea Eliade and Jonathan Z. Smith. I believe that they can be taken as relatively representative for how the relation between space, narrative and religion has been conceptualized in the study of religion.[1]

Eliade saw this myth as expressing the universal necessity of having an *axis mundi*, whereas Smith claimed that it expressed an etiology consisting of two parts: event and memorial. These views imply that narratives are necessary in constructing maps of the world. This entails that narratives map onto maps, and that human cognition makes use of maps in navigating the world. I will evaluate these assumptions in the light of research from neurology, robotics and the cognitive sciences, and conclude that these two assumptions cannot be verified. Based on this research and philosophy of mind, a different solution to the relation between space, narrative and religion is proposed.

## The Myth of Numbakulla

Numbakulla was a mythical ancestor whose doings were the center of attention in a myth recorded by Baldwin Spencer in 1896 at Alice Springs in connection with the Engwura ceremony. Numbakulla was the central figure of the Southern, Central and Northern Arrernte people (Spencer and Gillen 1927, 356).[2] I will first give a brief selective summary of the myth.

Numbakulla arose "out of nothing" at a place called Lamburkna. He created a great many central things, but most importantly he created the *Kauwa-auwa* pole at the center

1. I do not intend to discredit excellent anthropological work such as that of David Parkin (Parkin 1991), but for the sake of clarity and coherence, I have chosen to focus on Eliade's and Smith's work. I do, however, think that other work on space, narrative and religion essentially resembles Eliade's and Smith's work in their basic assumptions.

2. This is the linguistically correct orthography for what Spencer and Gillen spelled "Arunta" (Gill 1998).

of his camp in Lamburkna: "On the ground outside he…planted a pole called *kauwa-auwa* in the middle of it. A representation of this is now erected on the ceremonial ground at the close of the Engwura" (Spencer and Gillen 1927, 357). After having taught the first ancestor the ceremonies, "Numbakulla went on alone and drew up the pole after him and was never seen again." A small version was used in the Engwura ceremony: "You use a small Kauaua at the Engwura" (Spencer and Gillen 1927, 360).

Towards the end of a long and detailed myth we find the following, also concerned with the pole: Some descendants were walking with this pole the Kauaua now used regularly for ngwura:

> At a place called Okinyumpa an accident befell them which made them all feel very sad; as they were pulling up the kauaua which was very deeply implanted the old Oknirabata, who was leading them, broke it off just above the ground, and to the present day a tall stone standing above the ground at this spot represents the broken, and still implanted, end of the pole. Carrying the broken Kauaua they came to Unjiacherta, which means "the place of Unjiamba men" and lies near to the Hanson Creek. They arrived here utterly tired out, and found a number of Unjiamba men and women of all classes. They were too tired and sad to paint themselves; their Kauaua in its broken state was inferior to many of those which the Unjiamba people had, so they did not erect it, but, lying down together, died where they lay. A large hill, covered with big stones, arose to mark the spot.
>
> (Spencer and Gillen 1927, 388)

## The Eliadean View

The phenomenology of religion is, of course, a multifaceted discipline (cf. Jensen 2003). Here Mircea Eliade will be taken as representative of this. Eliade deserves credit for focusing on the spatial elements of religion, for example in *The Sacred and the Profane* (1957). He does this among other things through the Numbakulla myth of the pole recounted above. He notes that "This pole represented a cosmic axis, for it is around the sacred pole that territory becomes habitable, hence becomes transformed into a world" (Eliade 1957, 23–33). A more thorough analysis is found in *Australian Religions*:

> The pole is charged with important symbolism and plays an important role in ritual. The fact that Numbakulla disappeared into the sky after climbing it suggests that the *Kauwa-auwa* is somehow an *axis mundi* which unites heaven and earth…in other words the "center" imparts structure to the surrounding amorphous space. (Eliade 1973, 51–52)

Concerning the next element, Eliade writes:

> One day an accident befell one of these mythical groups: While pulling up the *kauwa-auwa*, which was deeply implanted, the old chief broke it just above ground. They carried the broken pole until they met another group. They were so tired and sad that they did not even try to erect their own *Kauwa-auwa* "but, lying down together, died where they lay. A large hill, covered with big stones, arose to mark the spot." Seldom do we find a more pathetic avowal that man cannot live without a "Sacred center" which permits him to "cosmicize" space and to communicate with the trans-human world of heaven.
>
> (Eliade 1973, 52–53)

In the Eliadean view this is but one more concrete manifestation of religion's trans-human form, particularly the form of the *axis mundi*. According to Eliade's theory, humans need a sacred center to give shape to the surrounding environment; this is imparted through religious narratives.

## A Critique by Jonathan Z. Smith

Jonathan Z. Smith has thoroughly criticized Eliade's appropriation of this myth in support of his theory. He has convincingly argued that Eliade's theory of the morphology of the sacred has several empirical problems which this myth illustrates. I will not go into detail with all the claims, but just provide a summary and touch on a few points.

Jonathan Z. Smith summarizes Eliade's exposition of the myth as consisting of nine elements, each of which is evaluated in terms of its relation to the original text. For each point, Smith examines Eliade's interpretation, and concludes that it is either verified, or that it should be rejected, or that it is in need of revision (Smith 1987, 6):

1. The pole was fashioned by Numbakulla from a gum tree—revision
2. After anointing the pole with blood, Numbakulla climbed up the pole and disappeared into the sky—revision
3. The pole is a "cosmic axis"—rejected
4. Following Numbakulla's withdrawal, the pole plays a ritual role—revision
5. The pole is always carried about by the ancestors in their wanderings—revision
6. The ancestors determine the direction in which they travel by the direction toward which the pole bends—rejected
7. The pole is broken by accident—verified
8. Ancestors die because the pole was broken—revision
9. This is because the breaking of the pole is like "the end of the world"—rejected

Smith's reading cannot be termed sympathetic to Eliade. It is a bit problematic as it does not allow Eliade the benefit of a doubt, whereas Smith allows himself "interpretive freedom." Overall, however, it is safe to say the criticism is justified. I will comment on a few central problems.

Point 2 does need revision. This point is crucial because it leads Smith to conclude that the ever present danger of the historian of religion might be looming—namely the contamination of the sources by Christian influence. I think Smith is too quick on the trigger.[3] Another reading is also possible. Smith fails to point out that the sky is never actually mentioned in the text. It is an inference from the fact that Numbakulla climbed up. True, the sky is up, but other things are too. In the text Numbakulla is inviting the first ancestor to come to his camp to get secret ceremonial objects from the *Pertalchera*. He then says:

> We two will go up to see my camp. Telling the Inkata [first ancestor] to follow him, he began to climb up the tall Kauwa-auwa, and reached the crosspiece. But the blood had made it too slippery for the Inkata, who slid down so Numbakulla went on alone, drew the pole up after him and was never seen again. (Spencer and Gillen 1927, 360)

I have not been able to localize the place of the camp called Lamburkna on any map, but the *Pertalchera*, the storage of the secret things, is a cave Numbakulla made in a rock (356). This could suggest an elevated position. He might simply have been climbing up to his camp. It is correct that the ancestor already has made it to the camp in the following paragraph. This might be a result of the mythical style, or that a possible alternative route has

3. The same conclusion is reached by Sam D. Gill after a consultation of the original fieldnotes, which is the basis for the text (Gill 1998, 17).

been left out. Anyhow it is not reasonable to assume any great continuity in the story since this part was concocted from two different stories, where Baldwin Spencer (who was the editor of Spencer and Gillen 1927) personally elaborates detail (Gill 1998, 16). From the field notes, on which this text is based, no mention is made of the ancestor climbing the pole. It could be that he actually did make it up the pole anyway. His failure to do so in the myth could thus be a product of Baldwin Spencer's imagination (Gill 1998, 16).

Point 4 cannot be said to be subject to revision. Here it is Smith who needs to be revised. It is too harsh to say that there is "no evidence"[4] of any connection to ritual. The crux of the matter is, as far as I can see, Eliade's and Spencer and Gillen's identification of the primordial mythical pole *Kauwa-auwa* created by Numbakulla, with the later mythical and contemporary *Kauaua*. The latter was used in the final stages of the men's final initiation ritual the Engwura (Spencer and Gillen 1927), and mentioned later in the myth. Here is a case where Smith's reading isn't the most obvious one. Numbakulla standing by the *Kauwa-auwa* together with the first ancestor says: "You use a small Kauaua at the Engwura." I fail to see how this constitutes "no evidence" for any ritual role. On the contrary, I will argue that the ritual role is quite central.

These two points seem to indicate that the connection between ancestors, territory, pole and ritual seems strong. Only the first two are emphasized in Jonathan Z. Smith's alternative interpretation of the myth. Let us take a closer look at this since it also focuses on the relation between space, narrative and religion.

## Tales of Wandering and Wondering

Instead of Eliade's cosmic outlook, Jonathan Z. Smith concludes that the myth is a type of itinerary consisting of 5 generative elements (Smith 1987, 7):

1. The ancestors journeyed from this place to that
2. Something happened
3. For this reason the place is called "so and so"
4. A feature in the present topography either was formed by or memorializes this event
5. The ancestors moved on to another place

Event and Memorial comprise the two central parts of the mythic narrative. Eliade's mistake was to ignore the memorial. Thus Eliade's points 7 and 8 are revised thus:

[EVENT] The pole is broken
[MEMORIAL] "and to this present day a tall stone standing above the ground at this spot represents the broken, and still implanted, end of the pole"
[EVENT] The Tjilpa ancestors die
[MEMORIAL] "A large hill covered with big stones arose to mark the spot," furthermore their *Tjurunga* "remained" behind and are "now" in their "Storehouse" accessible at ceremonial occasions" (Smith 1987, 9)

According to Smith, then, the myth: "...is above all an etiology for a topographical feature in the aboriginal environment. It is the memorial that has priority" (Smith 1987, 10). Smith then focuses on how humans are transformed in the narrative:

> ...through this very process of change...the ancestor achieves permanence.... [T]he central

4. "I can find no evidence that the poles in the Tjilpa wandering-ancestor myths or the poles in the various Tjilpa (or other Aranda) ceremonies are in any way identified with 'the pole' of Numbakulla" (Smith 1987, 126, n.4).

> mode of celebrating and signifying objectification is not dramatization, but recollection. Rupture occurs through the human act of forgetfulness. (Smith 1987, 12)

Everything fits nicely then. The myths are tales of wandering and wondering: while wandering, the Arrernte wonder at the strange shapes of the landscape, and these myths come in handy as explanations.

The pole's stated centrality in ritual is ignored. It is therefore all the more interesting to see how Smith exemplifies his point by quoting Strehlow who, in contrast to Spencer and Gillen, had intimate knowledge of Arrernte life and language, having growing up among them: "...the Northern native believes in the presences of his revered and dreaded ancestor wherever and whenever *ceremonies* are being given in honor of his remembrance" (quoted in Smith 1987, 13, my italics). The quote explicitly makes the opposite point: that the rituals are central. The ancestors' presence is thus not bound to the individual memorials, but to the rituals wherever they are performed.[5] It is ironic that Jonathan Z. Smith starts a book on ritual by ignoring ritual. I will return to the ritual aspect in the end, but for now let us focus on the relation between space, narrative and religion in Eliade's and Smith's account.

For Eliade the narrative produced a religious space, for Smith the space produced a religious narrative. Both assumed that this space is mapped in the narrative, and that these maps take the form of passive internal recapitulations of the external environment which serve to guide humans in navigating their environment. Neither had any clear solution for how acting in this environment was related to maps and narratives. But do narratives map onto maps? And do humans use internal maps to navigate their environment? And how does this relate to acting in an environment?

## *Chronicle of a Death Foretold*—Do Narratives Map onto Maps?

Not much research has treated the relation between narratives and maps, but Marie-Laure Ryan has provided an illustrative experiment (Ryan 2003). She asked 55 college students to draw a map of the city featured in the novel *Chronicle of a Death Foretold* by Gabriel Garcia Márquez. The novel is about a murder and takes place in a Caribbean city. The students had read and discussed the novel in the preceding weeks and were subsequently asked to draw a map of the city. The results are fascinating because they show how varied a map can be drawn from the same narrative even though it had been repeated and discussed in class.

Concerning the landmarks, almost all (98%) could remember the main character's house. The next landmarks were the river, the square and the docks remembered by respectively 78%, 67% and 58% (239). Thus only four out of 23 landmarks mentioned by the students show over fifty percent recall. Another interesting fact is that about half of the respondents (45%) also remembered the fountain on the square (225). The only problem is that this is never mentioned in the narrative. It was simply made up by half of the students.

Concerning the relation between these landmarks, only 17% were able to map the central landmark, the main character's house, as lying between the square and river (Ryan 2003, 227). So whatever mental models were responsible for organizing spatial relations in

5. Sam Gill's book on this myth and the scholarly treatment of it is a masterful examination of the case and the way it has been treated by different scholars, but it does not advance any new interpretation of the myth as such.

narratives, they are not map-like. Ryan concludes "If spatial imagination proceeds piecemeal, we will situate site *a* with respect to site *b*, and site *b* with respect to site *c* when a character is shown moving between these points, but we will not necessarily situate *c* with respect to *a* in our mental model" (Ryan 2003, 235).

So as far as we can see, narratives map very poorly onto maps; it is very difficult to produce maps from narrative input. Thus the assumption that narratives map directly unto maps can be rejected. But it might be that humans have an unconscious map which just has difficulty in being visualized. Humans may in actual practice, as opposed to mere reflection, make extensive use of internal maps. So let us turn to evidence for how humans navigate.

## Rats and Robots

A good place to start is to find a model of how navigation might possibly take place cognitively. For some reason, neuroscientists seem to think that humans have a deep affinity to rats. This is, of course, debatable, but I will here assume that there are good neurobiological and evolutionary reasons to assume some analogy to humans in simple sectors of knowledge, such as navigation, since both are mammals.

Research, primarily on rats, has shown that animals navigate using two basic systems: dead-reckoning or path integration, and landmark based cognition. Neurologically both seem to be rooted in the hippocampus (O'Keefe and Nadel 1978).

Dead-reckoning is a "continual updating of position by summing up small *displacements* (changes in position).... This process is equivalent to obtaining the position vector by integrating the velocity vector with respect to time" (Gallistel 2001, 24–25; Gallistel 1990, chapter 4). This in turn builds on an ability to navigate according to a framework. These frameworks utilize cognitive maps: "These cognitive maps are known as *Euclidean sense-preserving* representations of the environment, that is, they encode the metric relations (distance and angles) and the sense relation (right versus left)" (Gallistel 2001, 25). Research on rats has shown how this works. A rat was put in a rectangular room and shown in which corner a food item was hidden. It was then made disoriented and put back in the rectangular room. The results showed that rats searched in either the correct corner or the one diagonally opposed to it (Cheng 1996). This indicates that the information stored was something like "short wall as opposed to long wall (metric relation) to the right (sense relationship)".

But mammals also use landmark-based navigation. This works by storing and using information which determines the relations between landmarks in the environment, in much the same way: relations, such as direction, and the distance between them.

In rats something called "place cells" have been discovered (Tolman 1948; cf. Wang *et al.* 1999). They are located in an area of the hippocampus, and have a unique region of firing relative to one specific part of the environment. When moving through the environment "these cells seem to form a map-like representation, with each location represented by neural activity in a specific set of cells" (Sharp 2001, 26). These cells are used in both types of navigation.

Based on these insights into rat navigation, Maja Mataric has developed a Robot Rat, which successfully simulates real rat navigation. This Mataric robot employs very simple quasi-independent functions. For example it has a sonar and a function that keeps it at a given distance from the wall, another detects landmarks which are recorded as a combination of the robot's movement and the distance to the landmark. A third function makes a "map" of this information, but this map is not what we normally understand as a map. It is not a representation of all known landmarks and all their individual interrelations such as relative size, and distance from each other, which can be consulted and turned into action. Rather it is "a network of landmarks, each of which is...a combination of motor and sensory readings" (Clark 1997, 48). The map employed thus consists of landmarks and their interrelations as a sequence used in real-time navigation.

The philosopher Andy Clark forcefully sums up the implications: "The internal representations the mind uses to guide actions may thus be best understood as action-and-context-specific control structures rather than as passive recapitulations of external reality" (Clark 1997, 48).

### But Humans Are Neither Rats Nor Robots...

The skeptic will probably still have a feeling that he does not sufficiently resemble a rat or a robot for this evidence to be pertinent to our problem. It might very well be that humans could have evolved the capacity for rich inner maps beyond the rat stage, and do not use this form of navigation at all. This is a fair and not altogether implausible assumption.

As a first step of persuasion, let us try an exercise: try to imagine how you get from your house to your local supermarket. I suspect that you did not think of a richly detailed map where you could visualize the whole and then pick the shortest or most scenic route. Rather I suspect that you imagined going out of the door, taking out the keys, getting into the car, starting the car, and following the road until you reached a certain salient landmark (e.g. the great oak tree), three traffic lights, or a configuration of metric and sense-relations (a road crossing with one big and three small roads to the right and left respectively), where you would have to turn before or after, and so forth until your simulated trip ended at the supermarket. No map or model or system was most likely ever represented. You imagined going from point *a* to *b* then *b* to *c* and so on. At no point was any direct relation between *a* and *c* represented. You also included completely inconsequential details of no use to navigation, while omitting familiar movements essential to charting.

The hardcore skeptic will maybe still disagree, or claim that such small conjuring tricks are unscientific and do not impress him. He will still say that humans do in fact use maps all the time, also to solve navigational problems facing them in traversing their local environment. Actually he is right. Humans do, at least sometimes, use a map-like form of navigation. But first of all, they are very rarely internalized: the reason for making maps is exactly that they are hard to internalize. Secondly, in these cases it is a heavily culturally and educationally mediated task, as the following anecdote will illustrate.

Melanesian navigation, as used by the Caroline Islanders, is based on an extraordinary system of islands, ETAKs, which serve as reference points when traveling from one point to another. It is too complicated to give the details of the system here (cf. Hutchins 1995). During fieldwork, the anthropologist D. Lewis was trying to grasp the ideas of the navigation system, which he learned from master navigators on the Caroline Islands:

> On one occasion I was trying to determine the identity of an island called Ngatik—there were no charts to be consulted of course—that lay somewhere south-west of Ponape. It has not been visited by Central Carolinian canoes for several generations but was an ETAK reference island for the Oroluk-Ponape voyage.... I drew a diagram [that is, of a standard

> navigational type chart] to illustrate that Ngatik must necessarily lie where these ETAK bearings intersected. Hipour could not grasp the idea at all. His concept is the wholly dynamic one of moving Islands. (Lewis 1972, 142)

The problem was that he had not learnt that practical navigation could be understood by using a chart. Thus cases where maps are used as basis for finding our way in the world are most often artifacts of cultural learning practices.

## And Language…

Human navigation through the environment, however, does have different features than that of rats and robots. But the difference is not, as should be evident by now, any ability to form rich inner maps of the world. Instead humans can use language to solve navigational problems. Psychologist Linda Hermer-Vazquez and colleagues sum up two features of human navigation that differ from animals, and pre-linguistic children: 1) the process of dead-reckoning is highly variable across different cultures, and 2) that humans typically use verbal directions (Hermer-Vazquez *et al.* 1999, 4). Later research has shown that 1 can be subsumed under 2 in that this cultural specificity is linguistically mediated (Majid *et al.* 2004). So, human navigation has the extra feature of using language.

Hermer-Vasques performed a series of experiments comparing performance of a so-called flexible reorientation task. This was the task originally done with rats by Cheng (Cheng 1996). This showed that human adults performed at the same level as rats and pre-linguistic children when deprived of the use of language (Hermer and Spelke 1996). This condition was produced through verbal shadowing, in the form of word by word repetition of linguistic stimuli. The thesis was that:

> …the grammar of any language allows terms to be combined irrespective of their domain-specific content…. A natural language therefore could provide speakers of the language with a medium of representation in which multiple sources of information can be combined flexibly. (Hermer-Vazquez *et al.* 1999, 9)

They find good evidence to support this thesis. For example it seems that the linguistic ability is able to integrate the dead-reckoning and landmark based form of navigation. Human adults, for example, perform almost perfectly every time they have both landmark and geometric information. But these landmarks can be something as abstract as "The spot where a great injustice took place" (cf. Hermer-Vazquez *et al.* 2001, 266).

Thus, using dead-reckoning and geometrical information, humans—as opposed to rats and robots (and small children)—would be able to find the hidden treasure to the left of "where a great injustice took place." It can be seen that here the "landmark" concept has been greatly expanded to include knowledge from other domains.

They thus conclude that in addition to the basic spatial orientation systems that animals possess, humans have a further system:

> This system may connect to many other systems of representation, regardless of their domain-specific content. Its operation may be governed by rules and principles allowing the arbitrary combination of information from distinct, domain-specific sources. This uniquely human system may serve as a medium in which information captured from different encapsulated systems can be combined rapidly and with ease. The language faculty appears to have all the right properties to serve as this uniquely human combinatorial system of representation. (Hermer-Vazquez *et al.* 1999, 34)

Later corroborative evidence seems to support this conclusion (Carruthers 2002; Hermer-Vazquez *et al.* 2001; Majid *et al.* 2004). Further it has also been shown that there is cultural variability in how people perform in spatial problem-solving tests, which can be demonstrated to relate to characteristics in their language (Majid *et al.* 2004).

In addition to the basic animal system of navigation, humans have a linguistically mediated cognitive system which can integrate information from different domains. So far we have found no evidence that humans have any map-like representations which serve to guide action. But how do the navigational abilities relate to action? Or more precisely what is the relation between representation and action?

## Pushmi-Pullyu Representations

The American philosopher Ruth Millikan has proposed that humans show extensive use of a form of representation she calls Pushmi-Pullyu Representations (PPR). Consider a shopping list. This can be both descriptive, for example of what *is* in the shopping bag, or directive, for example of what *should* be in the bag. Millikan proposes that there exist more primitive representations that include both aspects. These are called Pushmi-Pullyu Representations. They are primitive in the sense that descriptive as well as directive representations use PPR. They are also primitive in the sense that these are the basic signals animals use among conspecifics, for example danger signals (Millikan 2006).

Millikan has a theory of representation which focuses on function; of particular importance is proper function (Millikan 1984). Here it is important to notice that proper function is not applied directly to representations, but "to the mechanisms whose function it is to produce and to use inner representations. When functioning properly, inner-representation-producing mechanisms produce representations in response to, and appropriate to, situations in which the individual organism finds itself" (Millikan 1984, 17–19; cf. Millikan 1996, 148). Here it is interesting to note that representations are byproducts of function. Further, she writes that:

> ...any state constituting a stage in a (proper-) functional process, when the shape of the process and hence of the state is determined by input from—and as a function of certain features of—the environment, can be viewed as itself having proper functions. The proper functions of this state are to help in the production of various further stages in the process to which this state will give rise if the whole system continues to function properly.
> (Millikan 1996, 149)

This means that a functional process can integrate the environment. It even implies some kind of loop with it. Humans as functionally integrated systems exist only insofar as they are able to survive, which Millikan herself is hinting at. And this involves the environment in the overall proper function of surviving.

Another important point is that there are sub-functions, which have proper functions, but these can also malfunction. Many functions don't even function properly most of the time. What is important is that they do function when it is necessary. One example is the eye-blink reflex, which is triggered uselessly many times. Millikan writes: "in order to proliferate, often a type needs to perform only in some critical but small amount of cases, a proportion that varies widely depending upon other factors that entered into the mechanisms of evolution by natural selection" (Millikan 1996, 147).

Let us return to the descriptive and directive representational functions. According to Millikan, "a representation is directive when it has a proper function to guide mechanisms that use it so that they produce its satisfaction condition. Like a blue print, it shows what is to be done" (Millikan 1996, 149). It amounts to a plan or a desire. Thus here the world can be at fault, in not (yet) fulfilling this desire or plan. In the example of the shopping list, the contents of the shopping bag may not have fulfilled the satisfaction conditions of the list.

This is opposed to descriptive representations: "The descriptive representation's truth condition is a condition to which it adapts its interpreters or users in the service of their proper function." This equals beliefs. Thus the representation can be at fault, because beliefs can be at a loss to describe the world adequately for proper function. If the shopping list, for example, is used as a descriptive representation, it can fail to do so adequately in case something is missing in the shopping bag.

We have now come full circle and a quite different picture of how humans navigate and act in their environment emerges: humans navigate their environment by the use of basic mammalian cognitive systems such as dead-reckoning and landmark based navigation. On top of this, language can assist further in landmark based navigation, since language can be used to integrate knowledge from other domains of knowledge to make symbolic or abstract "landmarks" which are not perceived. This functions in basically the same way, where the relations between individual landmarks are represented, but not all relations between all landmarks, as is the case with a map. These representations are PPR, which implies that they have a descriptive as well as a directive function. They are not passive recapitulations of reality; they are directly functionally integrated through humans with the environment, in the end serving the survival of the functional system: *homo navigans*—the navigating human, navigating the environment in search of attractive affordances.

## Another Look at the Map?

We return to the two assumptions governing previous research into the relations between space, narrative and religion: that narratives map onto maps, and that humans use internal maps to navigate their environment. And we conclude that they are not feasible.

It can now be seen that both Eliade's and Smith's theories of myth are inadequate, because they assume rich inner models (rather than locally integrated functions), and because their view of myth is a purely descriptive view, that does not explain the relation to action. This is like arguing that the shopping list can only be about what is in the bag, and not a list for what "should" be in the bag. However, we still need to propose a viable solution to how the narrative is related to space and cognition.

First of all we need to look critically at the narrative fragments under consideration. As Sam Gill has demonstrated, the story of Numbakulla has serious source critical problems. Let us recapitulate Gill's meticulous work on the sources. First of all Numbakulla (or Nambakala or Ngambakala), seems to be an adjective and not a noun or name as used by Spencer. The story of Numbakulla was elicited from an aboriginal, Charlie Cooper, in Alice Springs the 11th of June 1926. It was told in aboriginal English. Gill has juxtaposed the notes from Spencer's notebook used at Alice Springs that day with the version given in *The Arunta*. Let us look at the fragments from Spencer and Gillen used in the present exposition.[6]

6. Based on transcriptions by Sam Gill, taken from http://www.us.oup.com/us/companion.websites/0195115880/creation.story/?view=usa (September 1, 2006).

## Comparison of Sources

| Spencer and Gillen's The Arunta (356–360) | Spencer's field-notes June 11, 1926 |
|---|---|
| Numbakulla arose "out of nothing" at a place called Lamburkna | Nikata [i.e. inkata] Alchurinya Numbuka jump up self |
| On the ground outside he…planted a pole called *Kauwa-auwa* in the middle of it. A representation of this is now erected on the ceremonial ground at the close of the Engwura | He had big Kanana [in a second notebook from the same day, Spencer interprets this as kauwa-auwa] (special nurturinya [ibid., called nurtunja. The kauwa-auwa is a special kind of this] which he erected in a second ilpintira [ibid., described as a totem emblem drawn on rock or ground] outside the podalchura [ibid. described as a storehouse for the tjurunga]) |
| You use a small Kauaua at the Engwura | Kauaua at Engwura |
| Numbakulla went on alone and drew up the pole after him and was never seen again | Ngumbúkule pulled the kanana up. No trace left him. |

The language of Spencer's filed notes from June 11th seems to reflect aboriginal English. Thus we can be reasonably sure that this source is as close as we will ever get to the words of Charlie Cooper on June 11, 1926 in Alice Springs. It can be seen that what ended up in *The Arunta* of the fragments under examination here is reasonably close to what his informant can be thought to have said.

There is however reason to assume that what his informant said did not unequivocally reflect traditional Arrernte thought and mythology. Let us start with the name Numbakulla. The word Ingkaa Altjipa Ngambakala, which can be gathered from the notes, is an aboriginal translation of the Hermannsburg missionaries "Lord God Eternal." The mission station in Hermannsburg was established already in 1877 (Gill 1998, 220 n.25). This leaves about 50 years for the Arrernte to develop their thoughts on or facilitate syncretism with the Christian god before Spencer conducted his inquiry (cf. Gill 1998, 115–116). Thus the story may indeed be, as Gill claims, concocted for Spencer, but it is still based, in all other aspects than Numbakulla, on Arrernte concepts and interaction with their environment. Indeed, you could say that Numbakulla reflects the white man's god who is a recent addition to their environment. It may have roots in Arrernte mythology. Numbakulla was most likely ungambikula. This was an adjective used of the lizard ancestors who had the specific property that they were not created (Gill 1998, 13). This must originally have facilitated the syncretism with the Christian god. Behind the Numbakulla myth thus lies a special group of ancestors.

This passage has some descriptive elements, that is, what is the case: the existence of the ancestors as a category (since Numbakulla is a kin of the ancestors and the person he is talking to is the totemic ancestor, Achilpa), there are poles and there was and is a ceremonial ground and a ceremony called Engwura.[7] It also has directive elements, that is, what should

7. Spencer and Gillen observed the Engwura ceremony in Alice Springs in 1896. It is supposed to mark the end of male initiations, but Gillen had arranged for it to be performed (Gill 1998, 131). So their recordings may diverge from the actual performances. At the ceremony, the kauaua is painted with blood and decorated with eagle feathers (Gill 1998, 131). Poles were used in other initiation rituals, which is the case of the Arachita poles, which resemble descriptions given of the pole in the Engwura. They were poles also decorated with feathers, like the kauaua-auwa (Cf. Spencer and Gillen 1899,

be the case: a pole *should* be used, and Engwura *should* be held, and there *should* be a ceremonial ground. This double nature of descriptive and directive in the narrative can be compared to when adults tell their children: "we do not eat with our fingers" (Millikan's example). This does not mean that it always will be like this, or that everybody agrees, or even that it is necessarily important. The key issue is that the narrative is not *just* etiological, describing a place; or to put it another way: it is not a tale of wandering and wondering, or a myth of primordial times reflecting a universal morphology of the sacred, as was the interpretation of Smith and Eliade. There are some different types of information in the narrative: social information: the existence of ancestors (Numbakulla), artefacts: the pole, event: Engwura, places: Lamburkna/camp/ceremonial ground. Now let's look at the second fragment:[8]

| Spencer and Gillen's The Arunta (388) | Gillen's field-notes April-May 1897 |
|---|---|
| At a place called Okinyumpa an accident befell them which made them all feel very sad; as they were pulling up the *Kauaua*, which was very deeply implanted, the old *Oknirrabata*, who was leading them, broke it off just above the ground | and went on to Okinyumpa made quabara undattha and when the old man was trying to pull up the Kauaua which was planted deeply it broke off close to the ground and they all felt very sorry |
| and to the present day a tall stone standing up above the ground at this spot represents the broken and still implanted end of the pole | a tall stone now sticking up at the spot represents the broken Kauaua |
| Carrying the broken Kauaua they came to Unjiacherta, which means "the place of Unjiamba men" and lies near to the Hanson Creek. They arrived here utterly tired out, and found a number of Unjiamba men and women of all classes. | thence on carrying the broken Kauaua to Unjiacherta (which means the place of the Unjiamba men) near the Hanson (and about west of Sterling Creek Station) here the travellers arrived utterly knocked up and found a large mob of Unjiamba men and women of all classes |
| They were too tired and sad to paint themselves; | the travellers were too tired and sad to paint themselves, |
| their Kauaua in its broken state was inferior to many of those which the Unjiamba people had, So they did not erect it, but, lying down together, died where they lay. | their Kauaua in its broken condition was inferior to many in the possession of the unjiamba people So they did not erect it, but, lying down died where they lay |
| A large hill, covered with big stones, arose to mark the spot. | a large hill covered with big stones sprung up and marks the spot (where they died) |

Gillen was a telegraph operator in Alice Springs. He had been recording stories from the Arrernte since 1890. So at the time of recording this story he had some degree of command of Arrernte language and experience of their culture. Here we see that the correspondence between the notes of Gillen and the published results are almost verbatim. There is no reason in this case to seriously question the authenticity of the fragments under consideration.

This passage has descriptive elements: the two places, an accident with the pole and the death of the group. It also describes a headman of the group, who leads the rest. Again it also has directive elements: poles *should not* be broken, death *will* follow if it happens.

196–202). So, while the actual Engwura observed by Spencer and Gillen in 1896 may not be in all detail the normal Engwura ceremony, there is no reason not to suppose that there was such a ceremony involving a kauaua pole in a male initiation rite, and that poles in general were widely used in initiation rites among the Arrernte.

8. Based on transcripts by Sam Gill taken from http://www.us.oup.com/us/companion.websites/0195115880/wanderings/?view=usa (September 1, 2006).

Groups *should* have leaders who *should*, or who *will* from time to time, make decisions.

Again we have a few elements: places: Okinyumpa and the hill, events: breaking of the pole and dying, social information: *Oknirrabata* and the existence of groups.

Thus in our two fragments we have a list of salient elements:

| | |
|---|---|
| Places | Ceremonial ground |
| | Okinyumpa |
| | Hill |
| Events | Numbakulla "invents" pole |
| | Engwura (and other rituals involving poles) |
| | Breaking of pole |
| | Dying |
| Artefacts | Pole (*Kauaua*) |
| Social Categories | |
| Ancestors | *Oknirrabata* (leader) |
| Groups | |

For rats and robots, only the first category could possibly be a landmark for navigation involving Pushmi-Pullyu Representations. For humans, however, the last three categories are also landmarks in navigation through the environment. This is because, as we saw, humans can use language to integrate knowledge from different domains in problem-solving by juxtaposing them in a linguistic format. Narratives are such juxtapositions of information from these different domains of knowledge. This might be why myths frequently appear quite weird and counterintuitive. What is important in this connection, however, is also that it is not just passive information: there is a direct link between the representation and action that follows from this. I believe that this is why people bother to tell myths in the first place: they assist *homo navigans* in navigating his physical, psychological, and social environment in order to obtain a better life.

I would like to suggest a list of points concerning how this is achieved through the myth of Numbakulla. It is highly tentative and consists of loose ideas that might prove useful in further research. First let us take the obvious example: the groups travel from Okinyumpa to the hill, this is juxtaposed by the breaking of the pole and death—thus if you want to navigate away from death, do not break the pole. It is part of a larger pattern, though:

| | | |
|---|---|---|
| Numbakulla[9] "invents" pole | Breaking of pole | Dying |
| Pole-ritual (Engwura)[10] | No ritual | Death |
| The primordial camp—Lamburkna | Okinyumpa | Hill |
| Primordial ancestor | Oknirrabata | Group |

The first three rows describe a temporal relation from left to right, the last describes one of hierarchy. The columns describe a relation of contiguity, not causality. It could be phrased as things that go loosely together or mental landmarks in close proximity. The first two rows are sequences of events or narratives, the third is topography, and the last is concerned with social categories. Through the myth, these diverse mental landmarks can be

9. It may be more correct to say the ancestors.

10. I believe this to be the origin of the generic pole ritual, as there are different kinds of these. This renders it comparable with the last episode.

perceived as having a relation to each other. This juxtaposition in narrative format makes it possible to perceive the relation between breaking the pole, dying, ancestors and group.

It should now be quite clear what Eliade and Smith missed in their treatment of the sacred in narrative. They only focused on the descriptive elements and did not adequately see the complicated structure of the Arrernte myth. It is not just simply a pattern, *axis mundi*, from a universal morphology of the sacred, or a simple etiological description of a memorial to an event, it is an intricate network of landmarks, physical, psychological and social, that serve not as a map, but as a tool for navigating the environment (cf. Herman 2003).

The overall goal is, as is any other proper function in Millikan's framework, to provide for the continuing function of the function, that is, survival. The attractiveness of Millikan's framework is that proper functions might malfunction. They even might do so most of the time. This exempts us from taking the uncomfortable task of arguing for the adaptive value of religion, which has been convincingly refuted (Atran 2002). Religion, and therefore also religious narrative, might very well be the product of the "malfunction" of perfectly fine proper functions, which is what contemporary cognitive science of religion argues (Boyer 2001).

Thus we find that myths are products of functions that ensure that individuals can navigate properly in their environment, be it physical, social or mental (if there is any difference).

## A Surprise…

At the end of this mystery of the relations of space, narrative and religion we find the revelation that the age-old question of whether ritual reflects myth or whether myth is an explanation of ritual is ill-posed from the beginning. The answer is neither/nor. Instead there is a deeper relation between description and direction that makes the problem disappear, since both are from the start integrated, as a function of human/animal functioning in interaction with the environment. For different species the environment is made up of different things, this is what Jakob von Uexkül aimed at with the idea of *Umwelt* (Uexküll 1992). These things are the only ones it can perceive and directly relate to action in this *Umwelt*. The *Umwelt* of an amoeba, for example, consists of chemical substances. The *Umwelt* of humans consists of caves, rocks resembling broken poles, hills, ancestors, leaders and groups as well as initiation rites. Drawing on James J. Gibson, we could say that mythical narratives might be an instrument of direct perception of the environment (Gibson 1979),[11] designed to ascertain its affordances needed for navigating so as to stay alive to navigate another day, or to put it another way: to get "pole position" in the race for survival.

## References

Atran, Scott. 2002. *In Gods We Trust: The Evolutionary Landscape of Religion*. Oxford: Oxford University Press.

Boyer, Pascal. 2001. *Religion Explained: The Evolutionary Origins of Religious Thought*. London: Heinemann.

Carruthers, Peter. 2002. "The Cognitive Functions of Language." *Behavioral and Brain Sciences* 25: 657–726.

11. Millikan has said the same of language (Millikan 2002).

Cheng, K. 1996. "A Purely Geometric Module in the Rat's Spatial Representation." *Cognition* 23: 149–178.
Clark, Andy. 1997. *Being There: Putting Brain, Body, and World Together Again.* Cambridge, MA: MIT Press.
Eliade, Mircea. 1957. *The Sacred and the Profane.* New York: Harcourt, Brace and World.
———. 1973. *Australian Religions.* Ithaca, NY: Cornell University Press.
Gallistel, C.R. 1990. *The Organization of Learning.* Cambridge, MA: MIT Press.
———. 2001. "Animal Navigation." In *The MIT Encyclopedia of the Cognitive Sciences,* edited by R.A. Wilson and F.C. Keil, 24–26. Cambridge, MA: MIT Press.
Gibson, James Jerome. 1979. *The Ecological Approach to Visual Perception.* Boston, MA: Houghton Mifflin.
Gill, Sam D. 1998. *Storytracking: Text, Stories, and Histories in Central Australia.* New York: Oxford University Press.
Herman, D. 2003. "Stories as a Tool for Thinking." In *Narrative Theory and the Cognitive Sciences,* edited by D. Herman, 163–192. Stanford, CA: CSLI Publications.
Hermer, Linda and Elisabeth Spelke. 1996. "Modularity and Development: The Case of Spatial Reorientation." *Cognition* 61: 195–232.
Hermer-Vazquez, L., A. Moffet and P. Munkholm. 2001. "Language, Space, and the Development of Cognitive Flexibility in Humans: the Case of Two Spatial Memory Tasks." *Cognition* 79: 263–299.
Hermer-Vazquez, L., E.S. Spelke and A.S. Katsnelson. 1995. "Sources of Flexibility in Human Cognition: Dual-Task Studies of Space and Language." *Cognitive Psychology* 39: 3–36.
Hutchins, E. 1995. *Cognition in the Wild.* Cambridge, MA: MIT Press.
Jensen, J.S. 2003. *The Study of Religion in a New Key: Theoretical and Philosophical Soundings in the Comparative and General Study of Religion.* Aarhus: Aarhus University Press.
Lewis, D. 1972. *We, the Navigators: The Ancient Art of Landfinding in the Pacific.* Honolulu: University Press of Hawaii.
Majid, A., M. Bowerman, S. Kita, D.B.M. Haun, and S.C. Levinson. 2004. "Can Language Restructure Cognition? The Case for Space." *Trends in Cognitive Sciences* 8: 108–114.
Millikan, Ruth G. 1984. *Language, Thought, and Other Biological Categories.* Cambridge, MA: MIT Press.
———. 1996. "Pushmi-Pullyu Representations." In *Mind and Morals: Essays on Cognitive Science and Ethics,* edited by M. Friedman, L. May and A. Clark, 145–161. Cambridge, MA: MIT Press.
———. 2006. "Styles of Rationality." In *Rational Animals,* edited by M. Nudds and S. Hurley, 117–126. Oxford: Oxford University Press.
O'Keefe, John and Lynn Nadel. 1978. *The Hippocampus as a Cognitive Map.* Oxford: Clarendon Press.
Parkin, David. 1991. *Sacred Void: Spatial Images of Work and Ritual among the Giriama of Kenya.* Cambridge: Cambridge University Press.
Ryan, M.L. 2003. "Cognitive Maps and the Construction of Narrative Space." In *Narrative Theory and the Cognitive Sciences,* edited by D. Herman, 214–242. Stanford: CSLI Publications.
Sharp, P.E. 2001. "Animal Navigation: Neural Networks." In *The MIT Encyclopedia of the Cognitive Sciences,* edited by R.A. Wilson and F.C. Keil, 26–28. Cambridge, MA: MIT Press.
Smith, J.Z. 1987. *To Take Place: Toward Theory in Ritual.* Chicago, IL: University of Chicago Press.
Spencer, B. and F.J. Gillen. 1899. *The Native Tribes of Central Australia.* London: Macmillan.

———. 1927. *The Arunta: A Study of a Stone Age People*. London: Macmillan.

Tolman, E. 1948. "Cognitive Maps in Rats and Men." *The Psychological Review* 55: 189–208.

Uexküll J.von. 1992. "A Stroll Through the Worlds of Animals and Men: A Picture Book of Invisible Worlds." *Semiotica* 89: 319–391.

Wang, R.X.F., L. Hermer, and E.S. Spelke. 1999. "Mechanisms of Reorientation and Object Localization by Children: A Comparison with Rats." *Behavioral Neuroscience* 113: 475–485.

# — 15 —

# Narrativity in View of a Theory of Syncretism

Anita Maria Leopold

Once in the 1980s I saw a *real* American Indian sitting in a café in the middle of Copenhagen. With much appreciation in her voice my friend, who had accompanied me, told me that he was the person in charge of the "Indian courses." At that time these "Indian courses"—with an origin in the religion of the Hopi Indians—were the new trend in the therapeutic market in Denmark. Everybody with a taste for New Age and the markets in alternative healing and spirituality knew about them. Even my newborn daughter's visiting nurse recommended the Indian courses as a good after-birth therapy. The interest in Denmark, in particular of the Indian courses and more generally of Hopi religion, can be traced back to the 1960s American hippie movement and its enthusiasm for exotic and alternative religions, including the Hopi.

In his book, *The Invention of Prophecy: Continuity and Meaning in Hopi Indian Religion* (1992), Armin W. Geertz analyzed the complex history of contact between the Hopi Indians and various White American groups. He was observant of the role the interpreter of religion and his audience play in the process of change and invention in religion. He described the invention of new syncretistic forms of Hopi prophecy, and how they were initiated by different political interests of the involved parties (315). Geertz also discussed a particular Hopi petroglyph, called "Prophecy Rock" near the village of Oraibi and the many interpretations of the petroglyph that have evolved depending on the various interpreters' beliefs and political strategies (303–306, 327–329). The special interest in the petroglyph emerged due to internal Hopi politics and outside involvement by White Americans, recently from New Age circles. There is an extraordinary interest in the Hopi mythical character called Pahaana, a kind of "Long Lost White Brother." He plays a central prophetic role in the Hopi Emergence Myth. According to Geertz, Hopi mythographers postulate a series of worlds, which all end in cataclysm. Just before entering the "fourth world," mankind met their tutelary deity Maasaw, who, before he withdrew from sight, taught them the rules of how they should live:

> ...until the end of the fourth world is heralded by the return from the east of their White Brother (who had parted ways with them in primordial times). This return would be signalled by a variety of signs.... Maasaw will return once again to punish and purify

> humanity and the earth as a prelude to a new "fifth" world era. The White Brother, called Pahaana, will bring a new way of life with him, and the Hopis were commanded to fulfil the impossible task of simultaneously clinging to their ancient ways and following the new ways of their White Brother until Maasaw's apocryphal return. (Geertz 1992, 3)

It is this latter part which has played an important role in the development of syncretistic interpretations of Hopi prophecies. It has stimulated the imagination of Christian and New Age individuals even in the belief that the long awaited White Brother literally refers to them (336–337). Geertz called this the "Pahaana syndrome," thereby indicating how Euro-Americans "more or less adopt Hopi ideology as their own, enacting, as it were, their own fantasies about the Long Lost White Brother" (332–333).

The interest of this essay is the source of syncretistic interpretation. We can observe the formation of an analogy, however false it may be, of the apocalyptic vision of the White Brother in Hopi prophecies and Christian apocalyptic visions in addition to Christian messianic expectations. Nevertheless, an analogy is not necessarily innate to the object or phenomenon in question; in most cases it is a narrative construction. In this case, it is constructed through syncretism, and the following examples shall illustrate syncretistic narrative inventions.

## Syncretism as a Conceptual Model

For some scholars in the study of religion, the notion of syncretism is considered to be a critical category because, they claim, it denotes a sign of "impure" religiosity in contrast to "pure" religiosity. Following their line of discussion, in effect, every religion is syncretistic. Why then use such a concept that predisposes the mixing of pure objects? They are, however, partly right. Historically speaking there are no pure religions—all religions are the result of encounters and amalgamations of peoples and cultures. However, religious as well as cultural concepts are cognitively embedded as pure objects. Christians believe that their heavenly Christ is a purely divine figure although Christ's heavenly form emerged more or less from a fusion of the Jewish messiahs and Hellenistic beliefs about the heavenly state of man before the Fall (Martin 1987, 3–15, 120–126). In Latin American contexts, the identification of Catholic saints with African Orixás (Apter 2004, 160–184) does not make them less qualified deities in the minds and practices of the believers. Furthermore, Christian saints are no less Christian just because some of them began as pagan deities. The list of such cases is lengthy. To clarify my argument, allow me to make a comparison to language. Importing foreign words into Danish does not make Danish impure, because the syntax and practice are still Danish. Syncretism in religion follows a similar pattern as in language and culture—there is a particular *interpreter* or *syntax* that places elements of foreign origin into the particular social or semiotic context of the receiving tradition.

However, without theoretical consideration, syncretism remains a problematic category. Therefore in order to redefine the category, we must discuss the different issues and discourses connected with the category. The research history of syncretism displays a complex variation on definitions and many disagreements of usage. Basically, the notion of syncretism concerns the transmission of religion in general. Students of religion usually describe the category as "a generalization about diverse elements incorporated into some target religion from external religious or secular source or sources" (Martin and Leopold 2004, 93). The causes, as well as the effects of the historical processes that lead to syncretistic formations vary a great deal, depending on the historical, cultural, political and social climates. Most syncretistic formations go unnoticed since they appear as the natural results of interaction. Some scholars call this an "unconscious" mode of syncretism. Sometimes the source of syncretism is referred to as "the finding of one's own in the other." Such cases are considered to be adaptations, amalgamations or assimilations similar to the parallelization of Isis and Demeter in the Hellenistic age that resulted in an identification of the two goddesses. Other syncretistic formations have emerged as a consequence of a dominant culture's forcing or "interpenetration" of a new religion onto the dominated culture to such a degree that the original religion either became absorbed into the new official religion or its tradition prohibited. In such cases, syncretism is a result of a resistance to power and a means of preserving indigenous gods disguised as gods of the dominant culture. In certain cases, this form has been called "creolization" with reference to the dominance of the Christian culture on Afro-Caribbean or Afro-American cultures. Innovations are not necessarily the result of intercultural contact. They may arise independently as part of discursive differences in competing factions of shared religious and cultural backgrounds. New religious movements or new myth-making often incorporate "borrowed" elements from other religious (or secular) sources as a way of legitimating new contemporary values, often disguised as old religious teachings, as a means of contesting orthodox values. The break between early Christianity and Judaism produced new myth-making, together with the many factions of Christian movements, such as the Gnostics, which gave rise to a rich and innovative mythology disguised as revelations of earlier teachings. The New Age invention of Hopi prophecy can be compared to this type of syncretistic mythology. There are, of course, many more "syncretisms" than those mentioned here (see Berner 2004, 295–317).

Using syncretism taxonomically seems relatively unproblematical. However, many students of religion have rightly criticized the category for being so vague and indefinite as to be of no use in classifying its object (Leopold and Jensen 2004, Part II, 14–85). The criticism of the category has drawn attention to the lack of precision at the fundamental levels of explanation (2–12). For that reason I have constructed an analytical model that takes into consideration the different levels of discussion found in the research history, which consists of three theoretical levels: the *social* level, the *semiotic* level and the *cognitive* level (376–387). Each analytical level takes into consideration that there are divergent difficulties attached to the category. The first level, *the social,* to which we may assign issues of power to syncretism, also concerns the different politics and modes of religion with regard to the encounter of religions and religious innovation notwithstanding the antonym "anti-syncretism." At the second level, *the semiotic*, we must bear in mind the hermeneutical and transformative impact that syncretistic interactions have on religion. This has to do with the mechanisms of change and innovation in religious meaning that many scholars refer to as the dynamics of syncretism. The third level, *the cognitive*, is the basic level underlying the two other levels because it explains the nature of conceptual blending and religious categorizing from the structure of human cognition that may also help explain the transformative functions of "syncretistic semiotics," as well as the various constraints against syncretism in religious systems (Leopold and Jensen 2004, 378–387, Leopold 2004c).

In this essay I will focus on narrative as the link between the three levels,[1] because nar-

1. In the Laboratory on Theories of Religion at the Department of the Study of Religion in Aarhus, we have been working with theories on narrativity and cognition, and it is our conviction that narrative plays a major constitutional role in human cognition and social life. See http://www.teo.au.dk/en/research/current/cognition

rative is crucial to the dynamics of syncretism. I shall also draw attention to the fact that there are discrepancies on all three levels. This is due to the fact that we generally believe that we are living in coherent communities and cultures and speaking coherent languages, despite the fact that culture, language and lineage consist of "patchworks" of very different elements. Members of a village or a community, or even a nation, may believe that they share some essence that is the foundation of a social group, whereas much of it is "just" a narrative construction rather than a genealogical fact (Boyer 2001, 252–53; Leopold and Jensen 2004, 3–4).

## The Crisis of Change

The problem of syncretism can be located between the frontiers of the mental activity of adapting to new ideas and ways of life and the social control of traditional ways of life. In narrative we have the means to defend, refute and negotiate changes of "mind" and "lifestyle" in a society. Consequently, syncretistic transmission favours narratives. Religious "syncretism," however, and "anti-syncretism" for that matter, both define the social or religious constraints in society. The ability to adapt to new ways of life is basic to survival, being innate to human cognition as well as intrinsic to our social lives. We can therefore compare the evolution of cultures to a process of "trial and error" that from a multitude of possibilities seeks out the best workable solutions. As such, creative thinking, as part of the ability to adapt to new ways of life, is a dynamic factor in working through crises resulting from inter-religious or cultural contact. This also explains the patchwork effect of religious as well as cultural forms. The survival of religion depends on its ability to function as both a model for and of a community. Therefore every religion, even the most dogmatic ones, adapt to times and circumstances. Significantly, when a religious tradition takes an anti-syncretistic stand, it does so to maintain stability in the midst of changing circumstances.

No doubt the major issue of religious syncretism is the dilemma of the need to adapt to change and the need to maintain tradition. This dilemma comes into view at different levels of religious life, depending on how a religion administrates its power and politics in a community. Social control is more often than not embedded both in religious narratives as well as in ritual forms. Therefore, innovations will provoke some kind of authoritarian mechanism to deal with individual reactions. Defectors of a tradition are known to transform cosmological narratives to correspond to a new set of social order. For example the early Christian church recruited believers through a new mythology corresponding to a new set of social rules. Syncretism can therefore be studied in view of the dilemma of change according to the psychological and cognitive responses to change within a particular religious group.

## "Misrepresentation": The Dynamics of Syncretistic Invention

Looking at interactions between the Hopi Indians and New Agers, a web of various political interests and emotional aspects appear to be interwoven. This web, however, gave rise to new narratives and new meanings. Geertz claims that Euro-American interest groups have consistently misinterpreted and misrepresented Hopi religion primarily due to the influence of Frank Waters' *Book of the Hopi*—a book that initiated the hippie invasion of Hopiland (Geertz 1992, 333). He maintains that these groups used the Hopis "as an excuse for something else, as a focus for a movement bound up in its own voraciously apocalyptic vanities" (335). Furthermore, the flow of syncretistic interpretations of Hopi prophecy also came from some Hopis themselves. Thomas Banancya, for instance, entered the New Age market in order to gain support for the "Traditionalist Movement" in its battle against progressive elements in the Hopi community. His status in his home community, however, was marginal (306–307, 318–324). Banancya introduced a new set of interpretations of the prophecies that were collated by the Traditionalist Movement. Thus, for instance, the Hopi tutelary deity, Maasaw, became equated with the Biblical Messiah or the Great Spirit, neither of whom he is (318). In rephrasing Hopi tradition in more ecological terms, the Traditionalist Movement placed itself on a new and more universal spiritual path in terms of which they claimed to be appointed as care-takers of the North American Continent (318). The authority for this move was drawn from a number of drawings of the Prophecy Rock petroglyph, whereby the Hopis were given a vital role to play in the coming universal redemption that they believe will lead to a new world: "a paradisic life-everlasting, where the earth will be renewed and all people will speak one tongue and believe in one religion" (319). Geertz calls this type of syncretism a mutual "misrepresentation" (357). Power is also an inescapable effect of this type of syncretistic relationship (Leopold and Jensen 2004, 142–255). There are, furthermore, ethical problems involved in "borrowing" from or "shopping" with peoples' most sacred narratives.

## The Role of Power in Syncretism

Power and syncretism are related to issues of cross-cultural communication in general. Linguist Rukmini Bhaya Nair asks: to whom do stories belong? She recognizes the fundamental problem of ownership of stories, considering the changeable nature of narrative, and acknowledges the emotional impact it has on people:

> How is individual ownership of a story established and maintained, since it is necessarily a shared resource …? When stories are passed on as texts within, as well as across, communities, does not the moral issue of plagiarism, the case of "pretended" authorship, then inevitably arise? Are stories "fragile" in the same way that "selves" are fragile, subject to the same processes of mortality, and the same sorts of morality? Who or what are the subjects of narrative? (Nair 2002, 17)

People are generally protective of their narratives. Religious narratives, in particular, because they are considered to be sacred by their believers. As a consequence, "misrepresentation" can become a matter of sacrilege. Several Christian denominations practice a very strict control of their belief system. Unfortunately this has led them to resort to violent oppression of peoples with different or individual beliefs (Leopold 2004a, Whitehouse 2000 and 2002). According to Geertz, the role of symbolization has far more impact in social reality than, for example, military might (Geertz 1992, 373). Timothy Light reminds us that it is "symbols to which loyalty is sworn and which are used to define religious groups. It is symbols around which we unify and symbols over which we fight" (Light 2000, Leopold and Jensen 2004, 326). However, a symbol may well denote a kind of universality with the ability to prompt meaning across religious borders. Biologist and semiotician Terrence Deacon argues that learning a new symbol does not necessarily include the learning of the whole set of cultural indices that are otherwise included in the symbol. He claims that a symbol can easily be projected onto a whole new set of indices—even when stemming from a different culture (Deacon 1997, 95). Thus, when a symbol from one religious context is used in a new religious context, its success depends of how well it translates not only in meaning but also how new users respond to it emotionally. A symbolic "crossover" such as the "White Brother" suggests an emotionally powerful symbol, because it proposes an analogy between the mythological "White Brother" and White Americans. No completely alien symbol can survive—each symbol that successfully crosses over to a new culture must resemble something in that culture. The "White Brother" symbol qualifies as a symbolic crossover, because it permits white people to project themselves onto Hopi narrative and participate in it. Yet, resemblance is, according to Deacon, interpretive. He suggests that drawing an analogy is cognitively a process of *recognizing* a similarity (71). This he describes as being similar to when a bird mistakes a moth on a tree whose wings resemble the colour of the bark to just be more bark (75). He therefore claims that the interpretive step in making an analogy is negative—namely, the act of *not making* a distinction (Deacon 1997, 74). Part of a successful "syncretism" involves seeing resemblances where there were none before and to ignore differences in order for it to become "recognizable" in the minds of people. This may happen through a process of interpretation primarily on a discursive level.

## The Cross Field of Narrative

Banancya knew the symbolic features of both Hopi and White American worlds. This knowledge helped him to seek legitimization of his beliefs in appealing to New Age rhetoric and symbolization. As such, different groups could share their opinions in a new discourse that reworked the different religious paradigms to make them fit into a new "meta-paradigm" shared by both groups such as "ecology and mysticism" (Geertz 1992, 311–313). A published drawing of the petroglyph appeared in 1972 in an environmental monthly based in San Francisco. It called attention to the symbols carved on the rock as representing a warning against the devastation of the Earth because of mankind's interference (Anonymous 1972, 34 in Geertz 1992, 311).

As discourses meet, new interpretations of Hopi prophecies are invented, and, furthermore, change also occurs on the social level because people participate emotionally in narratives. The rhetoric and religious symbolization used in the discourses give Hopis and New Agers alike the option of constructing new identities by placing themselves within the narrative as if they were acting out the story. According to Nair, narrativity is fundamental in human cultures "because it underwrites our collective illusions of being unified selves with reliable social identities" (Nair 2002, 212). When traditional stability changes, fault lines may appear in people's self esteem as their tradition comes under challenge. The need for a new unifying symbolization may therefore arise to help maintain meaningful relationships with themselves, their traditions and the intruding other.

## The "Trespassing" of Narrative

Syncretism effectuates cognitive changes in such a way that it helps people form new identities. The opposite, anti-syncretism, is a means of protecting traditional ways from change and intrusive forces. Wherever there is syncretism, there is also some degree of anti-syncretism. Timothy Light argued that "blending" or "melding" first happens in peoples' minds, which is partly a psychological problem in group-identity and partly a phenomenon that emerges quite naturally from the ways we acquire religion (Light 2004, 318–324).

Cognitive linguists Gilles Fauconnier and Mark Turner claim that "blending" is a constant mental activity of the human mind. No matter how or what we blend, the mind uses the same skeletal mapping schemes and combines them in the same simple ways. "Meanings themselves are the imaginative products of blending, whether simple or complex, and are not predictable from the forms used to evoke them. The mapping schemes, by contrast, *are* predictable from the language forms used to evoke them" (Fauconnier and Turner 2002, 147). Like Light, Fauconnier and Turner underscore that the mind reformulates information rather than just replicating it. This may explain why there is the need of social control in most religions to constrain the individual from too much imaginative thinking that goes against authoritative interpretations (Leopold 2004a). In religion in general, a new and unexpected situation sets off precautionary rules such as a ritual of crisis to deal with it. Subsequently, we cannot escape the classical topic of *syncretism versus anti-syncretism* (Stewart and Shaw 1994). Trespassing others' sacred narratives can be as severe as trespassing a forbidden sacred area, or a foreign nation without a passport. Light claims that all religions have categories that are more exclusive than others (Light 2004, 338–341). Categories that are most sacred are fundamentally important to their respective cultural or religious systems. Therefore we will find resistance to innovation only in some areas of a religion, whereas in other areas, syncretistic elements will be accepted or go unnoticed.

## The Emotional Impact of Narrativity

Narrativity is meaningful at a fundamental level of cognition because it is capable of reforming people's feelings about other ethnic and religious groups. Daniel J. Siegel, a cognitive neuroscientist, sees emotion as integral to the process of interpersonal communication (Siegel 2001, 80). He claims that there is a convergence between social processing and the creation of meaning in emotional processing and cognition (81). We must therefore expect that of empathy plays a part in how we mirror ourselves in foreign peoples and their religions, such as in the case of the Hopis and New Age groups. Nair acknowledges that "Narrative creates emotional resonances through *purely linguistic means*" (Nair 2002, 11). In effect, she claims that meaningfulness emerges in the space of interpretation between teller and listener. This may explain the discursive effects of the "teller" and the "listener" and may be equalised in syncretistic discourses to the "narrator" and "listener" relationship (Leopold 2001). She further argues that story-telling permits a re-formation of experience that allows us to reform or to review ourselves when we share our stories (Nair 2002, 11). In my opinion, religious narrative may actually provoke a certain type of experience, such as to experience oneself as the "White Brother" of Hopi mythology. In the next section, I will discuss the impact of syncretistic narrativity on "identity" and "self."

## Rabbi Yonassan Gershom

Rabbi Yonassan Gershom is an American Jew, who describes himself as "neo-Hasidic." He has no affiliation in any branch of Judaism, nor does he have a regular congregation. In his book *Beyond the Ashes: Cases of Reincarnation from the Holocaust* (1992), he described the "neo-Hasidic" path as a combination of traditional Hasidism and new ideas from fields such as psychology, sociology and modern metaphysical philosophies. Most importantly, he believes in *reincarnation* (xv–xvi). The main purpose of the book is to provide hypnotherapists, parapsychologists and past-life readers with some theological and cultural contexts in order to help individuals with past-life experience of the Holocaust and to provide the general public with an overview of Jewish teachings on reincarnation (xix).

In many ways this case study is similar to the Hopi situation. It involves a member of a minority, who seeks to legitimate his beliefs through New Age. New Age is useful for that purpose precisely because it is eclectic and pro-syncretistic. This open-mindedness, nevertheless has a drawback: one must adhere to the New Age worldview to become accepted. The password is *reincarnation*. Historian of religions Ingvild Gilhus pointed out that the belief in reincarnation is a key symbol in New Age, and as such it serves many functions in today's western religiosity (Gilhus 1998, 157–158). Gershom knows how to appeal to New Age-inspired audiences through Jewish mysticism (Gershom 1992, 3–7). He described how in the beginning he was met with resistance by both the Jewish community and "alternative spirituality" because no one wanted to deal with the Holocaust (7). But he finally found a place within the New Age-inspired Association for Research and Enlightenment (A.R.E.) which is the international headquarters for the work of Edgar Cayce (1877–1945).

Gershom's article "Are Holocaust Victims Returning?", published in the A.R.E.-related magazine *Venture Inward* (1987), got him started. People contacted him because of the article, especially non-Jews, who claimed they had Holocaust reincarnation memories. He described how people started to call him. Some hung up two or three times before they worked up the courage to talk to him, while others again tested him out before explaining why they had called. He uses these kinds of incidents and many more as "proofs" of authenticity. His material is taken from 250 people who believe they were reincarnated from the Holocaust. Furthermore, he has the written records of 78 people: 56 women and 22 men, 16 were born Jewish, 7 converted to Judaism, and 55 were not Jewish "in this life" (Gershom 1992, xx).

## Narrativity and Experience

The subject under examination here is how does this idea of being a victim from a past life in the Holocaust get shaped in the minds of people? Some of the persons that Gershom mentions had established already known narratives about reincarnation and the Holocaust, others were invented through Gershom's communications with them. A further question is how beliefs become *memory*. To what extent do we literally *change our minds* when we invent new narratives about ourselves? Louis Cozolino, who is a psychologist with a clinical practice, presented neurological evidence indicating that neurons both change and grow in response to new experiences and learning (Cozolino 2002, 21). He also argued that storytelling plays a crucial role in handling new experiences and learning processes in general. Cozolino sees language as a key integrative mechanism in neural processing:

> Storytelling weaves together sensations, feelings, thoughts, and actions in ways that organize both one's internal and external worlds. These autobiographical memories are at the core of our sense of history, our conscious connections with others, and our sense of self in interpersonal and physical space. (Cozolino 2002, 34)

Narratives allow us to actively place ourselves within a story. Cozolino claims that through stories, "we can escape our bodies and the present moment in imagination to create other possible selves, ways of being, and worlds that have yet to be created" (35). Dominick LaCapra, scholar of intellectual history, focuses on the interaction of history, memory and ethico-political concerns in his work on the aftermath of the Holocaust (LaCapra 1998, 2). He has re-thought Freudian psychoanalysis on working through neuroses connected with mourning and the ability to remember the traumatic event of the Holocaust (LaCapra 1998, 6). He argued that it has become one of post-Nazi genocide's most sublime narratives of victimization, carrying religious overtones (37–38). As a result we see in the aftermath of the Holocaust a need of mourning, of "acting-out" a mimetic relation to the past which is regenerated or relived as if it were fully present:

> ...acting out may be a necessary condition of working-through, at least for victims and in certain ways, for all those directly involved in events. Even the secondary witness or empathetic observer who resists full identification with, and the dubious appropriation of the status of, victim may nonetheless undergo muted trauma. (LaCapra 1998, 45)

The symbolic power of the Holocaust fits into the Western apocalyptic paradigm, even in secularised form through filmmaking and art. Those born and raised in post Second World War societies all share this paradigm. Moreover, LaCapra suggests that there is a tendency to sacralize the traumatic experience of victimization, especially among American Jews (14). Interestingly, Gershom assumes that membership in the post-war "baby-boom-generation" is a sign of rebirth from the Holocaust (Gershom 1992, 44–45). His therapeutic counselling is based on the Holocaust as a "mythicized" history of evil understood as an eschatological sign of the wake of the "new age" (Gilhus 1998, 152).

Nair claims that a "good" story "is one where the listener is so completely hooked that he completely identifies with the imaginary characters and imaginary worlds that the teller conjures up" (Nair 2002, 247). Cozolino confirms Nair's claim that the process of listening to and telling stories brings together behaviour, affect, sensation, and conscious awareness in a way that maximizes the integration of a wide variety of neural networks. "Through stories, we connect with others...and provide the opportunity for moral lessons, catharsis, and self-reflection" (Cozolino 2002, 35). In the case of Gershom and his clients, the connection corresponds to such a "teller-listener" relationship. Gershom as therapist interferes with his clients' narratives, and vice versa. The clients' belief that they are reincarnated from the Holocaust reinforces Gershom's own beliefs. As such, the reincarnation of Holocaust victims is made probable through a verifying discourse of a "teller-listener" relationship.

## The Cases of Proof

A recurrent theme among his non-Jewish clients is that they have experiences, which are meaningless to them, but which Gershom recognizes as descending from Jewish customs. One of his callers told him how his young son refused to drink milk if there was meat on the table (Gershom 1992, 12).[2] Another theme is the dreams and visions that people have of being victims of the Holocaust (25–26). An example is Nancy's story. She was an American college student who, according to Gershom, was thinking about converting to Judaism, and on the invitation of a guest rabbi at the college, she decided to attend her first synagogue service in nearby Philadelphia. The worship was in Hebrew, and although she did not understand a word of it, it seemed strangely familiar to her. When the congregation reached the "Shema" prayer, she began to choke and cough. Unable to get her breath, she kept repeating *Adonai Echad* ("the Lord is One") from the first line of the prayer. "'Oh, what a nightmare that was!' she gasped and described a vision of dying in the gas cham-

2. Jewish dietary law doesn't allow milk with meat.

ber.... [H]ow could she have known that those same words are spoken by Jews on their deathbed, words that were doubtlessly gasped out by millions of Jews with their dying breaths?" (29–30).

Gershom interpreted physical signs as proof of past-life experiences, *per se*. For instance Cynthia, another non-Jew, also "for no apparent reason" developed asthma at the age of twenty. She contacted Gershom because she believed that her asthma was connected with her death in her past life as a Jew in the Holocaust. She believed that the reason for her asthma in this life was because she was buried alive in her past life. She claimed she had had a vision with dirt on her face in a pit with dead bodies. And upon looking up, she saw men with guns in hats and coats and this person (one from this life) on a rack or tree being tortured (33–34). Barbara, another client, who had suffered from chronic sinusitis and bronchitis all her life, wrote to him because she too believed that she had had a past-life experience. Having descended from a family of Gypsies on her grandmother's side, she believed that she had been a Gypsy during the Holocaust and was killed by the Nazis. As with Cynthia, Barbara also had a vision of being buried alive. She felt she knew things from childhood like the colour of German uniforms even though there were no colour TV's in her childhood (34). Gershom described this story as intriguing because there appears to be a tree in Barbara's story which is similar to Cynthia's story, which, according to Gershom, may confirm that they were "tuned in" on the same event (35).

Rabbi Gershom's somewhat circular reasoning becomes a narrative verification of the past lives from the Holocaust (237–396). LaCapra describes in his analysis of Claude Lanzmann's film *Shoah* (LaCapra 1998, 95–138), how Lanzmann's presentation of the Holocaust in the film is a quest for a "full presence" of the past that represents, on a religious level, the idea of the full incarnation of divinity in the world (104). We can compare Gershom's quest to heal his clients with the way the Holocaust is staged in Lanzmann's film. It is an "acting-out" that works because of its "full presence" of the past and its religious overtones. This form of acting-out, staging victimization and healing at the same time, is necessary to develop the correct emotional sensation of empathy between filmmaker and his audience, or in this case between healer and client. As such, Gershom is able to cross over religious boundaries because the scenario that Gershom and his clients create in common is very similar in its narrative compositions to a film-script.

The dreams and visions give the reader the feeling of "full presence" as if they were documentaries. But for that to happen, there must be some shared symbolic features to begin with. First of all, his clients already believed in reincarnation prior to their contact with him. They were likely to have been in contact with a New Age lifestyle in some way or another, since it was his article in a New Age Magazine that prompted the many callers in the first place. Secondly, his clients also seem to have been motivated *a priori* to believe that they had been victimized in a past life during the Holocaust. "Nancy" wanted to convert to Judaism, "Cynthia" had for several years believed that she was a Jew in the Holocaust, and "Barbara," partly descended from Gypsies, "knew" she had died as a Gypsy at the hand of the Nazis. Some sort of *empathy* towards victims of the Holocaust was already in their minds for different reasons.[3] Most of them suffered from chronic illnesses, which they claimed were caused by past-life experiences. Gershom gave them the opportunity to verify their own beliefs. He gave authority to their stories, and vice versa.

3. I find reason to believe that most of the non-Jewish people who sought contact with Gershom were motivated not by the fact that they were attracted to Judaism, rather, my guess is, by the mechanisms of victimization.

Gershom claimed that in believing the stories, it works like a "kind of catharsis" (Gershom 1992, 11, 22). Nair claimed that the catharsis is effectuated through the mutual acceptance between teller and listener, "evaluative responses to a story" that is formed between teller and listener (Nair 2002, 199), i.e. what she also described as the "tacit acceptance of the notion of multiple authorship" (22). That is also the very nature of syncretistic narrative. Both Nair and Cozolino stressed the significance of narratives providing the illusionary sense of "self" by pointing to narratives as vehicles of defining both social and private selves. As such, a new story can serve as a blueprint for new behaviour (Nair 2002, 205; Cozolino 2002, 166). The combination of the Holocaust and reincarnation beliefs is an effective narrative that Gershom and his clients mutually constructed in order to confirm/invent new selves and to heal traumas.

Can narrative *really* be so effective as to give the listener a sensation of "being present" in the same vivid way that Gershom and his clients experienced it? Nair claimed that a good story is one where the listener is so completely hooked that he completely identifies with the imaginary characters. Additionally, Cozolino described how our "malleability of memory" plays a role in how we compose memory. "A therapist's belief that her client has been abused may influence that patient to unconsciously fabricate a memory that they both then come to believe as true. This process is a clear demonstration of both the malleability of memory and the power of co-constructed narrative in shaping experience" (Cozolino 2002, 100). This suggests that memory is not just an innate cognitive matter, but that we make use of public symbolic memory systems, much like LaCapra described as memory that conflates with myth or ideology (LaCapra 1998, 17). He stated that memory is a crucial source for history which has complicated relations to documentary sources: "Even in its falsifications, repressions, displacements, and denials, memory may nonetheless be informative—not in terms of an accurate empirical representation of its object but in terms of that object's often anxiety-ridden reception and assimilation by both participants in events and those born later" (19).

This indicates that a symbolic representation which is used to prompt a memory allows a much more emotional engagement in historical events. Thus *teller* and *listener*, in Gershom's counselling sessions reciprocally verify their dreams and visions to be documentary and experienced memory because of the signs, the asthma, bronchitis and so on, become symbolic representations of memory from the Holocaust. Memory becomes the object of interpretation, very much like Hopi prophecy. It confirms analogous encoding as a phenomenon, similar to Deacon's description of iconic relationships, an interpretation of *failing to make a distinction* of things (Deacon 1997, 74).

## Souls with Memory

Basic to Gershom's belief in reincarnation is that we have souls with memory. He is, however, well aware of the lack of a scientific explanation of what a soul is, let alone how souls could have memories. For this purpose, he turned to both Jewish mysticism and new discoveries from cognitive studies to find confirmation of his beliefs (Gershom 1992, 178–179). Thus he combined Jewish mysticism, modern psychology, and his own personal observations on the Kabbalistic teaching of the soul's five levels: (1) *nefesh* becomes the biological life force of the body; (2) *ruach* becomes the lower emotional spirit or "ego"; (3) *neshamah* becomes the individual higher consciousness; (4) *chayah* becomes the collective unconscious of the group; (5) *yechidah* becomes the level of unity with creation and God (179). According to

Gershom, it is the level of *neshamah* that survives after death. He claimed that according to some kabbalists, the level of *neshamah* must be consciously developed, and links this with "recent discoveries about child psychology" (182). He confessed to integrating two systems of thought, kabbalah and psychology, in order to know more about what the earlier kabbalistic mystics were trying to say. His intention in interpreting kabbalistic mysticism in light of modern psychology was to make mysticism more relevant to modern readers. As such, he claimed that the brain is analogous to a very sophisticated computer which "is 'unprogrammed' at birth but which utilizes its childhood experiences to develop the 'software' which eventually enables the higher levels of the soul to manifest themselves on the earth plane" (184). This is the *neshamah* level that is also the level of the "eternal soul" which survives death.

His theory of soul and mind is a syncretism of Hasidic beliefs and scientific theory. We may claim it to be a "blended space" of systems of thought. Turner argued that blended spaces "can develop emergent structure of their own and can project structure *back* to their input spaces. Input spaces can be not only *providers* of projections to the blend, but also *receivers* of projections back from the developed blend" (Turner 1996, 60; cf. Fauconnier and Turner 2000, 39–57). This means that the blend composes a shared structure that constructs a relationship between the parts that is essentially of a different nature than the parts. Such relationships throw light on the nature of syncretism as a particular kind of analogy. In the analogy of a moth on the tree that the bird mistakes to be just more bark, the moth actually attains some of the bark-like nature of the tree when in relationship with the tree. But the moth is just a moth when *not sitting on the tree*. Similarly with analogy: analogous coding is fundamentally a social means to unite meanings with different semantic origins. Fauconnier and Turner's theory of blending seen in relation to my focus on syncretism may show us how cognition interrelates at the social level of communication.

Gershom used modern psychology to confirm or even authorize the kabbalistic standpoint of the soul by neglecting the fundamental epistemological differences between kabbalistic beliefs and modern psychology. The standard "architecture" of how to create a "syncretism," I believe, is by *not recognizing* the incompatibility of the input spaces in order to form a new and meaningful relationship between them. The paradigmatic motif that controls the blend, in this case reincarnation beliefs, gives, so to speak, room for an interpretation that legitimizes the analogy because it forms a hermeneutical circle. This is how new meaning emerges that may also secure a new social relationship between individuals and groups that normally are socially and mentally apart. Turner suggested how the "soul" is a result of conceptual blending composed of "animacy and agency":

> We see a small spatial story in which an actor other than ourselves behaves in certain ways, and we project features of animacy and agency onto it from stories in which we are the actor.... Objects that are prototypical actors are perceived as able to move themselves and able to move other objects. If actors move objects, what moves the actors? What is the source of their movement? One answer that has come up historically is *the soul*. The soul is what moves the body. The body is the object the soul moves as a consequence of its own self-movement. (Turner 1996, 21)

Turner's description of the soul as a conceptual blending suggests that it is basic to human perceptions of reality. If he is right, the basic nature of syncretism must be cognitive as long as we understand syncretism in the perspective of conceptual blending. The next example illustrates how a syncretistic blending is constructed by way of an analogy of "long-term" and "short-term" memory and the Hasidic notion of *neshamah*:

> We might say that the entire physical brain is the "short-term memory" of the soul for one lifetime, while the *neshamah* and higher levels serve as the long-term memory from one life to another. Unnecessary details are usually forgotten. However, if there is "unfinished business" left over from one or more incarnations, then the soul may preserve some specific details about that life. This type of carry-over is frequently connected with a tragedy that would generate very strong emotions, such as being tortured during the Holocaust.
> (Gershom 1992, 185)

The blending of science and religion justifies how a soul can have memory. This relation only works at the expense of the epistemological differences of the input spaces. Turner claimed that narrative imagining is fundamentally a cognitive instrument for evaluating a problem of mind (Turner 1996, 20). In cases of remembering traumatic events, LaCapra stressed that there exists a complicated set of relations between traumatic events, memory, and imagination because the empirical content is extremely difficult for the mind to handle (LaCapra 1998, 180). In Holocaust cases, he stressed that the complexity of memory may be *exaggerated* because of the difficulty of remembering events that challenge the imagination (182). Because of this fragility of memory, it may well need narrative in order to fill in the blank spaces that can never fully escape the use of imagination. Even empathy is an imaginative component within the scope of surrogate victimage (182). Indeed narrative emphasizes empathic behaviour in order for us to participate in other people's victimage. That again confirms why "surrogate victimization" in stories of reincarnation may have a cathartic effect, which Gershom claimed his clients experience from simply being *believed*. "Healing" as a form of New Age therapy basically works as a "narrative" collaboration between healer and client to relieve emotional or physical obstructions in the client that she believes stem from earlier reincarnations (Gilhus 1998, 178). In the case of Gershom's clients, they all suffered from chronic illnesses that both Gershom and the clients believed stemmed from their victimization in the Holocaust. Research in neuroscience and psychotherapy seem to indicate how imaginative healing actually works. Cozolino argued that rewriting narratives about the "self" can effectively change the organization and nature of memory:

> As such, narratives are a valuable tool in the organization and integration of neural networks throughout the nervous system. Memories are distributed throughout the brain and body and are subject to modification. Because we can write and rewrite the stories of ourselves, new stories hold the potential for new ways of experiencing ourselves and our lives. In editing our narratives, we change the organization and nature of our memories and, hence, reorganize our brains as well as our minds. (Cozolino 2002, 103)

If we can manipulate memory to fit emotional states of mind as well as physical states of being, then memory may also be changed to accommodate our social needs of sharing more general stories, including world history. In the case of "souls with memory," it is a syncretism of different types of narrative, i.e. Gershom and his clients' reincarnation beliefs and the historical documentary of the Holocaust that basically is formed through the interaction of the social level of the healer-client relationship and the cognitive level of empathetic and emotional reaction to narrative. Memory, then, is not just a cognitive system but also a social construction. Siegel pointed out that emotion plays a central role in creating and regulating mental life and is communicated among individuals in a way that sheds light on how the mind develops and functions within the social context of human relationships (Siegel 2001, 82).

## Gershom's Soteriological Program

Both Gershom's neo-Hasidic programme and Banancya's Traditionalist Movement have adapted New Age inspired universalism. They both have a soteriological programme that appeals to people with New Age tastes. Their projects, however, are also attempts to form alliances between minority groups/persons and people from the majority culture in order to gain a more powerful and estimated position in society. The soteriological paradigm of New Age serves that cause well because of its "holistic" ideology that allows minority religions/persons to play an important and necessary role for the sake of universal redemption (Gilhus 1998, 29–30, 152–153). New Age seems to have a capacity for collecting a plurality of minority cultures, religions, mythologies and misfits in a narrative whole that promises them an elect position in the coming utopia on planet Earth. Gershom therefore found parallels between Jewish history and Native Americans in order to appeal to New Age (Gershom 1992, 241). In that quest, he even mentioned Hopi prophecies, and Banancya's trip to the United Nations building and the Pentagon in Washington (Gershom 1992, 247). Finally, the Holocaust is a sign of the "Messianic Age" that he translates into the Jungian-inspired concept of a "collective Messiah" (247, 250–251). His hope was to reinstate Jews in the spiritual evolution of the planet, wherein both Jews as well as Gentiles must participate (251). Gershom's neo-Hasidism is a syncretistic construct in order to reinstate Judaism as an important religion in the reign of New Age universalism. He called this soteriological path "the pluralistic path," wherein each culture may work within their specific paths as a "unity in diversity" to save the world (259–260).

The cases of Gershom and the Hopis represent some of the many minority voices in the United States that use New Age as a platform to deal with their invisibility in a pluralistic world. In both cases, syncretism was used as a stepping-stone from isolation to social acceptance. Gershom's project also intended to gain "universal" acceptance of Jewish victimization in the Holocaust. LaCapra claimed that those born later tend to sacralize it and make it their master narrative, which evokes the desire to redeem the self (LaCapra 1998, 177). In Gershom's story, the suffering and redemption of the Holocaust became a shared responsibility through contact with New Age circles. That Gershom and Banancya used the New Age as a vehicle to "go popular," points to a mechanism in syncretism that enhanced their cultural, religious and personal positions and thus facilitated survival. This mechanism seems to interact between the social and the cognitive levels of contact.

## Conclusion

Thus I come round to my initial argument that syncretism is basically a cognitive as well as social ability to adapt to new ways of life. Paradoxically, the phenomenon of anti-syncretism is in essence also a "survival kit" to withstand possible annihilation in the crisis of change (Leopold 2004b, 123–134). Both "syncretism" and "anti-syncretism" are very different ways of responding to the dilemmas posed by crises of change: through the adaptation of the new, on the one hand, and the preservation of the traditional, on the other. The evolution of cultures depends on both approaches. Nevertheless, in order to communicate with other people, we must always enter "blended spaces" regardless of incompatibilities. Fauconnier and Turner claim that "mental spaces can exist routinely alongside incompatible mental spaces" (Fauconnier and Turner 2002, 29). Luckily, war is not the first thing that comes to mind when people meet with divergent opinions. We use creative thinking, instead, to convert incompatibility to compatibility. Entrenched in creative thinking, we find the analogous coding that allows us to connect emotionally and empathetically with strangers. As such, syncretism is basically grounded in a cognitive means of compromise: analogous coding allows us to *not* recognize essential differences. This doesn't mean, however, that we end up in a mental "melting pot." This would be to oversimplify the evolution of thought, culture and religion. We should remember that it is precisely in particular symbolic relationships that analogies work. Fauconnier and Turner suggest that matching is not obvious and pre-constructed. It is driven by the intended blend, not by any obvious analogy (Fauconnier and Turner 2002, 29). To recognize an analogy, then, is always an interpretive process that, in Deacon's words, never interprets a symbol identically alike. "[T]here is something more," he claims, "some additional interpretants that are crucial to their symbolic capacity, and a great deal of additional learning has intervened to make their production possible" (Deacon 1997, 64). In other words, a symbol is never perceived in exactly the same way by two different people.

What sort of phenomenon is syncretism besides being a cognitive one? The drama of syncretism happens as well on the social level. Let me repeat Light's both wise and precise description of the problem of "melding" religion: "It is symbols to which loyalty is sworn and which are used to define religious groups. It is symbols around which we unify and symbols over which we fight" (Light 2000, Leopold and Jensen 2004, 326). We cannot ignore the issue of the misrepresentation of people's most sacred symbols. The moral issues of syncretism, such as Geertz implied, appear to be inescapable. It goes beyond the question of dogma and various modes of religiosity. Nair observed that narrative is an important instrument of self-protection; to protect the fragile self's claim to identity within society is to protect itself against the danger of mental death (Nair 2002, 18–19). Our needs to draw borderlines around our fragile selves are sorely tested, however, by the mind's flexibility in creating and re-creating self and society. Thus the borderline question in cases of peoples and cultures in contact is to find out how to balance this delicate form of communication and avoid the clash of cultures.

As such we find syncretism situated in between cultural imperialism and cultural appeasement. I have examined how syncretism interacts with narrativity from different cognitive angles. One of these angles is, the "malleability of memory" together with the "plasticity of memory" which are the main composers of narrative memory. Another aspect is the influence of "empathy" on interpersonal and intercultural communication. LaCapra's "acting-out" works because it represents "full presence" which is essential for a successful syncretistic story. Gershom heals his clients, just like a good filmmaker, by a ritualized and narrated presence of the Holocaust. Furthermore, blending is in my opinion innate to syncretistic architecture as a way to find unity behind diversity (Fauconnier and Turner 2002, 137). What makes a "syncretistic blending" different to blending in general, however, is that it figures as a general cross-cultural phenomenon. Therefore it has and will always have, I am afraid, the paradoxical imbalance of power. Whether we refer to the notion or the phenomenon, syncretism never remains neutral because it is fundamentally human. But that should not prevent us from studying this phenomenon even though the category of syncretism is controversial in the study of religion.

## References

Apter, Andrew. 2004. "Herskovits's heritage: Rethinking syncretism in the African Diaspora." In *Syncretism in the Study of Religion*, edited by Anita Maria Leopold and Jeppe Sinding Jensen, 160–184. London: Equinox.

Berner, Ulrich. 2004. "The Concept of 'Syncretism': An instrument of Historical Insight/Discovery?" In *Syncretism in the Study of Religion*, edited by Anita Maria Leopold and Jeppe Sinding Jensen, 295–317. London: Equinox.

Boyer Pascal. 2001. *Religion Explained: The Evolutionary Origins of Religious Thought*. New York: Basic Books.

Cozolino, Louis. 2002. *The Neuroscience of Psychotherapy: Building and Rebuilding the Human Brain*. New York: W.W. Norton and Company.

Deacon, Terrence. 1997. *The Symbolic Species: The Co-evolution of Language and the Human Brain*. London: Penguin Books.

Fauconnier, Gilles and Mark Turner. 2002. *The Way We Think. Conceptual Blending and the Mind's Hidden Complexities*. New York: Basic Books.

Geertz, Armin W. 1992. *The Invention of Prophecy: Continuity and Meaning in Hopi Indian Religion*. Knebel: Brunbakke Publications.

Gershom, Yonassan. 1992. *Beyond the Ashes: Cases of Reincarnation from the Holocaust*. Virginia Beach, VA: A.R.E. Press.

Gilhus, Ingvild. 1998. *Kulturens Refortrylling. Nyreligiøsitet i moderne samfund*. Oslo: Universitetsforlaget.

LaCapra, Dominick. 1998. *The History and Memory after Auschwitz*. Ithaca, NY: Cornell University Press.

Leopold, Anita Maria. 2001. "The Architecture of Syncretism: A Methodological Illustration of the Dynamics of Syncretism." *Historical Reflections / Réflexions Historiques* 27 (3): 401–424.

———. 2002. "Syncretism and Transformation in the Gospel of Truth." In *The Nag Hammadi Texts in the History of Religions*, edited by Søren Giversen, Tage Petersen and Jørgen Podemann Sørensen, 46–53. Copenhagen: The Royal Danish Academy of Sciences and Letters.

———. 2004a. "Syncretism and the Interaction of Modes of Religiosity: a Formative Perspective in 'Gnostic-Christian' Movements in Late Antiquity." In *Theorizing Religion's Past: Archaeology, History, and Cognition*, edited by Luther H. Martin and Harvey Whitehouse, 105–124. Walnut Creek: Altamira Press.

———. 2004b. "En strid om synkretisme i nutidens USA." In *Det Brede and Det Skarpe. Religionsvidenskabelige studier. En gave til Per Bilde på 65 års dagen*, edited by Armin W. Geertz, Hans J. Lundager Jensen and Jens Peter Schjødt, 123–134. Copenhagen: Forlaget Anis.

———. 2004c. *Syncretism: The Problem of Defining a Category*. Ph.D. dissertation, Aarhus University.

Leopold, Anita Maria and Jeppe Sinding Jensen, eds. 2004. *Syncretism in the Study of Religion*. London: Equinox.

Light, Timothy. 2004. "Orthosyncretism: An Account of Melding in Religion." In *Syncretism in the Study of Religion*, edited by Anita Maria Leopold and Jeppe Sinding Jensen, 318–324. London: Equinox.

Martin, Luther H. and Anita Maria Leopold. 2004. "New Approaches in the Study of Syncretism." In *New Approaches to the Study of Religion*, vol. 2, edited by Peter Antes, Armin W. Geertz and Randi R. Warne, 93–108. Berlin: Walter de Gruyter.

Martin, Luther H. 1987. *Hellenistic religions. An Introduction*. Oxford: Oxford University Press.

Nair, Rukmini Bhaya. 2002. *Narrative Gravity: Conversation, Cognition, Culture*. Oxford: Oxford

University Press.

Siegel, Daniel J. 2001. "Toward an Interpersonal Neurobiology of the Developing Mind: Attachment Relationships, 'Mindsight,' and Neural Integration." In *Infant Mental Health Journal* 22(1-2): 67–94.

Stewart, Charles and Rosalind Shaw, eds. 1994. *Syncretism / Anti-Syncretism: The Politics of Religious Synthesis*. London: Routledge.

Turner, Mark. 1996. *The Literary Mind: The Origins of Thought and Language*. Oxford: Oxford University Press.

Whitehouse, Harvey. 2000. *Arguments and Icons: Divergent Modes of Religiosity*. Oxford: Oxford University Press.

———. 2002. "Modes of Religiosity: Towards a Cognitive Explanation of the Sociopolitical Dynamics of Religion." *Method and Theory in the Study of Religion* 14(3-4): 293–315.

# — 16 —

# The Salience and Relevance of Modern Icelandic Ásatrú: A Preliminary Case Study in the Immunology of Culture

Guðmundur Ingi Markússon

The concern of this preliminary investigation is an extraordinary increase in the membership of the officially recognized neo-pagan denomination in Iceland, the Ásatrú Association (*Ásatrúarfélagið*).[1] It is suggested that the Icelandic "national narrative" is the variable of interest to explain this increase. The results are discussed in the context of the cognitivist critique of the concept of "culture" (e.g. Pascal Boyer) and Jesper Sørensen's "immunology of cultural systems." It is argued that a cognitive account of the salience and relevance of cultural phenomena is insufficient to account for a bias found in the data and that a theoretical account of culturally-based salience and relevance is needed.

## A Variation in Need of an Explanation

The present study has its point of departure in a simple statistical fact: after having "flat lined" with less than one hundred members for about two decades after its official recognition as a denomination in 1973, the membership curve of the Ásatrú Association (here after the Ásatrú) rose by more than 420% during the 1990s—from 98 members in 1990 to 512 in 2000 (see Figure 16.1).[2] This increase is extraordinary considering the stable his-

1. The Ásatrú Association was founded in 1972 with the aim of furthering a modern "renaissance" of the Old Norse, polytheistic religion, commonly referred to as *ásatrú* ("*ása* faith" or "faith in the *æsir*," i.e. the chief group of Old Norse gods), brought to Iceland from Scandinavia by its ninth century settlers (Beinteinsson and Gunnarsdóttir 1992, 132ff; Pétursson 1985). The central inspiration is Old Norse mythology, preserved in the Prose Edda of Snorri Sturluson (1178–1241) and the Poetic Edda. Icelandic folklore, some of which has survived to the present day, is also important. The Association was recognized by the state as a religious denomination in 1973. Such recognition grants the denomination's head (the "High Priest," *Allsherjargoði*) the right to perform legally binding acts, such as marriage, baptism, confirmation and burial. The current High Priest is film score composer Hilmar Örn Hilmarsson. His predecessors were Jörmundur Ingi Hansen, in office 1994-2003, and Sveinbjörn Beinteinsson (1924–1993), presiding from 1972–1993, both founding members.

2. Statistics and figures must be viewed in the context of the small population of Iceland which counts approximately 300,000 people (01/12/2000 about 283,000). All statistics are from Hagstofa Íslands (Statistical Bureau of Iceland) 1974/1984, 1991–2001.

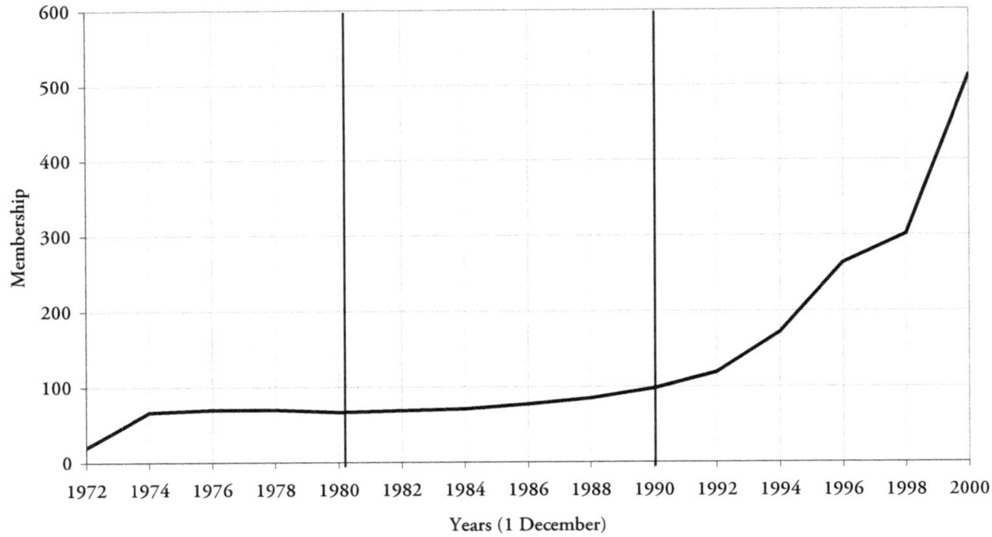


Figure 16.1 Ásatrú membership development from foundation (1972-2000)

tory of membership for the first two decades of the Ásatrú's existence. Also, as I will return to below, the Ásatrú is the only marginal religious group in Iceland that shows such dramatic change in membership during this period. Seeking an explanation for this statistical anomaly is the purpose of this study.

## Ecclesiastical Image Crisis

The search for an explanation soon focused on the Lutheran National Church (hereafter the Church). The Church's dominance of the Icelandic religious landscape finds expression in the fact that it counts close to 90% of the population as its membership (01/12/2000, about 88%). Given this dominance, could it be that events involving the Church somehow stimulated the rapid rise in Ásatrú membership in the 1990s? In this context, it was noted that the Church suffered negative discourse in the Icelandic media in the 1990s, especially in 1996 and 2000.

In early 1996, a woman came forward accusing the Church's head, the bishop of Iceland, of sexual misconduct years before, something which he denied. At the time, the matter ended unresolved as word against word. Understandably, this did not prevent the media from giving it their full attention. Only recently has the bishop's guilt become evident. The close media attention tarnished the Church's public image, and by the end of 1997, the bishop was forced to resign before finishing his appointed term.

In the summer of 2000, Church and state decided to celebrate a millennium of Christianity on the plain of Thingvellir, where the ancient Althing ("Parliament" founded around 930 CE) was situated. There in 999/1000 CE, according to written sources, Icelanders made a peaceful transition to Christianity (on the condition that they might still practice the old religion secretly if they wanted to). These official celebrations which were to last a whole weekend (July 1–2), led to less-than-celebratory discursive battles in the press. Some complained of Christian triumphalism, that the transition to Christianity was not so peaceful after all, and so on. A respected former bishop ended up before an ecclesiastical disciplinary committee after comparing the critics' statements to those of the Nazis and communists.

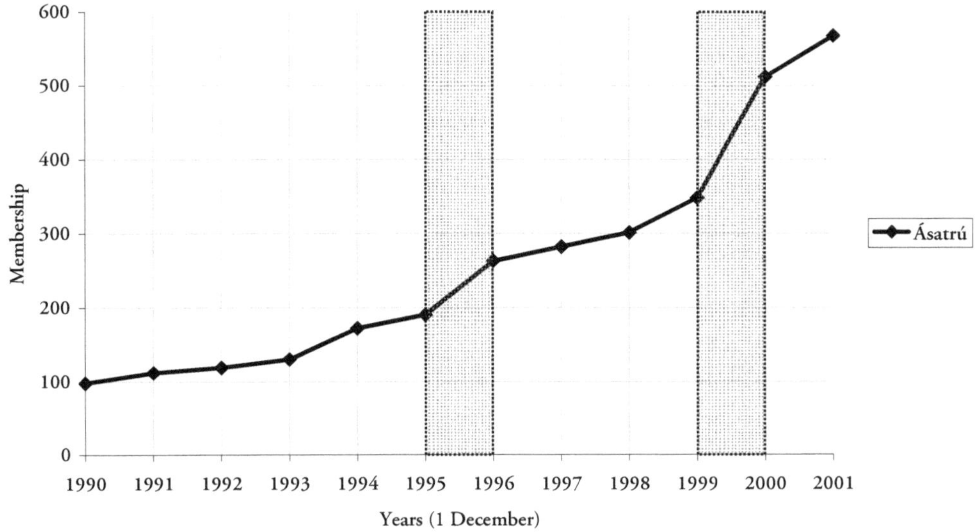


Figure 16.2 Correlation of ecclesiastical bad press and Ásatrú membership increase in 1996 and 2000

To add insult to injury, fewer people turned up than expected. Judging from the media coverage and the differences in opinion, before and after, about their merit and success, the celebrations of 2000 were not the whole-heartedly happy affair they were meant to be.

From these considerations, a hypothesis emerged to the effect that the Ásatrú had benefited from the "bad ecclesiastical press" the Church suffered in 1996 and 2000. If there is anything to this hypothesis, a positive correlation should be discernible; viz. the rise in membership should be significantly more accentuated in 1996 and 2000. As can be seen in Figure 16.2, this appears to be the case.

## Common Denominator: National Narrative

But why should negative media discourse about the Church positively stimulate the membership of a marginal new religious group like the Ásatrú? Is there something that links these very different religious denominations? As it turns out, there is: both denominations claim a share in what I will term the "Icelandic national narrative."

While keeping firmly in mind the pitfalls of summarizing the complexities of the past, whether real or imaginary, the broad—and admittedly *simplified*—strokes of Icelandic history can be broken down into these main points (I should like to emphasize that I do not have in mind "scientific" historical accounts of academics, but rather common interpretations made in popular and day-to-day discourse):

1. Iceland is settled in the ninth century CE mainly by Vikings arriving directly from Norway or by way of the British Isles. The settler's religious outlook is the polytheistic *ásatrú*.
2. With the foundation of Althing ("Parliament") around 930, Iceland becomes an independent commonwealth.
3. A transition to Christianity on Althing in 999/1000 by peaceful means.
4. Internal warfare results in loss of independence to Norway in the thirteenth century (later, Iceland comes under Danish rule).

5. The Lutheran reformation in the sixteenth century.
6. General misery at the hands of nature and the Danes for centuries.
7. Independence from Denmark in 1944.
8. Rapid modernization in the twentieth century.

It is important to emphasize that this basic narrative is to all intents and purposes *shared*, viz. its elementary points are familiar to more or less everyone due to the various means by which it is maintained. These means include the educational system (e.g. courses in Icelandic history, language and literature), the wide availability of Old Icelandic sources (such as the Sagas, Eddas, and so on), and the stability of the language (Old and Modern Icelandic being very close).[3]

As already mentioned, both the Church and Ásatrú claim a share in the national narrative. An important element in the Church's rhetoric when defending its privileged status as national Church (which is protected by the constitution) is that Icelandic culture, through the influence of the Church (first Catholic, then Lutheran), is inherently Christian. The Ásatrú maintain that they represent a renaissance of the Viking settlers' religion within a modern framework and, further, that pagan attitudes never really disappeared in Iceland. In addition, the mere application of the term *ásatrú*, commonly used to cover the pre-Christian religion, creates a reference to Icelandic pre-history. In this way, the Icelandic national narrative is the common denominator linking both denominations.

On the basis of these considerations, the hypothesis emerged that the Ásatrú benefited from the image crisis of the Church in the media due to the fact that both denominations "fit into" the national narrative of Iceland. To evaluate this hypothesis two predictions are advanced:

1. Since the Ásatrú is the only non-Christian or marginal religious group that fits into the national narrative, other marginal denominations offer ideal possibilities for comparison. Those that have a sufficiently long statistical history (i.e., recognized well before the 1996) and which can readily be defined as marginal and outside mainstream Christianity (which in the Icelandic context is primarily the Lutheran and, to some extent, the Roman Catholic varieties) are the Church of Jesus Christ of Latter-Day Saints (Mormons), Jehovah's Witnesses and Baha'i. In light of the above hypothesis, changes in the membership of these denominations should *not* show any correlation with the image-crisis of the Church in 1996 and 2000. This is indeed the case (see Figure 16.3).
2. Since the Church's image-crisis of 2000, unlike that of 1996, was openly about the national narrative itself (the Christian millennium celebrations taking place at Thingvellir, the site of the Althing, the "sanctuary" of Icelandic nationalism, where Iceland became a Christian nation, and so on), one should expect the Ásatrú membership increase to have been significantly more accentuated in 2000 than 1996. As is illustrated in Figure 16.4, this is the case: in 1996: about a 38.4% increase; in 2000: about 47%.

## The Salience and Relevance of the Ásatrú

Why did the Ásatrú's claim to a share in the Icelandic national narrative give rise to such membership dynamics? To put it differently, what are the workings of a "narrative fit"? My suggestion is that since its inception in 1972, the implicit and explicit references to the national narrative—via its designation (*ásatrú*) and its self-representation (as a modern renewal of the ancient religion)—made the Ásatrú and what it represents into a salient

3. For a concise introduction to the background of the Icelandic language see Karlsson 2004.

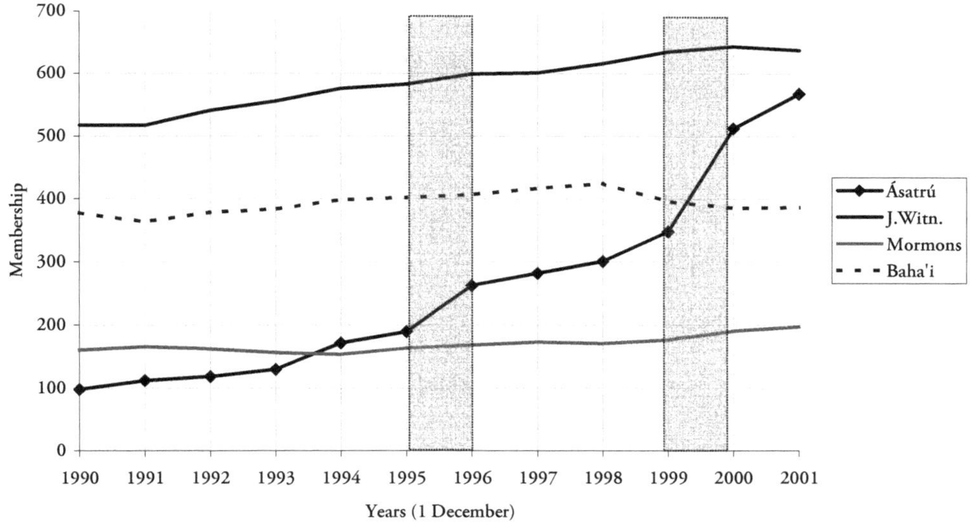


Figure 16.3 Comparison with other marginal denominations

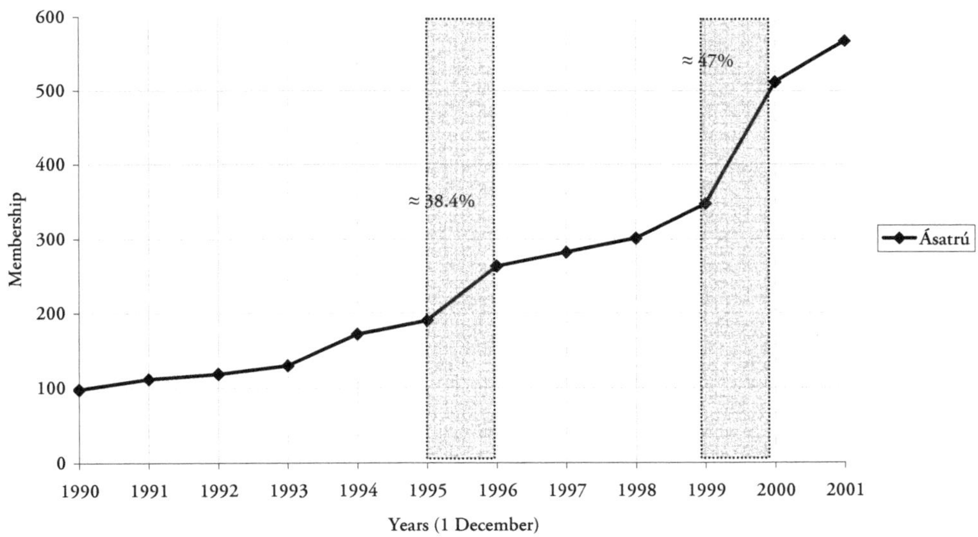


Figure 16.4 Comparison of increase in 1996 (≈ 38.4%) and 2000 (≈ 47%)

cultural phenomenon.[4] Thus, the Ásatrú was a more salient and relevant possibility for those disaffected enough with the Church to leave, but also for those already outside the Church who felt compelled to express their disagreement. It should be possible to evaluate this hypothesis by interviews and/or surveys.

4. That the Ásatrú has enjoyed cultural salience in Iceland from its foundation is difficult to dispute. The mass media, whether domestic or international, has never lacked interest in the group, a circumstance Icelandic sociologist of religion Pétur Pétursson has made the subject of research (1985, 1987). Frequent media appearances can be taken as a measure of the Ásatrú's salience as a cultural phenomenon.

One way in which the "narrative fit" makes the Ásatrú more salient and relevant is that it provides for various "ready made" inferences about the Ásatrú made in popular discourse. For example, the media frequently represents the Ásatrú as directly descendant from the ancient religion and its members as performing ancient rituals, or, in keeping with their own self-representation, that they are in some sense "keepers of the faith."[5] Thus the distinction between the ancient *ásatrú* and its modern counterpart (which is by all standards a new religion) is frequently blurred. In fact, the mere designation *ásatrú* is enough to conjure up various inferences of this sort, inferences that in the last analysis depend on elementary familiarity with the national narrative of Iceland.

I will close this section by returning to the Christian millennium celebrations in 2000 with an illustrative example of how invoking the national narrative can make the Ásatrú salient almost by default. Celebrating a historical event like the transition to Christianity is impossible without making references to the old religion which was pushed aside—the ancient *ásatrú*. As part of the celebrations, a "ritual theatre" was organized where idols of the Old Norse gods were thrown down a waterfall. The performance was intended as a reenactment of what a major player in the conversion, Þorgeir Ljósvetningagoði, supposedly did when he returned home after the conversion, thus demonstrating his rejection of the old religion. The Ásatrú Association protested what it saw as an official desecration of its religion, to no avail. The same morning before the "ritual theatre" took place, Ásatrú members held their own ritual at the same waterfall where they sacrificed an idol of the god Freyr, claiming that if there was anything to the legend, Þorgeir would not have been rejecting the old gods, but making a sacrifice to the waterfall. Thus, they both protested the "ritual theatre" and claimed an alternative interpretation of the act in question. To all intents and purposes, this was a ritual struggle about points in narrative space.

## Discussion: Salience, Relevance and Cultural Immunology

The first thing to notice is an interesting bias in the data: in 1996 and 2000, when the Church was suffering negative discourse in the press, a significant number of people joined a marginal, non-Christian denomination. The important point is that these people were *only* joining *one particular* non-mainstream religious group, viz. the Ásatrú. This represents a significant bias. If there was no bias, the distribution among marginal groups should be even—people should have been joining the Mormons, Jehovah's Witnesses and Baha'i as well. As can be seen in Figure 16.3, this is not the case—the Ásatrú is the *only* denomination that benefited from the Church's image crisis. It is my claim, with reference to the above, that the most likely variable implicated in this bias is the socially shared, Icelandic national narrative. Hence, the narrative has to be seen as a significant, *causal* factor.

The wider, theoretical implication is that something as "abstract" as narrative (a construct of symbolic culture)[6] can enter significantly into causal relations, and thus cannot be

5. An excellent example of this is a theme issue of *National Geographic* about the Vikings from 2000. In the theme article, a large colour photograph with text, covering the greater part of two pages, shows the Ásatrú High Priest walking in the group's graveyard in Reykjavík. A headline in boldface under the photo states "Keeper of the faith" (Vesilind 2000, 26f.). One can wonder why so much space was given to a leader of a marginal new religious group in a historical and sociological issue about the Vikings and their modern descendants?

6. *Symbolic* being a reference to Peirce's notion of *symbols* (as opposed to *icons* and *indices*) as the basis of linguistic reference (Deacon 1997).

ignored as an epiphenomenon, contrary to the views of some major players in the cognitive field, such as Pascal Boyer and Ilkka Pyysiäinen. Boyer has stated that the term "culture" only denotes a similarity between mental representations or ideas in people's heads, and that similarity as such cannot cause anything (2001, 35f.). Pyysiäinen, who shares a view similar to Boyer's, stated that "culture-concepts...cannot be employed as causal factors in explaining individual thought and behavior" (2002, 177; cf. 2003, 25ff.). This epiphenomenal view of culture cannot account for the bias pointed out previously. The reason hinges on the distinction between the Ásatrú and the other marginal denominations. What distinguishes the Ásatrú is *not* its ability to excite deep cognitive mechanisms (such as those which help us navigate and make sense of the physical and social world around us) but its "narrative fit." And the narrative in question is undoubtedly a cultural construct dependent on the human capacity for symbolic/linguistic reference. To further relate this to cognitive theorizing, the *salience* (attention-grabbing quality) and *relevance* (potential for generating inferences) of the Ásatrú depends on cultural, not cognitive factors.[7]

While we certainly need to acknowledge the importance of cognition, ignoring culture is misguided for it means turning a blind eye to a range of issues. In this context, Peter J. Richerson and Robert Boyd have stated that:

> the single most important adaptive feature of culture is that it allows the gradual, cumulative assembly of adaptations over many generations, adaptations that no single individual could evoke on his or her own. Cumulative cultural adaptation cannot be based directly, or in detail, on innate, genetically encoded information. (2005, 45)

What is needed is an approach that *complements* the emphasis on cognition. This is what Jesper Sørensen attempts with his so-called *immunology* of cultural systems (with reference to Dan Sperber's *epidemiology* of cultural representations). In addition to cognitively based selection mechanisms in the maintenance and transmission of culture, Sørensen's immunology proposes a selective principle:

> based on the degree of fitness of...new cultural representations into...the existing ecology of representations..., in short, a selection based on how compatible new representations are with already existing public and mental representations organised in systems of mutual reference. (2004, 62)

I find Sørensen's proposal promising, and the notion that cultural phenomena are organized in systems of mutual reference is especially relevant to the present study, as such referential arrangements (here, implicit and explicit references to the Icelandic national narrative) could provide a theoretical grounding for *culturally-dependent* principles of salience and relevance.

Among the areas to which Sørensen turns in building his case are cognitive linguistics and related fields of cognitive science where human cognition is considered in a socio-cultural context. This, I believe, is the right direction if we are to develop a theoretical description of the "stuff" of cultural systems and gain a better understanding of the interdependence of culture and cognition. In this regard I find Terrence Deacon's work on emergence, sym-

7. The terms *salience* and *relevance* are applied by Boyer in his cognitive theory of religion. *Salience* denotes the memorability or attention-grabbing quality of a cultural concept while *relevance* refers to its inferential potentiality, i.e., the more inferences a concept generates the more relevant it is (this concept is originally developed by Sperber and Wilson 1995). In Boyer's cognitive theory, contrary to the view expressed above, the *salience* and *relevance* of cultural concepts is entirely dependent on cognitive architecture (1992, 1994, 2001).

bolic reference and "evolutionary semiotics" particularly interesting as a possible theoretical ground on which an ontology of symbolic culture could be based (1997, 1999, 2003a, 2003b).

## Concluding Remarks

As is emphasized in the title, this paper is preliminary in character and very much a work in progress. Media discourse must be quantified, interviews and/or surveys taken, and hypothesis and predictions put to statistical tests. The present results must be viewed with this in mind. These limitations notwithstanding, this case study is suggestive as to the necessity of an approach that takes the reality of symbolic culture (a "cultural-immunology") seriously, to *complement* cognitive approaches and thereby reach a better account of human culture.

## Acknowledgements

I am grateful to Bob McCauley for generous and useful comments which have helped me gather my thoughts in a more organized manner. Needless to say, any lapses are due to my own cognitive and cultural shortcomings.

## References

Beinteinsson, Sveinbjörn and Berglind Gunnarsdóttir. 1992. *Allsherjargoðinn*. Akranes: Hörpuútgáfan.

Boyer, Pascal. 1992. "Explaining Religious Ideas: Elements of a Cognitive Approach." *Numen* 39: 27–57.

Boyer, Pascal. 1994. *The Naturalness of Religious Ideas: A Cognitive Theory of Religion*. Berkeley: University of California Press.

Boyer, Pascal. 2001. *Religion Explained: The Evolutionary Origins of Religious Thought*. New York: Basic Books.

Deacon, Terrence. 1997. *The Symbolic Species: The Co-Evolution of Language and the Human Brain*. London: Penguin Books.

———. 1999. "The Trouble with Memes (and What to Do about It)." *The Semiotic Review of Books* 10(3): 1–3. [www.chass.utoronto.ca/epc/srb/srb/10-3edit.html]

———. 2003a. "Multilevel Selection in a Complex Adaptive System: The Problem of Language Origins." In *Evolution and Learning: The Baldwin Effect Reconsidered*, edited by Bruce H. Weber and David J. Depew, 81–106. Cambridge, MA: MIT Press.

———. 2003b. "The Hierarchic Logic of Emergence: Untangling the Interdependence of Evolution and Self-Organization." In *Evolution and Learning: The Baldwin Effect Reconsidered*, edited by Bruce H. Weber and David J. Depew, 273–308. Cambridge, MA: MIT Press.

Hagstofa Íslands (Statistical Bureau of Iceland). 1974–1984. *Tölfræðihandbókin*. Reykjavík: Hagstofa Íslands.

Hagstofa Íslands (Statistical Bureau of Iceland). 1991–2001. *Landshagir: Statistical Abstract of Iceland*. Reykjavík: Hagstofa Íslands.

Karlsson, Stefán. 2004. *The Icelandic Language*. London: Viking Society for Northern Research.

Pétursson, Pétur. 1985. "Asasamfundet på Island och massmedia." *Forskningsrapport* 1985:1 nr. 185, Stockholm: Religionssociologiska Institutet.

———. 1987. "Religiösa minoriteter på Island i samspel med majoriteten." In *Religiøse Minoriteter*, edited by Ragnhild Kristensen and Ole Riis, 63–83. Aarhus: Aarhus University Press.

Pyysiäinen, Ilkka. 2002. "Ontology of Culture and the Study of Human Behavior," *Journal of Cognition and Culture* 2 (3), 167-181.

———. 2003. *How Religion Works: Towards a New Cognitive Science of Religion*. Leiden: Brill.

Richerson, Peter J. and Robert Boyd. 2005. *Not by Genes Alone: How Culture Transformed Human Evolution*. Chicago, IL: The University of Chicago Press.

Sperber, Dan and Deirdre Wilson. 1995. *Relevance: Communication and Cognition*, 2nd ed. Oxford: Blackwell Publishers.

Sørensen, Jesper. 2004. "Religion, Evolution, and an Immunology of Cultural Systems." *Evolution and Cognition* 10 (1): 61–73.

Vesilind, Priit J. 2000. "In Search of Vikings." *National Geographic* 197(5): 2–27.

# — 17 —

# "Baby in a Bowl" and Other Stories: Socialization in Astrological Narrative

Kirstine Munk

> Man has not one and the same life. He has many lives, placed end to end, and that is the cause of his misery. (Chateaubriand)

Thus begins a fabulous novel by Paul Auster (2002). With a quotation by the romanticist writer Chateaubriand, we embark in *Book of Illusions* on a wild journey in the story of a man whose life circumstances change dramatically time and again, bringing an almost surreal quality to a common psychological reality in contemporary life. As readers, we ask ourselves: is this person *really* the same person as he learns the ropes, managing to navigate in many different social contexts? From being a poor immigrant to a famous silent comedian, to murderer on the run, harbour worker, salesman, actor in a touring peep show, local hero in Sandusky, Ohio etc., until he ends his days in a desert in New Mexico with the one love of his life and lots of secretly directed movies. How much of this life story is his own making, and to what degree can one say that he is already being narrated as with determination he repeatedly musters all his courage, exploiting creative intuition to place himself somewhere else in order to become somebody else?

With this book, Paul Auster plays with a longstanding problem in contemporary philosophy, namely: to what extent is man already narrated through language, myth, and culture, and to what extent is he the creator of his own life, though invariably perceived and performed in the form of narrative? It is also a marvellous story about the incredible capacity of humans to actively re-create themselves in new social contexts and to use available cultural knowledge to form a new sense of self and to make decisions on how to act in the world. These are the questions at the core of this chapter.[1]

From the beginning of life, humans organized their perception and their understanding of the world by means of stories. Research in early child development has pointed to

1. This article is an extended version of a paper presented at the conference "Religious Narrative, Cognition and Culture" at Aarhus University, 2–4 December 2004. I am grateful for the feedback I received for this paper, especially from Merlin Donald and Anita Leopold whose ideas have been included in the revised article. My warm thanks also goes to Lars Dencik, who read parts of a draft version of the article.

the fact that the acquisition of language and narrative competence creates new subjective perspectives for the child. The verbal self begins to evolve when the child is 15–18 months old and permits increasing self-reflection (Stern 1985). Sometime after the second year, a new sense of self emerges: the narrative self. Each new sense of self opens a new domain of experience, but does not necessarily encompass the previous senses of self. According to child psychologist Daniel Stern, all senses of self are perceived as coexisting together (Nelson 1989, 318). These new narrative capacities force the child to reorganize her subjective perspective about who she is, and how she is, in relations with others. The social world largely comes to the child with language, and the early narratives that children tell, often to themselves, are mainly aimed at world-making and problem-solving. Significantly, it is through these processes that a firm sense of self develops (Nelson 1989).

Research by Katherine Nelson and her colleagues at the New York Language Acquisition Group in analyzing a toddler's own bedtime stories suggested that children create world order by creating real stories. Sequencing, distinguishing the ordinary from the extraordinary, employing intentionality, causality, and perspective are narrative measures used by the child to make sense of the world. Personal perspective in the stories emerged as the child managed to get her world structured into necessary, appropriate, and causal sequences (Nelson 1989, 89). Before the development of these narrative competences—when children are between two and three years of age—memory is impossible to organize. The development of this cognitive apparatus simultaneously produces "world," "self" and "other," but the building blocks consist mainly of cultural symbols employed from early childhood to deal with "trouble" and to make up one's mind.[2]

The work done by Katherine Nelson and her colleagues showed in minute detail how the child makes up her mind about the world and her place in it by using the parental input in creating little stories. Similar studies analysing how grown-ups actually make decisions, create narratives and make up their minds in the narrative mode have yet to be done. It is well known that we constantly construct and reconstruct our "selves" to meet the needs of the situations we encounter (Bruner 2002). It is also a well-known fact that the everyday mind is essentially literary. Story is a basic principle of thought, and narrative imagining is our chief means of looking into the future, of planning and of explaining (Turner 1996). These narrative acts of "self-making" are guided by cultural models of what is possible and agreeable, and what is not (Bruner 2002).

In this chapter, I present some of my findings from my ethnographic field studies on

2. "Trouble" is the narrative fuel. As pointed out by Jerome Bruner, it is trouble rather than drama, that is always implicit or explicit in narrative, and when we encounter "trouble" in life, we primarily seek a solution in the narrative mode (Bruner 1986; Nelson 1989, 76). In early life stories, "trouble" is about ordering and trying to come to terms with the socio-cultural world. Problems involving only physical objects, space and causality hardly play a role in early narratives (Nelson 1989, 284). Nor are early narratives clearly autobiographical in the sense that the stories deal with major life events—events that are presumably very affective, rare, novel, unexpected and transformative such as having a baby brother. Instead, what occupies the small child is the actual process of life in the midst of important events (Stern 1985). In later years, however, roughly from the age of four, children start developing an active autobiographical account of their lives (Donald 2001, 296). From this time on, autobiographical narrative plays an important part in the child's efforts to make sense of the world and regulate emotional states. Furthermore, these narratives often try to solve the question of what goes on in other people's minds. In this continuing process of creating autobiographical understanding of oneself, one's relations, and one's place in the world, the parents play an important role. They are the scaffolding of the story, and they validate and rehearse it (Siegel 1999; Nelson 1989).

how people in the economically and technologically advanced parts of the world use Western astrology. Astrology is a largely neglected, but very interesting and rewarding field of research, because it is a popular tool in people's narrative constructions of self specifically designed to deal with trouble and to help people make up their minds.[3] The reasons why people use astrology, and how they do so, have not been thoroughly investigated before.

What kinds of questions and issues are raised in an astrological consultation? What impact does it have on the way people make up their minds and deal with the world?[4] In this chapter, I will present some of my empirical findings in order to shed light on how various kinds of cultural knowledge, mainly from astrological symbols and narratives, may affect the ways our minds work.

So, please join me on a mental voyage to Anaheim (California) where I found myself at the beginning of my fieldwork, as a participant observer at a huge international conference for astrologers. One of the speakers was the Danish astrologer Christian Borup. He is well known to the Danish public, because he is the astrologer in a very popular prime time TV show called *The 6th Sense*.[5] His performance in this show has had a huge impact on astrology in Denmark, giving it momentum, and this astrological revival is termed "The Christian Borup effect" by his colleagues, who are of course very pleased with this development. Christian Borup has also managed to become quite a name abroad, not least because of his performance at the conference in Anaheim, where he spoke on "The birth myth."

The birth myth is a new narrative-based introduction to astrology that Christian Borup has developed. Over the more than 25 years of his professional career as a consulting astrologer, he has always asked his clients to provide him with a *narrative* of their birth as well their birth data, namely the exact time and the place of birth provided by the birth certificate. From an astrological viewpoint, the natal chart provides a symbolic image of a person's character as well as of the actual birth situation, and this narrative, "the birth myth," is the story the child has been told about the birth. Christian Borup explained: "It is the child's 'first religion.' It is the first story about why you are here. It doesn't have to be a true story, often it is a story that the child makes up," and he then presented the audience with an illustrative example of this: one of Christian Borup's clients is a man named Bjørn, meaning "Bear" in English. Bjørn wrote this brief but concise story:

> My name is Bjørn, because I was born close to the royal castle. And just as I was born, the royal parade passed by [the soldiers wear huge hats of bearskin] so this is why my name is Bjørn.[6]

The birth certificate, however, showed that Bjørn was born in a hospital 15 km away from the royal castle, and there was no way anybody could have heard the royal parade from

3. As pointed out by linguist Charlotte Linde (1993, 127, 166 ff.), narrative order is the basis for the two major coherence principles of life stories, namely causality and continuity. According to her research on life stories, astrology is a very popular cultural "coherence system" used to establish narrative order.

4. Given this lack of available material, I have had to supply the empirical basis for reflection. My fieldwork (in Denmark, England, the USA, and Japan) consists of interviews with astrologers and their clients, as well as participant observation at astrological courses, conferences and consultations in the period between March 2003 and April 2005. Unless otherwise specified, the references to astrology in the main text are based on my fieldwork.

5. This entertainment program was invented by *Mastiff Media* in Denmark, but is presently being sold to other countries. It started in spring 2002 and has according to them close to one million viewers each week, i.e. a fifth of the total Danish population (Source: Mastiff Media 2005).

6. Bjørn is not an unusual Nordic name, but it is unusual to associate it with royalty, as narrated here.

there. It was a story that had been made up. But as Christian Borup remarked, Bjørn behaved as if he was royal: he was very well dressed, had exquisite manners and the fact that he was just an accountant in a very small firm didn't affect him at all. Christian Borup explained: "That's the whole idea about the birth myth. It is something the child makes up to explain why it is here, a *raison d'être*, and it has deep psychological implications for the rest of that person's life because this story makes a *script* for how you react in certain situations."

Christian Borup uses the birth myth as a key to the chart in order to establish a rapport with the client. It is a tool to get a breakthrough, to open up and make the client see connections she hasn't thought of before. To illustrate this, he asked the audience if anyone would provide a story to be used as an example in the session, and a woman in her late fifties immediately raised her hand and told this story:[7]

> I was born in the middle of Poland at the end of World War Two when my parents were on the escape path from Eastern Europe. My mother was sixteen, and my entire family had been killed in the Holocaust, so I was the only living child. So, yeah, there was a lot of "this child *has* to live." And I was born a "preemie" without any incubators or anything in a Russian army camp. And my mom—an old woman told her to put me in a bowl of water, so she kept me in this warm bowl of water, and I was such a preemie—I was five month prem—I was such a preemie that I still had all the laguna hair [sic]. She said, she thought she gave birth to this little monkey.

This is a very short and compact story, yet ripe with emotion. When the woman phrased the sentence "this child *has* to live," she was saying it with great intensity and emotion, and when the story rang out with "she thought she gave birth to this little monkey," it was said with ease and humour. And the audience laughed relieved at the end of this story, and many of the listeners spontaneously exclaimed, "Now, that was a good story!"

Christian Borup's feedback immediately engaged the storyteller in a dialogue:

> CB: There are several things in this story. There is a Uranian element because of the hair, – because you were born a freak.[8]
>
> Storyteller: And we came to the United States and the United States is a country that is very Uranian.
>
> CB: Yeah, I know. And you've probably moved a lot of times in your life.
>
> Storyteller: Yes. (Very firm.)
>
> CB: Because it is in your birth myth that your mother was on the move while she was having you. You are a survivor. You probably tried to live in cold states but didn't like it, because you need the warmth which is symbolized by the bath.
>
> Storyteller: Oh,—tried to live in cold places....oh, okay.
>
> CB: Yeah, you probably... where are you living now? California?
>
> Storyteller: I am living in California. I like it warm in winter. I dare say this is my favourite place to be. (Audience laughs.)

The conversation went on to other topics such as the storyteller's relationship with her son and her survival instinct, and she readily made the connections between the story and her life. The interesting thing is that when Christian Borup explained how the birth myth works astrologically—the theme of the story being expressed by the latest aspect the moon makes before the birth takes place—the planet in question, Jupiter, didn't really fit the

7. Author's fieldwork in Anaheim 2003.

8. Uranus is a planet that, among other things, signifies eccentricity, inventiveness and originality.

story. Nevertheless, it didn't matter in the situation. Neither the storyteller nor the audience chose to question that too much, because the story in itself obviously worked. It was a *good story* in the sense that the storyteller was the hero in her own narrative. The story had drama, emotion, and plot, and the story conveyed a well-defined purpose in life, namely to survive and to help her only son, who is a mellow type, to become a survivor too in his own right. So she was content with this story. She liked the identity it gave her. There was power and a strong symbolism of transformation in it, and these are keywords in modern astrology: to mould your life-world according to your personality in creative and satisfying ways is what modern astrology is about.

It seems from Christian Borup's account that the birth myth is a static story, once created it continues to work in its own subtle ways. Yet, I would imagine that it actually does change over time. A sound story about identity and life purpose has to be dynamic in the meeting with new circumstances. As Anthony Giddens has pointed out, a biographic narrative will change in the course of a lifetime due to our continuous self-reflection, so that the story retold will make a better fit with our "colonization of the future" (Giddens 1991).

A personal myth is created from life circumstances, but it could be told in a number of ways, and the way it is told forms, reinforces and reflects cognitive dispositions in terms of perception and memory (Schank 1990). It is well known that biographic narratives are socially produced. They have to make sense to others (Gergen 1997). And it is therefore not surprising that a person who is stuck with a narrative that is too static and too difficult to fit into new future scenarios can find help to change it through the dialogue in the astrological consultation. This is an important part of an astrologer's job.[9]

An astrological consultation takes about 1½ hours. The horoscope is always on the table for the client to see, although she usually doesn't understand anything of it: just looking at the chart engages her imaginatively. The horoscope is a symbolic image of self and cosmos. It forms a round circle comprising the twelve signs of the zodiac with the planets and other heavenly bodies marked in the middle by symbols, each representing a variety of meanings. Thus, the planet Neptune may refer to idealism as well as alcohol, ministers and make-up artists, dreams, deceptions, allergies, and many other things that in some way refers to illusions, water and the hidden. Like all other symbols, astrological symbols cannot refer to everything. The cognitive strength of symbols lies not only in their multivocality but also in the limitations of their meanings.

The planets may be connected to each other in aspects which in specific ways add to the symbolic meaning. The horoscope circle is divided by twelve dividing lines marking the astrological "houses" or areas in life such as health, money, marriage, siblings, career etc. Each astrological house has sub-layers of meaning. Thus, the fourth house could be the inquirer's home, her father, the psychological atmosphere in her childhood and her general upbringing, her partner's paternal grandmother or even her best friend's sister's money. Usually the astrologer starts explaining the chart, often by beginning to tell her about her childhood and her relations to her parents as these are shown in the chart, and the client will relatively quickly pick something up, and the dialogue begins.

It should follow from this very brief description, that astrology is a cultural system that demands and reinforces analogical thinking which is the core of creative thought. The many layers of symbolic meaning in a chart are mainly conveyed to the client by means of

9. Interestingly, there seems here to be some common ground between astrology as it is practised in the contemporary Western world and constructive narrative therapies (Hoyt 1998).

parables, myths and metaphors. This type of conversation engages the client in a particular kind of thinking that can have both cognitive and emotional effects. The mental leaps provided by these analogical processes can be used by human beings to create new narrative understandings which form the world as we experience and understand it (Holyoak and Thagard 1996).

According to Christian Borup, many people, especially women, come to see an astrologer when their kids move away from home, or when they are facing a divorce or other kinds of so-called "biographic situations" that may have potentially huge effects on their future life-course.[10] However, it doesn't have to be that dramatic. The turning points or crises in life that prompt people to seek astrological advice are many, and the decision to turn to astrologers does not always correlate with major occurrences in the outer world. Many women just feel stuck in their lives. They may have been brought up accustomed to thinking of being something for others, and suddenly there is nobody there to be something for. The client has to find a completely new way of life and in this process, she tells her story, and the astrologer helps her understand it in new ways and to mould it anew by means of a dialogue and the available symbolic input. The aim is that by creating a slightly different biographic story, the client perceives herself as being powerful and in the centre of her own life-world. The many symbols and the richness of symbolic meaning in the chart bring the needed flexibility into play. Any planet or planetary configuration will bring a different background theme into the story, and ideally this continuing process brings her to perceive her story from different angles, thereby loosening it up and making it dynamic.

This process of creating or re-creating a story in a dialogue is initiated in the ritual. The astrological input helps the client to scaffold a new story, but the process doesn't end there. The client will return home and reflect on the things said. New elements will tentatively be put into the foreground of the story, then into the background, when she thinks about it herself and when she talks with her friends afterwards about her experience at the astrologer's house.

According to cognitive narratologist Manfred Jahn, imaginary perception is a crucial concept in our understanding of stories and storytelling as psychological and cognitive forces, rather than mere forms of communication or entertainment. He distinguishes between external stories, namely stories as they are performed, written, or told, and internal stories, which are the stories "stored in memory and performed in the mental theatre of recollection, imagination, and dream" (Jahn 2003, 195). When a mother tells her child a fairytale, the child may internalize it, script her life on it and rely on it as one of her personalized internal stories (Berne 1972, 39 ff.; Jahn 2003, 200). A part of the procedure of internalization is the "indexing" of stories in memory. When the story is retold, it becomes external again. Imaginary perception is closely bound up with the process of externalization and associates with three major types of events in storytelling, namely with the drafting of a story for production; with remembering, daydreaming, and similar mental activities; as well as dreaming and, one may add, with the kind of dialogue present in and related to an astrological consultation. The actual composition of a story is understood to cycle be-

10. Due to a lack of statistical data, it is difficult to assess how many people use astrology. According to my interviews with astrologers, their clients are mainly women, and they are usually in their twenties when they see the astrologer for the first time. They are relatively well-educated and live mainly in urban areas. This is corroborated by statistics of Danish astrology users on the internet who share this profile. But there is actually a great deal of variety. On the internet, where astrology is an enormous success, astrology websites are used by roughly a quarter of the total population in Denmark (Source: Jubii.com 2004).

tween the processes of imagining and remembering, and the process of external reception in close or near-simultaneous succession. As Henry James put it, "The teller of a story is primarily, none the less, the listener to it, the reader of it too" (quoted in Jahn 2003, 202). Reception and production are closely related as the story unfolds in the loops by which we perceive, remember, imagine and tell stories in a process which is "like life itself" in that it transcends substance, media and form (Jahn 2003, 198, 212). Therefore, it is a process of which we are not totally in charge.

Many clients present biographical narratives in which the theme of socialization seems unresolved, either because they feel they don't fit into the immediate environment they were brought up in, or because they feel a mismatch between themselves and general cultural and social expectations. It seems as if most westerners, regardless of their educational background, use the scientific myth of early socialization as a narrative structure to give direction and purpose to their life story.

When referring to socialization as a myth, I am thinking of it in the same way as philosopher Mary Midgley would use the word: not as something *opposed* to science, but as a central part of science, namely the part that decides its significance in our lives and gives meaning. These kinds of myths form imaginative patterns that suggest particular ways of interpreting the world (Midgley 2003). And the theories of socialization seem very successful in this respect, even outside academia. Socialization is a modern myth in the sense that people don't think about it, they think with it in their self-understanding. The apparent determination in many theories of early socialization at least in their popular versions is not really questioned in itself, but in the telling of their life stories the astrologer's clients will often say something like, "Isn't it odd that I became a classical composer and my parents had a bridal store in Copenhagen." People who feel that their socialization should have brought them to something else sometimes turn to astrology for an alternative explanation.

One of my informants is a former child star with celebrity parents. Her mother was very ambitious on her behalf, and my informant acted in a number of films from the age of 3 to 14. Then it was suddenly all over, and it has been difficult for her later on to find her path in life. She has had a number of odd jobs, and her family background has been difficult for her too. Her parents divorced early, and she hasn't seen her father much, which has been an immense source of grief for her. This was only emphasized by the fact that people always associated him with her because of their common surname. My informant told me that she felt she had done everything she could to strengthen their relationship, but in vain. All the other kids had ordinary families, and in her mind, she carried a model image that a good family consisted of a mother, father and children. She missed her father a lot even though he didn't show much interest in her, and she still hoped for the day when she would get married, and her father would accompany her down the church aisle.

She got married, but her father never showed up. He didn't even respond to the letter she sent him, and my informant said:[11]

> When the astrologer told me that I had a strange chart in that there *really* was no father in my horoscope, then I understood that this is the way it is supposed to be. So I have learned to live with it now.

Socialization can be defined as the process in which the individual internalizes values, judgements, and behavioural patterns that are part of society and thereby come to be considered natural, normal and evident. But many socialization theories are challenged

11. Author's fieldwork in Denmark 2004.

by postmodern conditions (Dencik 2002). The stability and continuity of cultural life patterns doesn't exist to the same degree anymore, and socialization has to be seen as a life project and not as something that is more or less completed by the end of adolescence. Also, in a dynamic world, socialization is not something that happens to the individual. It is very much something the individual does herself by *choosing* diverse social situations.

The active involvement of the child in early socialization is a more recent discovery (Stern 1985; Dencik 1999). Formerly, popular psychodynamic theories emphasized the influence of the parents and other primary "socialization agents" in forming the child's character, but most children today move between many different social arenas (Dencik 2002).[12] A major investigation carried out among five-year-old children in Scandinavian countries concluded that in reality, socialization takes place in many contexts, and peer group socialization in institutions and elsewhere is a very important factor.[13] These investigations also highlighted the strong need in individuals to belong with others in the social settings in which they take part. By engaging in a "gazing behaviour" like an anthropologist doing fieldwork, children actively try to figure out what is happening in the social arena before they try to become part of a group, by imitating significant others. Social groups are dynamic, of course, and modern children—as well as modern adults—have to navigate in many relational settings. Studies in socialization processes have therefore changed, now focusing on how children manage the shifts between their "worlds" and how they actively participate in constructing them (Dencik 2005). For modern adults and people in early adolescence, the situation is somewhat similar.

From the viewpoint of social psychology, human beings become what they are through their relations with others, and the conditions of postmodernization therefore inevitably challenge traditional concepts of identity. Class, gender, family background and nationality have become less important for privileged groups in the Western world. While even a generation ago these parameters could have a determining effect on a person's self-identity and life path, the self has increasingly become a reflexive and open-ended project. Today, man not only has 'many lives, placed end to end' although the ever-present possibility of creating a completely new life is a strikingly late modern feature. Rather than the film metaphor where different pictures are placed in succession, the Hullabaloo House in Tivoli seems to be a more fitting image for life in today's shifting social arenas. As one continuously jumps between many moving platforms that turn around in unexpected ways, one can easily lose balance, focus and track of the people one entered the Hullabaloo House with—but then again: it's very easy to jump into new acquaintances.

One of the characteristics of contemporary Western society is that social relations have become unglued. We experience a massive increment of social stimulation and navigate through multiple disconnected relationships (Gergen 1991). Expanded networks of mediated communication provide individuals with access to what we might in general term 'non-local' knowledge, and individuals are constantly confronted with alternative visions

12. This shift away from a relatively fixed environment of socialization to a patchwork of socialization contexts is connected to the escalating transformations of modernity from the 1970s onwards. Industrialization and changes in the labour force created an institutionalization of childhood that demanded the development of particular social skills in children such as the ability to leap from one social context to another.

13. This investigation (BASUN) on Childhood, Society and Development in the Nordic countries was carried out in the late 1980s and comprised studies from Denmark, Sweden, Norway, Finland and Iceland (Dencik 1999).

of the world and countless narratives of self-formation (Thompson 1999). Out of the myriads of emotionally compelling input, the many social roles we are invited to play, and the symbolic materials available, the individual weaves a biographical narrative. For it is always by constructing a story that we are able to form a sense of who we are. Yet, like Penelope, individuals today have to un-weave their biographical work repeatedly in order to reinvent their understanding of their own self within the dynamic conditions of their ever-changing worlds.

The feeling that in important ways one doesn't fit into one's original social milieu (the micro system of socialization) or meet general cultural and social expectations (the macro system of socialization) is very common today. "Narratives of disaffiliation" form a major part of the narratives gathered during my fieldwork. Other researchers before me have singled out the narratives of disaffiliation as a common narrative category (Eakin 1999) and have pointed to the fact that narratives are based on a "breach" of the commonplace (Bruner 1986). As individuals *qua* human beings have an urge to belong somewhere with someone (Dencik 1999, 2001), disaffiliation and the way to dissolve it, could generally provide a very good narrative theme. Yet I suggest that the extreme commonality of narratives of disaffiliation in contemporary times is predominantly connected to the present conditions of postmodernization. While classical explanation models gathered from popular psychology and socialization theory do not fit because many people experience a wide gap between the person they were brought up to be and the person(s) they have come to be, they often try to explain their identity structure and important life decisions by other explanatory devices, one of which is modern astrology.

The ambiguity of a horoscope provides its users with an identity-making toolkit that enables people to hold on to a personal identity and to maintain their integrity while shifting between various social contexts. The many layers in a chart create the image of a "kaleidoscopic self."[14] It may be split into different facets, but the messiness of the multiple symbols that form the immediate impression when looking at a chart is elevated to a unified entity both at a visual level and through the conversation with the astrologer.

The self in modern astrology is perceived in a strange mixture of essentialism and hermeneutics. Despite postmodern conditions, astrology promises to formulate an authentic self, a "sameness" that runs through our lives as we experience them. Major crises are often triggered by the fact that we sometimes act in unexpected ways in unexpected situations thereby challenging the self-image we may have. Yet in accordance with postmodern conditions, the astrological self is also a relational self. One's significant others—friends, lovers, family and foes—are all represented in the chart, and the symmetric relation between self and other is highly pronounced. Furthermore, as exemplified above, the self in astrology is seen as a narrative construction. Through astrological interpretation, the individual can better use the potential of the horoscope to create a better life story—better in the sense that it enables the client to create a better life. Astrologers acknowledge the formative and redemptive aspects of storytelling. Mythology, anecdotes and other kinds of narratives play an important part of any successful astrological consultation.

The plurality of life styles that characterizes contemporary life in the privileged parts of the West, replace the kind of continuous normal biography that was common fifty years ago with discontinuing biographies that cannot be predicted in the same way (Kupferberg

14. The "kaleidoscopic self" is a term coined by social psychologist Lars Dencik to denote the self in postmodern conditions (Dencik 2001). The term describes the self perfectly in modern western astrology.

1995). The individual does have more options open, but it includes more risks, and the multirelational complexity that is pertinent to life in contemporary times constantly challenges the individual's biographical project, thus making narrative reflexivity so much more important (Giddens 1990).

This is also reflected in the way astrology is considered to work today. Formerly—in Babylon some 3,000 to 4,000 years ago—the astrological entities, the planets, could be described as *culturally postulated superhuman agents,* following the terminology of McCauley and Lawson (McCauley and Lawson 2002). There was an immediate authoritativeness in the system. If a planet was seen in the sky in a particular position then a particular action should take place: a sacrifice should be made, or the king should remain at home or whatever (Koch-Westenholz 1995; Stuckrad 2003). However, today, people don't follow the advice of astrologers.[15] It is considered a valuable input, but it is only treated as one input among many others. Even when the client *believes* she follows the astrologer's advice something else usually happens:[16]

> Mariann, a woman in her mid-thirties with a Master's degree in public relations, went to see an astrologer because she felt stuck at a point in her life when she had been recently divorced. She was unemployed, and being a mother of two, desperately needed a job. I accompanied her to the consultation, and in the car, we talked about her expectations. She felt certain that the astrologer could tell her what kind of job she should go for and hopefully also when she would meet a new man. She told me that she had thought about starting her own business with special services and care for elderly people, but she wasn't sure if this really was the thing to go for, or if she should put her efforts in a different direction. In the consultation, which was almost two hours long—it started with her relationship with her parents and proceeded onwards—the astrologer told her that she had many talents, good communication skills, and that she could have her own company, but she would also be good in jobs where she had to take care of others. She could be a nurse, a psychologist, even an astrologer. Working as a therapist would be good. When the consultation was over, and we stepped outside, I asked Mariann about her immediate reaction, and she said:
>
>> "Everything fitted so well, and the most amazing thing was when the astrologer said that I should become an NLP-therapist, and I always knew that. It makes so much sense!"

I concluded that a conceptual blending took place between "communication skills" and "therapist," in a mental process described by Mark Turner and Gilles Fauconnier (Fauconnier and Turner 2002), and Mariann was determined to follow the advice of the astrologer...until the following week, where she was re-deciding after conversations with somebody else. She is now, finally, trying to become a therapist, but of a different sort.[17]

Let me end this chapter with an extract of an interview I conducted with one of the lay people attending the astrology conference in Anaheim: Patrick, a Canadian in his mid-thirties. He is a former soldier who has used astrology in retrospect in order to understand

15. One logical implication here is of course that either astrology is not religion, or perhaps that Lawson and McCauley's theory is a bit too narrow to encapsulate all the varieties of modern religious life.

16. Author's fieldwork in Denmark 2003.

17. Historian of religions Anita Leopold has suggested that the meaning-making processes we find in the astrological divination rituals are similar to those found in syncretism. That being the case, the astrologer prompts what is in the mind of the client (Leopold 2004). Sometimes, however, ideas that are genuinely new to the client also emerge in the consultation. This issue, which relates further to the ritual properties of the consultation, is being treated in Munk (forthcoming).

the decisions he has made. When I met him, he worked independently as a wild life photographer, but he had also been an English teacher in Brazil, a farmer, a factory worker and a lumberjack in Canada. When I met him, he was considering whether he should stay in the U.S. and expand his photography business or do something else instead. He was still using intense mental efforts on processing his years in the army.

Nothing in Patrick's background suggested that he should have become a professional soldier, and suddenly after 14 years in the army, he felt that he should move on and do something entirely different. If you are a professional soldier, leaving the army is even more difficult than leaving a marriage. You are socialized into particular thought-patterns and ways of relating, and the process of leaving took years. He used astrology in that process to retell his life and formulate his identity in ways that made new symbolic connections. To paraphrase the philosopher G.W.F. Hegel, *Wesen ist was gewesen ist.*[18] Yet, even though we understand ourselves by means of our narratives of the past, we need some help from other people and from myths, rituals, symbol, and metaphor to make the cognitive leaps that we actually make all the time, and more so now than ever. Patrick said at the end of an interview that lasted four days and comprised his whole life story:[19]

> I won't be looking at my chart anymore. It validates everything that I am, so there is no reason for it anymore. Because once you are in touch with your inner voice, everything else is extra. Coming here has been the end of the process that has been going on for the last three years. It was an important tool, but actually mythology played a greater role.... When I came to the conference in Anaheim, I didn't expect this to be the finale. The astrologers at the ISAR conference were the cream of the crop, and I wasn't that impressed. There were too many big egos there. I am ending a cycle. I am ready for a new beginning now.

## References


Auster, Paul. 2002. *Book of Illusions*. New York: Faber and Faber.

Berne, Eric. 1972. *What Do You Say after You Say Hello? The Psychology of Human Destiny*. New York: Grove Press.

Bruner, Jerome. 1986. *Actual Minds. Possible Worlds*. Cambridge: Harvard University Press.

———. 1991. "The Narrative Construction of Reality," *Critical Inquiry* 18: 1–21.

———. 2002. *Making Stories: Law, Literature, Life*. Cambridge, MA: Harvard University Press.

Dencik, Lars. 1999. "Fremtidens børn—om postmodernisering og socialisering" [Children of the Future—on Postmodernization and Socialization]. In *Børn og familie i det postmoderne samfund [Children and Family in the Postmodern Society]*, edited by Lars Dencik and Per Schultz Jørgensen, 19–44. København: Hans Reitzels Forlag.

———. 2001. "Transformations of Identities in Rapidly Changing Societies." In *The Transformation of Modernity: Aspects of the Past, Present and Future of an Era*, edited by Mikael Carleheden and Michael Hviid Jacobsen, 183–221. Aldershot: Ashgate.

———. 2002. "Socialization." In *Psykologiske grundtemaer [Psychological Themes]*, edited by O. Løw and E. Svejgaard, 212–228. Aarhus: Kvan.

———. 2005. *Mennesket i postmoderniseringen—Om barndom, familie og identiteter i opbrud [The Human Being in Postmodernization: On Childhood, Family and Identities in Transformation]*. Værløse: Billesøe and Baltzer.


18. "Being is what has been" (Hegel 1972).

19. Author's fieldwork in Anaheim and Los Angeles 2003.

Donald, Merlin. 2001. *A Mind So Rare: The Evolution of Human Consciousness*. New York: W.W. Norton and Company.

Eakin, Paul John. 1999. *How Our Lives Become Stories Making Selves*. Ithaca, NY: Cornell University Press.

Fauconnier, Gilles and Mark Turner. 2002. *The Way We Think: Conceptual Blending and the Mind's Hidden Complexities*. New York: Basic Books.

Gergen, Kenneth J. 1991. *The Saturated Self: Dilemmas of Identity in Contemporary Life*. New York: Basic Books.

———. 1997. *Realities and Relationships*. Cambridge, MA: Harvard University Press.

Giddens, Anthony. 1990. *The Consequences of Modernity*. Cambridge: Polity Press.

———. 1991. *Modernity and Self-Identity: Self and Society in the Late Modern Age*. Cambridge: Polity Press.

Hegel, Friedrich. 1841. *Wissenschaft der Logik*. Vol. 2. Frankfurt am Main: Suhrkamp Verlag, 1972.

Holyoak, Keith J. and Paul Thagard, 1996. *Mental Leaps: Analogy in Creative Thought*. Cambridge: The MIT Press.

Hoyt, Michael F., ed. 1998. *The Handbook of Constructive Therapies: Innovative Approaches from Leading Practitioners*. San Francisco: Jossey-Bass Publishers.

Jahn, Manfred. 2003. "'Awake! Open Your Eyes!' The Cognitive Logic of External and Internal Stories." In *Narrative Theory and the Cognitive Sciences*, edited by David Herman, 195–213. Stanford: CSLI Publications.

Koch-Westenholz, Ulla. 1995. *Mesopotamian Astrology: An Introduction to Babylonian and Assyrian Celestial Divination*. Copenhagen: Museum Tusculanum Press.

Kupferberg, Feiwel. 1995. "Biografisk Självgestaltning." *Sociologisk Forskning* 31(3): 32–57.

Leopold, Anita Maria. 2004. *Syncretism: The Problem of Defining a Category*. Unpublished Ph.D. dissertation, University of Aarhus.

Leopold, Anita Maria and Jeppe Sinding Jensen, eds. 2004. *Syncretism in Religion: A Reader*. London: Equinox.

Linde, Charlotte. 1993. *Life Stories: The Creation of Coherence*. New York: Oxford University Press.

McCauley, Robert N. and E. Thomas Lawson. 2002. *Bringing Ritual to Mind: Psychological Foundations of Cultural Forms*. Cambridge: Cambridge University Press.

Midgley, Mary. 2003. *The Myths We Live By*. London: Routledge.

Munk, Kirstine (forthcoming). *Signs of the Times: Why do Modern People Use Astrology?* London: Equinox.

Nelson, Katherine, ed. 1989. *Narratives from the Crib*. Cambridge, MA: Harvard University Press.

Schank, Roger D. 1990. *Tell Me A Story: A New Look at Real and Artificial Memory*. New York: Charles Scribner's Sons.

Siegel, Daniel J. 1999. *The Developing Mind*. New York: Guilford Press.

Stern, Daniel N. 1985. *The Interpersonal World of the Infant*. New York: Basic Books.

Stuckrad, Kocku von. 2003. *Geschichte der Astrologie von den Anfängen bis zur Gegenwart*. Munich: C.H. Beck.

Thompson, John B. 1995. *The Modern Media and Modernity: A Social Theory of the Media*. Cambridge: Polity Press, 1999.

Turner, Mark. 1999. *The Literary Mind*. Oxford: Oxford University Press.

# — 18 —

# To Meet without Actually Meeting: Cultural Models of Virtual Rituals in 3D Cyberspace

Kenneth Hansen

## Introduction

The study of cultural practice supported by the Internet could benefit from a cognitive anthropological description of the cultural models in use. This chapter suggests that such a description could be achieved by the application of a generative theory, i.e. Lawson and McCauley's theory of religious rituals. The subject of this study is two cultural practices supported by the 3D cyberspace system "Active Worlds." The chapter briefly outlines the development of the system, the ritualized interaction structures, and the emergence of a central ritual: a wedding. Following this, Lawson and McCauley's theory is presented, and a comparative analysis of virtual rituals in the two cultural practices is carried out. Finally, a cultural model of the virtual rituals is outlined and discussed.

## A Cultural Model

A model is, according to cognitive anthropologist Roy D'Andrade, "an interrelated set of cognitive schemas used to represent something, to reason with or to calculate from by mentally manipulating the parts of the model to solve some problem" (1995, 180). A cultural model is typically a larger number of interrelated schemata used in a cultural practice, for example a model of navigation. The difference between the theory and the model is that the knowledge that makes up a cultural model most often is procedural in character: a cultural theory is made up primarily of declarative knowledge (1995, 73). Cultural models are good *to think with*, cultural theories are good *to think about* (1995, 178).

A cultural model is assumed to be part of a larger whole, for example a "culture." But, as D'Andrade has stressed (1995, 244–252), no such integrated whole seems to exist:

> The empirical fact is that culture looks more like the collected denizens of a tide pool than a single octopus. Empirical work on plant taxonomies, color terms, models of the mind, navigation, land tenure schemas, kin terms, etc., reveals a world of independent mental

> representations. Each cultural model is "thing like," but all the models together do not form any kind of thing (1995, 249).

Consequently, it is a mistake to look for a "virtual culture." Such a "thing" seems only to exist in fleeting performative moments. It is foamy and volatile. Similarly, it could seem futile to look for "universal principles" such as those Lawson and McCauley mention in the theory used here. However, it is possible to describe the cultural models that people use to do something within the cultural practices mediated by a 3D cyberspace system. Such a description might prove useful, it is my hope, in further work on computer supported cultural practice, for example the development of computer mediated learning, or the study of social computing in general.

## The Beginning: Active Worlds

The 3D cyberspace system "Active Worlds" has a rather chaotic history, and its development and design involves contributions from many people (see Hansen 2001 and 2002). In the beginning of 1994, three small software developers in the San Diego area were merged under the name "Knowledge Adventure Worlds." Their goal was to develop avatar-based multi-user 3D cyberspace systems for the Internet. The first attempt was made by the programmer Ron Britvich, and, because it was the first un-tested version, it was called "Alpha World." The first registered user was Britvich himself, who had set out alone in this venture. The company changed its name shortly afterwards to "Worlds Inc." and, after some discussion, decided to support the development. Initially, the designers Lynne Ann and Danny Viscas set to work on the design of avatars and building objects. The musician David Toley was asked to design background MIDI music and sound effects, and a dancer, Pamela Forth, later designed the movements of the avatars. The system was officially released June 27, 1995. It consisted at that time of a chat window and a window where 3D graphics were rendered as one walked through a virtual city. It looked like a combination of a web-based chat system and a 3D computer game. The heart of the system was a "3D engine" from the company "RenderWare," and the 3D files were either downloaded from the net or stored on a CD-ROM, which could be obtained by mail. In the first accessible 3D world, it was possible to use a single motionless avatar named "Cy," short for "Cyborg," and it was possible to build simple constructions by piling up basic Euclidean objects one by one.

The "Alpha World" soon got several thousand users, primarily from the curious developer community, who eagerly monitored the works of their colleagues on the net. The way to get started was the usual one on the Internet. The user had to download the AW browser, complete the installation and launch the program. After some marketing efforts, Worlds Inc. succeeded in getting enough interest to justify the hiring of another developer, the programmer Roland Vilet, to support Britvich. The influence of Vilet on the design of technology and social structures was significant. With a special understanding of the cultural aspects of the functions of the system, he made a crucial contribution to its continuing growth. Vilet's most important contribution was doubtless the development of a means to stream the descriptions of the 3D objects directly into the 3D window. This technique made the program a de facto 3D browser and meant that users did not have to store large amounts of 3D files on their computers in advance. Instead it was now possible to continuously send and update descriptions of 3D worlds to the user and combine these with cached files from earlier world visits (Figure 18.1).

Figure 18.1 Alpha World, the first world in Active Worlds. Avatars looking at poster at "Ground Zero." Ron Britvich is posing as "Protagonist" on the right. In the back a black "teleport" is visible.

The development was driven by an open dialogue with the users. Often this happened as a process in which the often very committed users drew the attention of the programmers to technical aspects of the system that prevented the on-going development of the culture; the programmers then concentrated on finding solutions to these issues. Several fundamental issues emerged in this way. First of all, it soon became obvious that the users had too many problems of orientation and navigation. Alpha World was, and still is, a very large place, and people were building anywhere they saw fit. Even though the system came with a horizontal system of coordinates, it was almost impossible to find your way. The system of coordinates was too abstract. The situation for the user could be compared to being put down by a helicopter in a surrealistic city. Buildings and other obstacles didn't appear on the screen until a certain approximation was reached, and they disappeared again when passed. Furthermore what could be traversed and what had to be bypassed was not obvious since the "collision detect" function for the 3D objects was not yet implemented in the browser. This situation was only remedied when Roland Vilet came into the project. The general consequence of this was that people had great difficulties in coordinating their actions. Put simply: it was hard to meet, and hard to meet again.

But Alpha World did not only seem enormous from the perspective of a single avatar. To save time on downloading and rendering, the visibility in the browser was limited to no more than 120 meters. Furthermore, the number of fellow avatars one could see was restricted to 12 at a time.[1] To find each other in such an environment was, obviously, almost impossible. Roland Vilet began therefore to develop a system of public transporta-

1. After 20 May 1998, the number rose to 50.

tion. With the teleportation feature, objects in Alpha World could be linked to a coordinate somewhere else in the world. Small black teleport booths soon began to mushroom everywhere. Vilet limited the number of these so that the impression of a world with a finite size would not be lost and gave permission to begin the construction of an official "metro net." Users now had a common transportation system, which they could use to move around more precisely (Figure 18.2). They could also use the official teleports as points of reference, for example the largest teleport "Worlds Transit Authority Teleport Station." The avatars "Dataman," "Little Bull," "Desi" and "Laura" built this "main station." Together they made up "The Robocob Team." This team was the first example of a more formal group of builders and laid the foundation for the later establishment of the first building "guild," the so-called "Titans." Such a group can be said to form a distributed functional system. It evolves slowly and provides a basis for complex actions later, for example a wedding.

The "Titans" consisted of Russell Freemant, known as the avatar "Dataman;" the avatar "Razzle," who was actually two persons, the Australians Liz and Dave Rasmussen; "Little Bull," whose real identity is not known; and "HenrikG," the Norwegian engineer, Henrik Gudbrandsen, who later constructed "Patagonia" (see below). The group constructed a handful of worlds, the most successful perhaps "Winter Wonderland." Originally conceptualized only as a Christmas world, it quickly became so popular that it was kept open for everybody all the year round. In "Winter" Henrik Gudbrandsen built a large traditional church, which was later used as the scene for the most significant interaction structure in the development of the AW universe, the wedding of "Trinket" and "Midnight Madness."

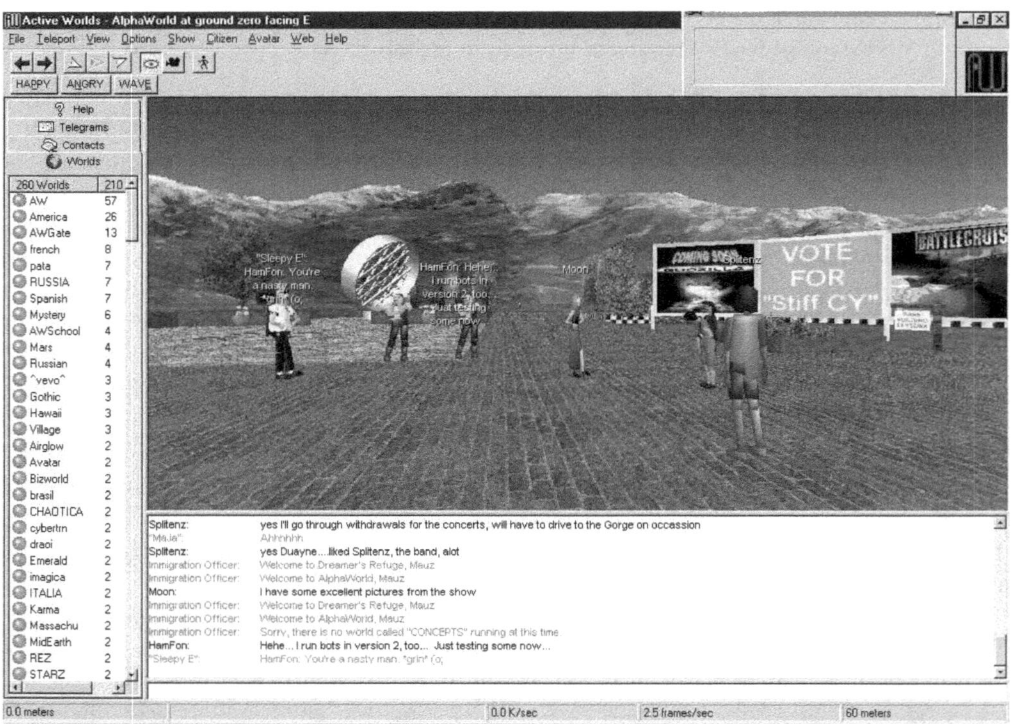


Figure 18.2 Active Worlds. The browser with a chain of worlds to the left, various communication options and a menu of avatar expressions.

## Virtual Weddings

Weddings have been part of the activities in Alpha World since the very start. The first wedding, between the avatars "Starshaddow" and "WetnWild," took place May 8, 1996. Following this event, weddings have been a recurrent happening. The most elaborate wedding to date, however, was the wedding between "Trinket" and "Midnight Madness" that took place May 16, 1998. The wedding was a carefully planned event that had been prepared for a long time. Not only the church had to be built, but also decorations and effects—flowers, tuxedoes, invitations, a wedding cake etc.—had to be produced. The design of each of these items had to be discussed and negotiated by the actors and the "Titans," who for the most part were in charge of the actual constructions. The final designs made up the contextual frame for the event, but the coordinated interactions in the design process itself also created and reproduced a basic social structure—a kind of interactional infrastructure—which the event then consolidated. The design framework still stands visible in "Winter Wonderland" and is reproduced every time an avatar walks through, rendering it all in the browser. The same goes for the corresponding social infrastructure. Thus the wedding of "Trinket" and "Midnight Madness" illustrates how the design of the tools has evolved to sustain a highly complicated task. The ritualized structures of interactions in the ongoing design process interlock and create a large system of coordinated actions.

## A Distributed System

The system might be further characterized from the perspective of cognitive anthropology. In this tradition, Edwin Hutchins has recently suggested that coordinated actions might be described as generally involving the processing of information using representational media: "The conduct of the activity proceeds by the operation of functional systems that bring representational media into coordination with one another. The representations may be inside as well as outside the individuals involved" (Hutchins 1995, 372).

The representational media can be "external," for example tools, maps, computer monitors, program windows etc. or "internal," for example mental models or schemata (see also Hansen 1999). In Alpha World, if one is alone in carrying out a task, for example building a church, one has to bring an array of these representational media into coordination with each other. Internally stored information on the locality: where suitable building materials can be located—and navigation: how is walking, flying, warping or teleporting performed—is brought into coordination with information represented on the browser: where are we, the time, the direction, the perspective, the avatar, and which options are used, etc. All this is used to carry out the task at hand using the appropriate building tools. A person doing this is thus what Hutchins calls a "functional system" for the coordination of internal and external representational media. If more than one person is part of the process, this is done in a similar way. The participants then form a kind of distributed functional system that propagates information by bringing representational media into coordination with each other.[2]

With this analytical frame in mind, it would seem reasonable to assume that a large event such as a wedding demands the existence of a comparable distributed system. A distributed system thus appears very early in the development with the birth of "The Titan Guild."

2. The concept of distributed cognition is discussed in greater detail in Cole and Engeström 1993. See also Hansen 2002 for a short discussion of definitions.

The wedding illustrates how the basic mechanisms in this distributed system are used to create an even larger functional system: the wedding. The wedding event is closely connected to the evolution of the building guild, which is closely connected to the development and design of the basic representational media and tools. It is thus no coincidence that the Titan Henrik Gudbrandsen constructed the church in which the wedding took place, and that the "virtual father" of the bride was the Titan "Razzle."

The participants in the wedding are members of this distributed functional system. They too have to be socially competent and must control their avatars in coordination with the other avatars. They must present their roles in a way that makes sense to the audience—a diverse group consisting of the other users present as avatars, users logged-in and present but not visible as avatars, the programmers who control everything and might be invisibly present, and a variety of bystanders following logged in users—and they have to make and continuously support a shared representation of the situation which is adequate to maintain their interactions and the progress of the event. The audience, for their part, has to understand the ongoing performance, its goal and processes, and must know how to play their parts when called on. Finally, the external representational media (the church, the avatars, the wedding cake, etc.) have to contain enough information to support the whole event and keep it moving towards its goal: the marriage.

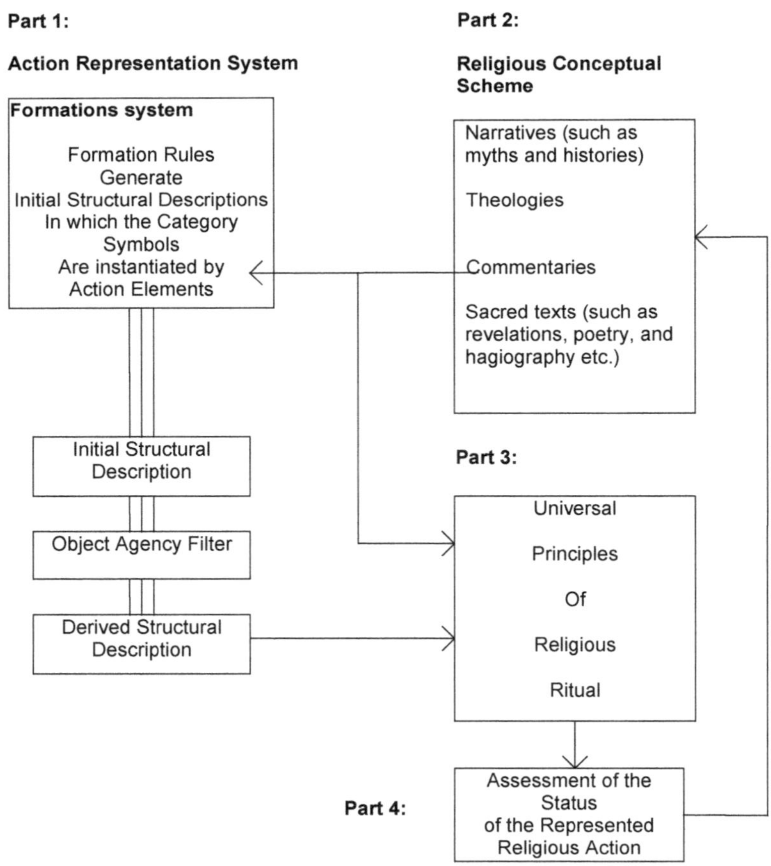


Figure 18.3 Lawson and McCauley's theory of religious rituals.

This system can now be placed in a cultural model by applying a generative theory of religious ritual to the description of "virtual rituals."

## A Generative Theory of Religious Rituals

E.T. Lawson and R.N. McCauley presented their theory on religious rituals in the book *Rethinking Religion: Connecting Cognition and Culture* from 1990. From the outset, it must be reiterated that the generative theory of Lawson and McCauley is used in this context not as a kind of "cognitive" theory, but merely as a distributed cultural model. This places it within the traditions of ethnomethodology (Garfinkel 1967, Sacks, Scheegloff and Jefferson 1974 or, for example, Cicourel 1973) and cognitive anthropology (D'Andrade 1995, Hutchins 1995). The work of Lawson and McCauley raises many important questions concerning, for instance, the relation between cognition, language and structural descriptions of culture and action. This fundamental discussion is, however, not within the scope of this chapter.

Lawson and McCauley's theory is based on the fundamental observation that all religious rituals involve supernatural agents. This typically happens through a ritually constructed relation between participants and a god or several gods. This fundamental relation is established either through the main ritual itself or has been presumed in a series of earlier minor rituals. Hence, the structural analysis of a religious ritual must include, somewhere, a supernatural agent. Likewise, "virtual rituals"[3] in the 3D cyberspace system "Active Worlds" always involve the representation of the user as a kind of "supernatural" agent, i.e. as an avatar. The original theory can now be outlined as follows.

The theory of religious rituals comprises four main parts, which together generate ritual religious actions and action complexes (Figure 18.3):

> The first part—*The Action Representation System*—is the basic syntactic component which implements formation rules on the category symbols from the conceptual scheme in order to generate structural descriptions. These are displayed in a formation tree as known from traditional generative linguistics (see examples later). The second part—*The Religious Conceptual Scheme*—is the semantic component, which connects the highly general formation system with, for example, a cultural system. The third part is the set of universal principles. These are generated through induction as the whole system runs. Every time it runs, it regulates itself through the fourth and last assessment part, a control mechanism for feedback and evaluation.[4]

From a large number of descriptions, several so-called "universal principles" are derived. Referring to generative linguistics, Lawson and McCauley designate these as "functional universals." Functional universals concern relations between category elements in the basic syntactic part of the theory; the two others relate to the combination of category symbols and action elements. These are, for example, represented in the "object agency filter," a mechanism that rearranges the structural description to ensure that objects do not appear, counter-intuitively, to act (see later). Lawson and McCauley define the universals as:

> ...a partial list of higher-order universal principles that will systematize participants' judgments about the relationships of various religious actions to the overall religious system in which they figure (201).

3. This concept will be defined later.
4. For an account of the fundamentals of generative linguistic theory, see for example Lyons 1986 and Radford 1981.

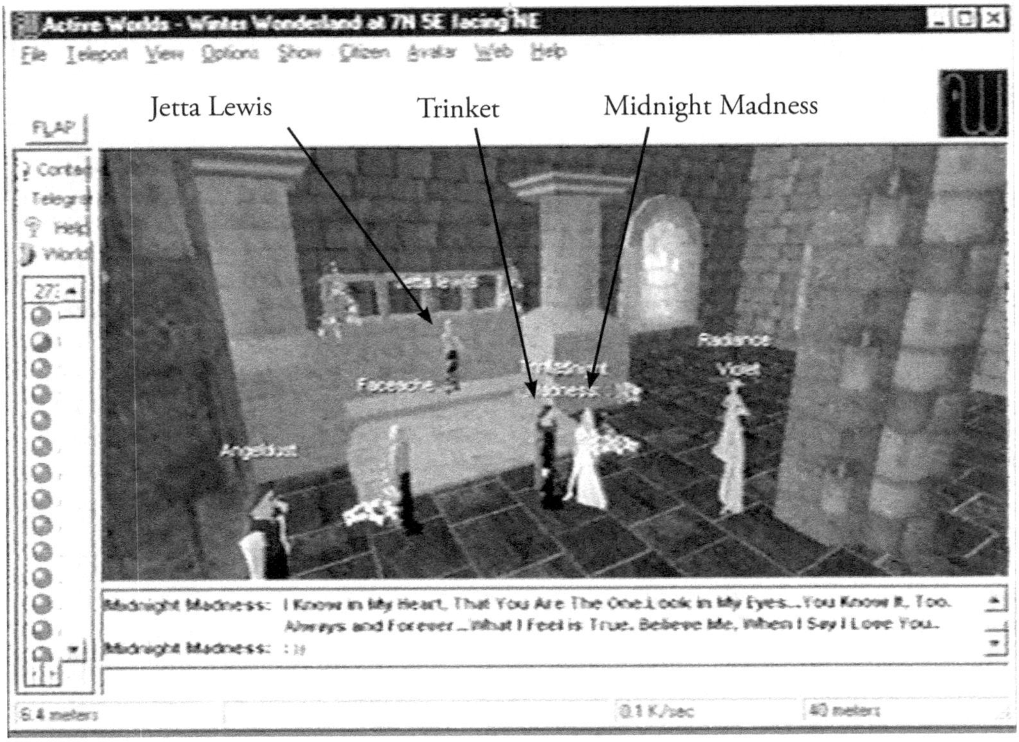


Figure 18.4 Virtual Wedding. Midnight Madness and Trinket, two avatars in Active Worlds, marry on May 5, 1997 in the large church of Winter Wonderland. Here, at the beginning of the ceremony, the bride recites a short poem. Above is the reverend, Jetta Lewis, and around them are gathered a selection of bridesmaids and servants. Approximately 80 users of the system are present but the software at this point of its development is only able to display 12 avatars at a time. The scene is recorded through the eyes of an avatar bird, and hence the poor quality. The text box reads:

Midnight Madness: I Know My Heart, That You Are The One. Look in My Eyes... You Know it, Too. Always and Forever...What I Feel is True, When I Say I Love You.
Midnight Madness: :))

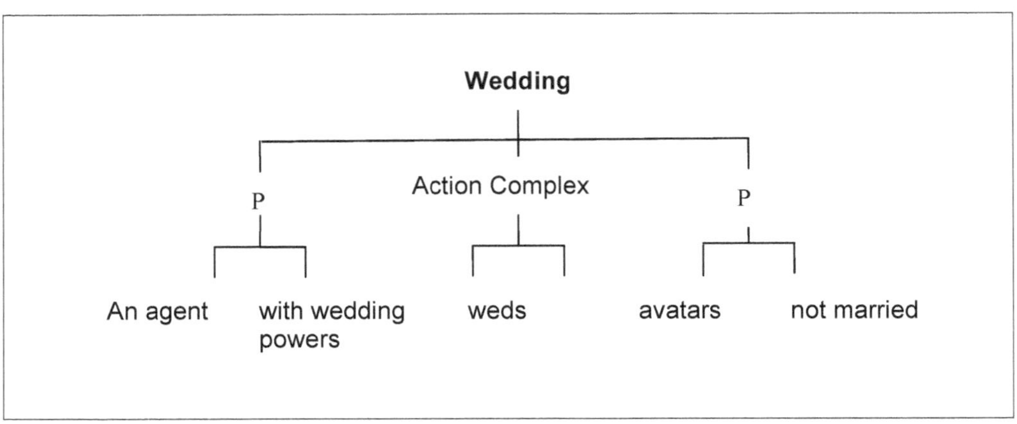


Figure 18.5 A first description of the event would look like this (P designates "participant").

The most important of these functional universals is, of course, what Lawson and McCauley call "The Principle of Superhuman Agency." This principle implies that all religious rituals somehow involve supernatural agents.[5]

Following this, a distinction of five ritual types shows how those rituals where the superhuman agent functions as the agent always prove more central to a religious system than those where the superhuman agents serve some other role. Thus an ordering of ritual types reflects the second universal principle, which is called "The Principle of Superhuman Immediacy."

Lawson and McCauley suggest a third and last universal principle. It prescribes that religious actions always involve an object. Rituals always do something to something or someone, i.e. rituals are performed in order to bring about a change within the religious context. Thus, for example, it is not possible to bless without blessing someone or something. Religious actions always have to have an object or they will become meaningless.

As stated in the introduction to this article, it may be futile to look for such "universal principles." Instead, Lawson and McCauley's theory is proposed as a cultural model in the tradition of ethnomethodology and cognitive anthropology. In this tradition, rules are seen as "methods" governing the creation of common social situations. The "universal" rules of the cultural model might be viewed as such rules.

## A Generative Theory of Virtual Rituals

In order to apply the theory to the study of virtual rituals of a cultural practice supported by a 3D cyberspace system, a way to start is to maintain the basic generative part and to attempt a description of the conceptual system, and then run the model and derive filters and universal principles. The first question, hence, is whether the basic generative "action engine" is able to generate the action complexes of virtual rituals. A typical example of such a virtual ritual is a virtual wedding (Figure 18.4).[6] A first description is depicted in Figure 18.5.

This preliminary and very limited description—which is expanded below—shows, not surprisingly, that the wedding has an active agent that acts, and that the acting is done in relation to an object, the wedding couple. This is true, of course, for both virtual and more normal weddings. Thus there does not seem to be a reason to assume that ritual actions in 3D cyberspace differ structurally from actions elsewhere. The only difference at this stage of the description is the fact that the agent—and the people appearing as object in the structural description—are only visible as avatars. Rituals that can be generated by the basic generative component in this way and carry this feature might thus be termed *virtual rituals* (see the final section of the article for a definition of this term). Other examples from the cultural practice of "Alpha World" include the "AW Bingo," the "Cy Awards" and the "A'tuin Treasure Hunt."

From this structural description it follows that the virtual ritual implies that an agent represented as an avatar gives something to someone who is represented as an avatar. This regularity gives reason to assume the existence of a distributed interpretational rule. This

5. It would seem problematic for the theory that it in this way apparently produces its own basic assumption, i.e. that it deals with religious rituals. However, since the output is only derived through processing and thus not formally proved, this does not violate the laws of formal systems. What it does do, though, is illustrate the limitations of such laws.

6. The wedding is described and analyzed in Hansen 1999, Barbatsis, Fegan and Hansen 1999, and Hansen 2002. The picture is a digitized still from a now rather ancient analog video recording, hence the poor quality.

rule is a feature of the distributed cultural model the participants use in performing virtual rituals. This first rule could be named "The Avatar Rule." It might be formulated like this:

***The Avatar Rule***
Virtual rituals are performed by agents represented as avatars.

In the ritual of the virtual wedding, however, one sees how this is not just a basic action complex with agents and other involved individuals represented as avatars. Following the theory of religious ritual, it is necessary to extend the syntactic part of the model with a conceptual system in order to reach a more thorough formal description. This conceptual system, however, is somewhat different from that of normal weddings. Compared, for example, to the official wedding procedures in a Protestant church in the United States, this becomes clear. The start of the ceremony shows this most clearly. Here the reverend declares (Figure 18.6):

> Midnight Madness and Trinket, today you begin your life journey together. The journey starts in Alphaworlds and has no destination in particular. Along the paths you follow, you will see many wonders and a few obstacles, but you embark upon your journey with a strong love, pure joy and a deep commitment to each other. With the following vows, you pledge to help fill this journey with ever growing love and affection, fierce passion, quiet tenderness, unanswering faithfulness, patience, understanding, and a love that will fill your heart and soul.

This introduction, of course, does not exist in a standard wedding ceremony. The so-called "life journey" of Midnight Madness and Trinket begins in "Alpha World"[7] and has no destination "in particular." The rather vague statement reflects a basic structural problem with these kinds of weddings, to which I will return to shortly.

The following exchange is also unusual:

| | |
|---|---|
| JL: | Who gives this woman to be married to this man? |
| Razzle: | I do |

The person who normally gives the bride away is the father of the bride, who on behalf of the family gives the daughter away to another man. In this case, it is the Titan "Razzle," the head of the Peacekeepers, the police force of Alpha World, and the owner of the world in which the wedding takes place, who acts as "father." The giving of gifts is, according to Lawson and McCauley, fundamentally a non-reversible ritual. Thus, gifts can be repaid but not returned. So in the case of traditional weddings, the groom has to find other ways to repay this generosity, for example by taking care of the family in the future. This is obviously a mechanism to secure relations and cultural formations. However, if the donator is a somewhat diffuse agent, the mechanism will not be able to operate, and the consolidating effect of the ritual will be limited. This is the case of the virtual wedding since the virtual father is both an avatar and two people, a married couple participating from Australia. The absence of the option of divorce, an act that could reverse the first ritual, also points in this direction. The marriage will most likely not last and a reversal is hardly needed.

The linguistic illocutionary force of the speech act also seems problematic. The person performing the wedding declares: "what God has joined, let no man put asunder," thereby

7. This could appear rather imprecise. Does it mean the whole Active Worlds system, the Alpha World part (the first world of Active Worlds) or the cultural practice of Alpha World (using several AW worlds, including Alpha World)? In fact it is a highly precise designation of the cultural practice being reproduced in the ceremony, i.e. a cultural practice that now incorporates the use of several AW worlds.

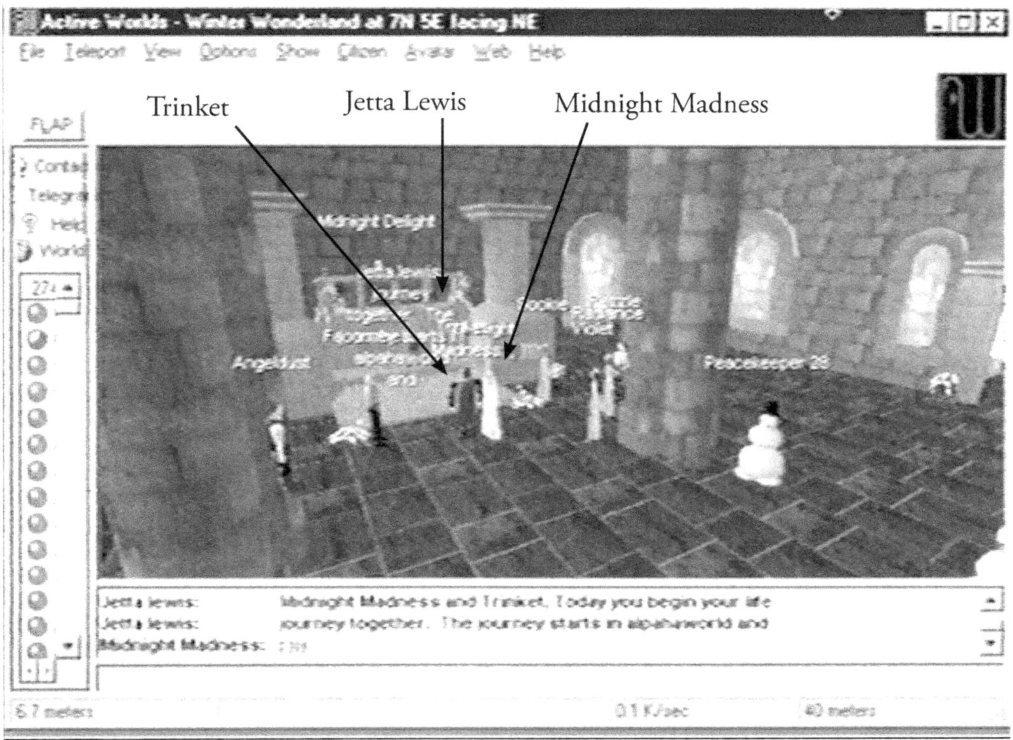


Figure 18.6 The ceremony starts with a brief speech by the reverend, Jetta Lewis. Text box reads:

Jetta lewis: Midnight Madness and Trinket, Today you begin your life
Jetta lewis: journey together. The Journey starts in Alphaworld
Midnight Madness: :))

acting as the representative of a superhuman agent. Similarly, category symbols such as altars, rings, churches, etc. are used in reference to traditional religious rituals that presume a superhuman agent. But the person performing the wedding has not been ordained in a previously performed ritual; she only has a license to perform marriages. Thus no superhuman agent is represented in the structural description. This is even more problematic as the wedding is not even legally valid, as people in the U.S. have to be physically co-present in order to be lawfully married. The final structural description can now be shown (Figure 18.7, next page).

This structural description shows that the cultural model is able to generate the same action complex in two co-existing ways: with or without the action element of **e**. This symbol represents a special quality, i.e. that the meeting is mediated by some kind of "transcendental instance," and that this is a quality of the action in itself. Two descriptions of the complex are, of course, intuitively wrong and formally not acceptable. This fact reflects a turn in the development of the original theory:

> The rules that McCauley and I have proposed initially permit not only descriptions of priests drinking wine (agents acting upon objects) but also permit wine to consecrate ritual participants (objects acting as agent upon objects). In this latter instance we clearly have a situation in which commonsense assumptions are violated. This is an infelicitous result not only in our ordinary understanding of action but also in some sense in religious understandings as well. (197)

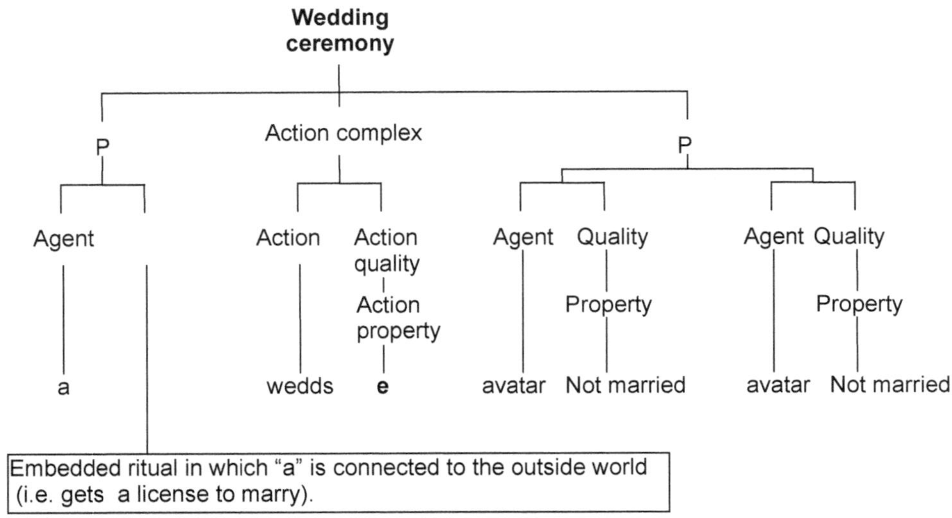


Figure 18.7 The final structural description of the wedding.

The answer to these problems is, like in the transformative turn of generative linguistics, to propose a transformative element that can prevent objects from appearing as agents. This is called the "object agency filter":

> The object agency filter is a principle which either eliminates such infelicitous descriptions (in which objects assume the role of agents acting upon objects, for example wine consecrating ritual participants) or re-arranges the description. For example, the object agency filter permits a redescription in which a permissible agent, a parishioner, consecrates herself by means of drinking consecrated wine. (198)[8]

From this perspective it would seem obvious to apply a filter in a similar fashion. But that would imply that we intuitively could decide which of the two descriptions was the right one. This is, however, not an easy task. Which one is the right one in this case? The situation appears absurd: how can the meeting happen as a consequence of the action in itself? Without a person with a sanctioned ability to wed, it is not normally possible to imagine a successful marriage act (Austin 1962). But as the wedding in an American context is not legally acceptable, the final description has to be the one where the action is special in itself. This is the description that the participants must accept in order to get married. If they did not do this, there would be no reason to go through the ritual. Thus no "object agency filter" or transformation rule is applied.

That the transformation rule is lacking points towards a general condition for the generation of descriptions. This can be seen as a second interpretational rule in the cultural model. This rule calls for a suspension of the formally necessary transformation rule, which demands that the quality of an action is part of the one who is acting. The transformation rule is suspended, or put in brackets. The transformation rule is necessary to describe

8. It has to be stressed that this example is taken from the ceremony of the Catholic church. Here the transformation of wine to blood is a ritual miracle. In the Christian Lutheran church, the act is only viewed as a symbol.

intuitively acceptable action complexes. This could be said to take place in what phenomenologists call "the natural attitude": the state of mind we normally use to perform in social situations. To put the natural attitude in brackets is what one of the founders of modern phenomenology, Edmund Husserl, has termed the phenomenological reduction: "The Epoché" (Husserl 1907/1997). I therefore propose to call this rule *The Epoché Rule.* This rule can be outlined like this:

***The Epoché Rule***
The agents must activate and maintain a special Epoché.

This formulation, however, needs some elaboration, to which I will return shortly.

Why do people choose to get married in this ambivalent way? From the descriptions of the cultural practice in Active Worlds (Hansen 2001 and 2002), it is possible to argue that they get married because it is the most symbolically elaborate expression of the meaning and function of these kinds of cultural practices. The goal and meaning of the system is to give people a way to meet and continue to meet.

This is initially done by developing and using a distributed functional system. This system is then used and reproduced by creating and maintaining a ritualized action structure, where the performance is reflected against a kind of "other place." From a system theoretical and phenomenological perspective, the "other place" is just that: something we refer to which is outside the world that surrounds us. In this way, it cannot be transcended. God is an example of this, so is a Utopia.

Such a performance can be regarded as what the French philosopher Michel Foucault has called a heterotopia (Foucault 1986). The role of the transcendent reference is, of course, most obvious in the marriage act as described here, where people come together "through" God (Figure 18.8, next page). But the virtual rituals previously described, all share the same feature, i.e. they all imply that agents' attributes have been given and are sanctioned by an initial God-like position somewhere outside the universe. This means that somewhere in the structural description of the virtual ritual there will always be a reference to the "other place." In developing the concept of heterotopia, Foucault used the term "Utopia" to designate this "other place," thus the third rule or feature of the cultural model might be named *The Utopia Rule.*

***The Utopia Rule***
The performance implies the reference to a place outside the world.

## Another Structure

As we have seen, the virtual wedding is a problematic mechanism if its purpose is to support the survival of the cultural practice. Could another structure serve this function in a more optimal way? If we turn to the cultural practice of Norwegian "Patagonia," the most important ritual is not a virtual wedding but the inauguration of a virtual parliament (Figure 18.9, next page) (For a thorough description of the situation and the history behind it, see Hansen 2001 and 2002).

Prior to the inauguration, the participants were elected in a elaborate election procedure, which can be seen as a set of embedded virtual rituals. The structure of the inauguration is depicted in Figure 18.10 (page 317).

This description shows how the users of Patagonia have solved the fundamental problem as depicted in the analysis of the virtual wedding. In this case it is not possible to gener-

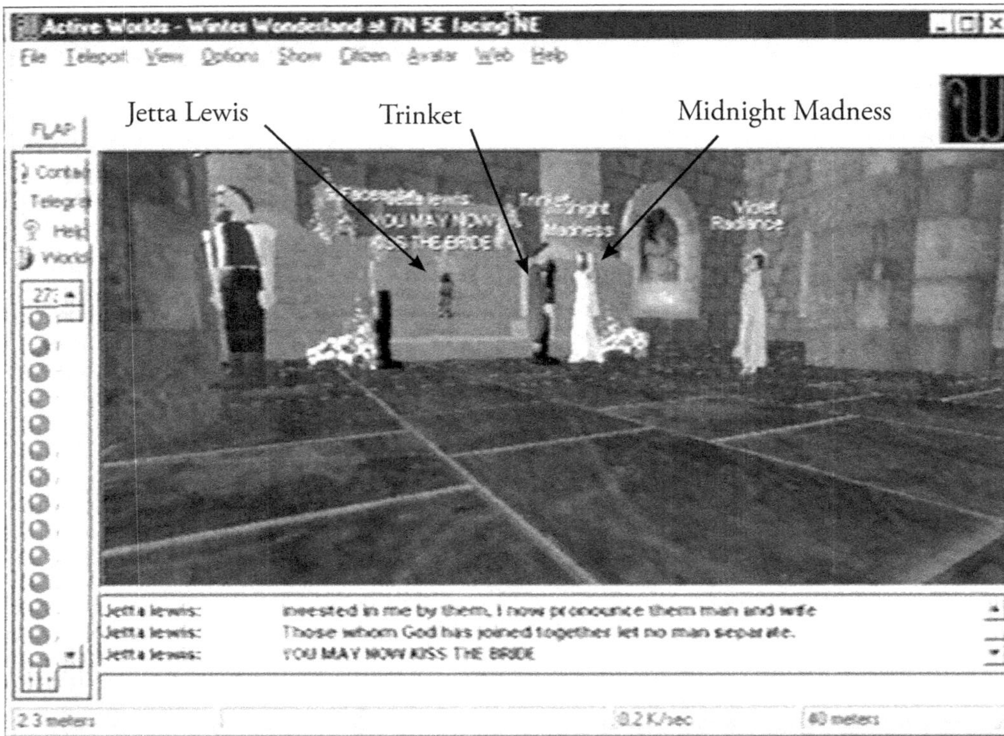


Figure 18.8 The wedding ends with the performative act of marrying. This is done with an explicit reference to God. Finally the groom may kiss the bride. Text box reads:
Jetta Lewis: invested in me by them, I now pronounce them man and wife
Jetta Lewis:Those whom God has joined together let no man separate.
Jetta Lewis: YOU MAY NOW KISS THE BRIDE

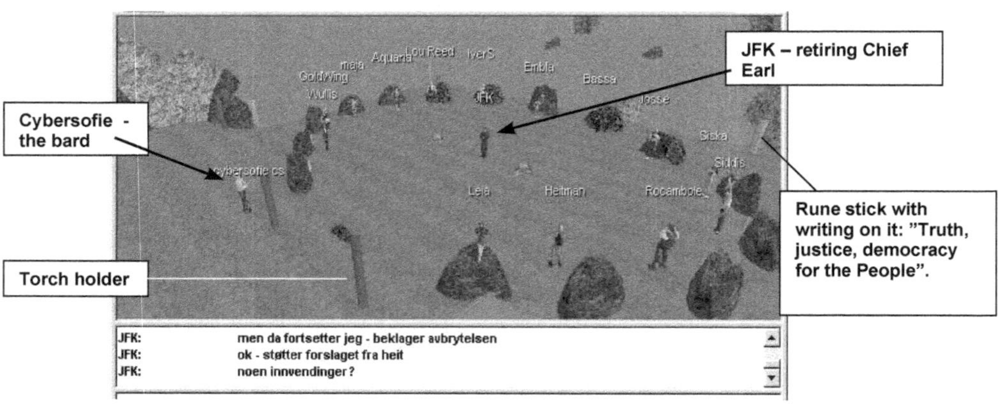


Figure 18.9 The virtual parliament of Patagonia has been inaugurated. Now the new Parliament performs their first meeting. It is led by the retiring Chief Earl, "JFK," in the middle. Each of the members, called Earls, stands in front of a rune stone, each with one rune on it. Together the runes form the word rettferdighet, Norwegian for "justice." Outside the Parliament circle, the "Rune Ring," "Cybersofie" awaits. She serves as the bard of the inauguration leading it from point to point in the ritual represented as verses in the "The Lay of the Bard."

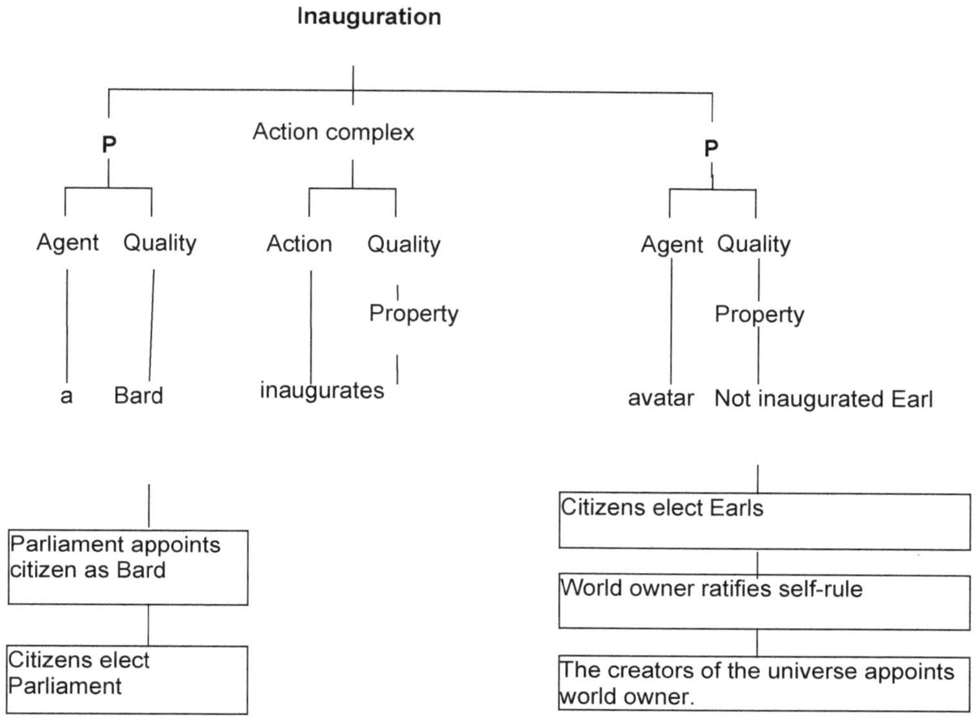


Figure 18.10 The structure of the inauguration.

ate two structural descriptions of the same virtual ritual. In addition, it is clear that the transcendental reference, the creators of the universe and the owner of the world have also essentially disappeared. Instead we find the Parliament. Consequently, there seems to be no reason to activate *The Epoché Rule*. This is, however, problematic. Phenomenologically, the users of a 3D cyberspace system have to handle a set of different social "worlds."[9]

This juggling of social worlds is perhaps most obvious in the case of the virtual wedding, where all the participants have to play their parts for hours in order to make the marriage happen successfully (the whole event lasted approx. 6 hours). This is a complicated task, as most of them probably sit at home and have to participate in the wedding—or example by taking up the right position, typing polite contributions, admiring the wedding dress, etc.—while shifting between operating the computer (for example the 3D browser and secondary chat programs) and being physically present in the physical context (for example interacting with family members). The participant is not "here" in the living room, but neither is he or she "there," at the wedding. Thus, the mental mechanism underlying *The*

9. "World" is here used in the social constructionist tradition. It designates what Schutz, with reference to Henri Bergson and Edmund Husserl, calls "finite provinces of meaning." "We speak of provinces of meaning and not of sub-universes because it is the meaning of our experiences and not the ontological structure of the objects which constitute reality" (Schutz 1973, 229). Erwin Goffman in particular has chosen not to speak about "worlds." Instead he uses the concept of "frames" (Goffman 1986, 6). When dealing with virtual reality, however, it would seem more appropriate to use the concept of "worlds" with the spatiality it implies (see also Welton 1997).

*Epoché Rule* can be understood as the ability to suspend, concurrently, belief in two social worlds, to be in a doubly reduced state. This is only possible as an oscillation between two reduced states of consciousness over time. It is this oscillation that creates the perception of "cyberspace"—like a mirror reflecting a mirror. This crucial point is treated more thoroughly in the article "The Performance of Cyberspace: An Exploration in Computer Mediated Reality" (Barbatsis, Fegan and Hansen 1999). Following this, *The Epoché Rule* can now be given a final formulation:

***The Epoché Rule***
The agents must activate and maintain an oscillation between two reduced states of consciousness over time.

If then, as in the case of Patagonia, this phenomenological oscillation disappears, it would seem problematic to construct the virtual at all. Furthermore, the lack of a suspension rule will allow the reappearance of the "object agency filter" in the model, making it impossible in the description for actions to be transcendentally sanctioned, i.e. special in themselves.

Thus it would seem that the users of Patagonia have exchanged freedom for stability. On the one hand, the more easily understandable election rituals in fact replace the ambivalent and paradoxical wedding rituals, giving the cultural practice more solidity. But on the other hand, the lack of ambivalence prevents the participants from meeting in a kind of performative space represented in the oscillation between two reduced states of consciousness. Thus it seems as if some of the liberating features of the system have vanished and the citizens have created a new paradox. In the effort to create a free democratic system, a "free space" independent from external forces, they have built a structure which will prevent the construction of the "free space." Thus, the more free they become, the less free they become. If this tendency continues, it will then, surely, lead to the collapse of this cultural practice. To counteract this tendency, the users of Patagonia have therefore developed another elaborate

Figure 18.11 The Pata Gathering took place on a camping lot one hour's drive from Oslo. In this picture the author raises his tent while people slowly arrive.

Figure 18.12 The three participants. Note the historical tattoos and the mobile phone, both common cultural items. On the left is Henrik Guldbrandsen, caretaker of Patagonia and creator of the virtual church in which the virtual wedding took place.

virtual ritual: "The Pata Gathering" (Figures 18.11, 18.12; 18.14 and 18.15, next page).

The gathering, which took place 3–5 September 1999, and which is described in more detail in Hansen 2001, reflects the desire to halt the deteriroiation of the cultural practice and keep the "free" performative space open. Almost all interactions were therefore affected by the need to reinforce *The Epoché Rule*. As stated above, it is through the mental mechanism underlying this rule that people are able to meet at all. The agents are thus still represented as avatars. That is the main imperative. So the ritual is, according to the definition stated above, still a virtual one, even though the representation is not directly mediated by the technology. Thus, people were only allowed to use their avatar names, and physical activities not previously represented in the 3D cyberspace system were frowned upon. People kept each other in check, so the cultural practice could survive and people could continue to meet.

The structure of this virtual ritual can thus be illustrated as below (Figure 18.13).

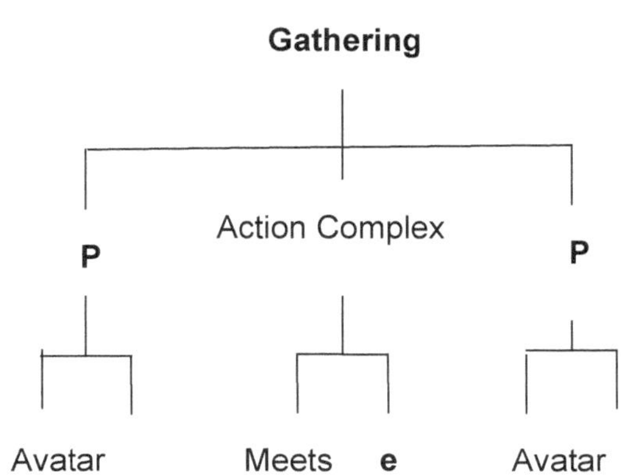


Figure 18.13 The structure of the ritual.

Figure 18.14 Two participants hanging out at night.

Figure 18.15 One participant brought a dog, another a red pick-up truck. The log cabin in the background was used for meals, a bar and bathing facilities.

Here we find again, as in the case of the wedding, two descriptions: one where the meeting takes place, because the individuals behind the avatars have moved physically and geographically, and one where the meeting is a function of the situation in itself. Thus, it is not the case that *The Epoché Rule* is not active any longer. It has just, to a certain extent, moved to other interaction structures.[10]

## Types of Virtual Rituals

From the structural descriptions presented here and in Hansen 2001, it is now possible to make a rough categorization. Three types of virtual rituals can be identified:

**Type 1:** Virtual rituals in which a meeting between two avatars is represented in the structural description as a quality of the action appended to the initial participant node of the target ritual. Examples of these kinds of virtual rituals are weddings and gatherings.

**Type 2:** Virtual rituals in which a meeting between two avatars is represented in the structural description somewhere other than under the initial participant node of the target ritual. Examples of these kinds of virtual rituals are the inauguration of the Parliament and other official meetings in Patagonia.

**Type 3:** Virtual rituals in which a meeting between two avatars is represented in the structural description as a quality of the action of an agent in a ritual embedded under the initial participant node of the target ritual. Examples are the Cy Awards ceremony and the AW Bingo.

From this a fourth interpretational rule can be described. This reflects the centrality of the meeting:

***The Meeting Rule***
The most central virtual rituals are those in which a meeting between two avatars is a direct part of the action complex.

This rule reflects how weddings and physical meetings are the most central of the cultural practices of Alpha World and Patagonia. After these, rituals like the inauguration of the Valhalla parliament occur, and, only after this, rituals like the Cy Awards and the AW Bingo.

## Conclusion

Virtual rituals in the 3D cyberspace system "Active Worlds" can, as we have seen, be described as action complexes generated by a theory such as the one presented by Lawson and McCauley. This description is here used as a distributed cultural model. The model can be represented as shown on the following page (Figure 18.16).

It follows from the descriptions and from the development of the cultural model that no transformation rule is used. This is different from Lawson and McCauley's theory. Instead, we find more weight placed on the feedback part, that is, the interpretational rules.

The rules are all derived from the first distinction: *The Avatar Rule*. This rule states that an avatar always has to be used in virtual rituals and thus distinguishes between virtual rituals and other rituals. Since this is the initial condition for choosing the object of description, this is of course always true. *The Epoché Rule* states that a certain oscillatory state of mind

10. It can also be located in "back-stage" interactions (Goffman 1963), for example on the chat service "ICQ."

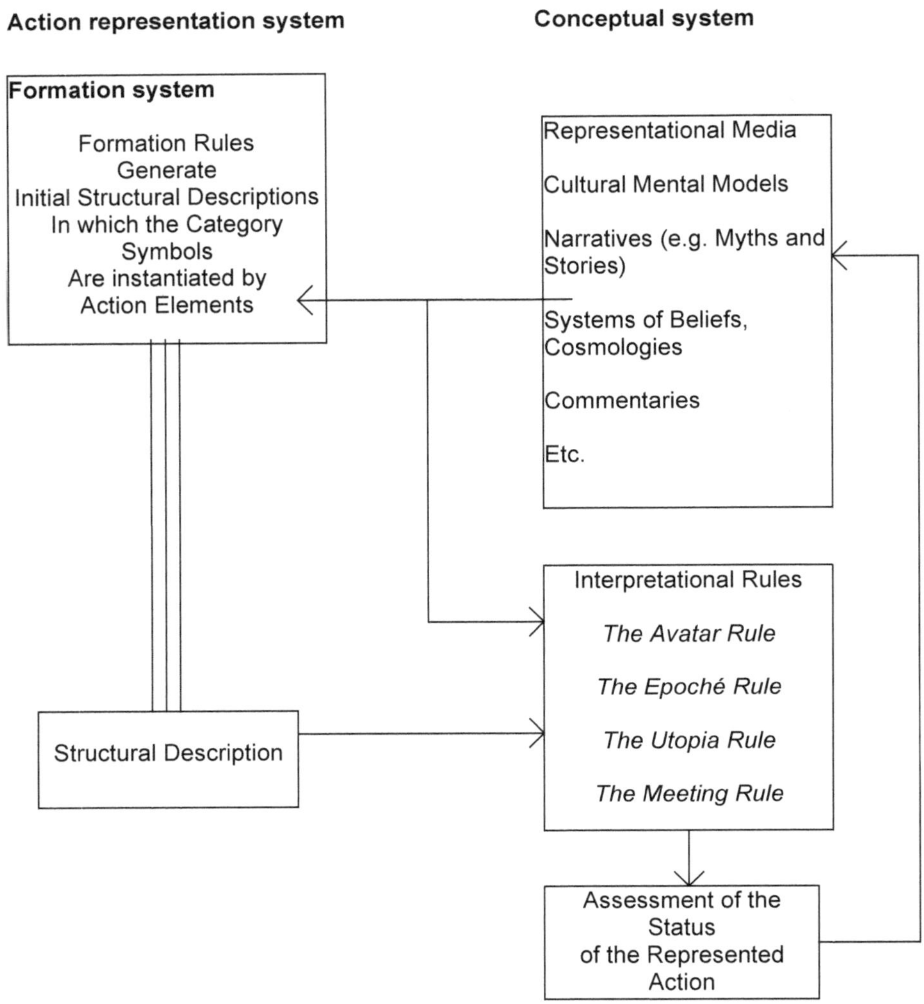


Figure 18.16 A cultural model of virtual rituals.

has to be maintained. The use of an avatar presumes the ability to perform this oscillation. *The Meeting Rule* states that virtual rituals in which a meeting between two avatars is a direct part of the action complex are the most central. This indicates that the participants want their avatars to meet. This intention is necessary in order to start the oscillation required by *The Epoché Rule*. Finally, *The Utopia Rule* states that virtual rituals always have a transcendental reference outside the social worlds. This is presumed in *The Meeting Rule* as it implies that the participants continuously "bend" their oscillations toward each other, aiming to transcend their own individual social worlds in a Utopian meeting. This is expressed in *The Utopia Rule*.

The cultural model generates social actions. As such it could be said to refer to the tradition of ethnomethodology, as mentioned above. As in this approach, the model in many ways reflects the fundamental phenomenological work of Edmund Husserl and Alfred Schutz. In their work, we find the constant struggle with the concept of transcendental intersubjectivity, i.e. the problem of how we construct a common ground for mutual understanding when the fact is that we are fundamentally separated from one another's experiences (see for example Schutz 1945 or 1957). This basic problem is also at the heart of this study.

People sitting in front of their computer screens in Norway, Denmark and the U.S. and meeting in 3D cyberspace as avatars in a virtual wedding or in a virtual inauguration do so by performing a paradoxical oscillation between two reduced states of consciousness. This might be the closest they will ever come to experiencing the promised land of "cyberspace."

## References


Austin, J.L. 1962. *How to Do Things with Words*. Oxford: Oxford University Press.

Barbatsis, G., M. Fegan and K. Hansen. 1999. "The Performance of Cyberspace: An Exploration into Computer Mediated Reality." *Electronic Journal of Computermediated Communication* 5 (1). http://www.ascusc.org/jcmc/vol5/issue1/barbatsis.html

Cicourel, A.V. 1973. *Cognitive Sociology*. Harmondsworth: Penguin Press.

Cole, M. and Y. Engeström. 1993. "A Cultural-Historical Approach to Distributed Cognition." In *Distributed Cognitions: Psychological and Educational Considerations*, edited by G. Salomon, 1–46. Cambridge: Cambridge University Press. .

D'Andrade, R. 1995. *The Development of Cognitive Anthropology*. Cambridge: Cambridge University Press.

Embre, L., E.A. Behnke, D. Carr, J.C. Evans, J. Huertas-Jourda, J.J. Kockelmans, W.R. McKenna, A. Mickunas, J.N. Mohanty, T.M. Seebohm, R.M. Zaner, eds. 1997. *Encyclopedia of Phenomenology*. Dordrecht: Kluwer Academic Publishers.

Foucault, M. 1986. "Of Other Spaces." *Diacritics* 16(1): 22–27.

Garfinkel, H. 1967. *Studies in Ethnomethodology*. Englewood Cliffs: Prentice-Hall.

Goffman, E. 1963. *Behavior in Public Places: Notes on the Social Organization of Gatherings*. New York: Free Press.

———. 1986. *Frame Analysis: An Essay on the Organization of Experience*. Boston: North Eastern Press.

Hansen, K. 1999. "Trinket and Midnight Madness Got Married—Or Did They?" Paper presented at the conference "Cyberculture and Technology," May 1999. Department of Literature, University of Copenhagen. Not published. Available in Danish at http://www.ruc.dk/~keha

———. 2001. *Virtual Culture*. Ph.D. Dissertation. Aalborg University/VR-Media Lab. Available with a summary in English at http://www.ruc.dk/~keha

———. 2002. "The Design of Public Virtual Spaces in 3D Virtual Worlds on the Internet." In *Virtual Space: Spaciality in Inhabited 3D Worlds*, edited by L. Qvortrup, 145–170. London: Springer.

Husserl, E. 1907. *Fænomenologiens idé* ("The Idea of Phenomenology"). København: Hans Reitzels Forlag, 1997.

Hutchins, E. 1995. *Cognition in the Wild*. Cambridge, MA: The MIT Press.

Lawson, E.T. and R.N. McCauley. 1990. *Rethinking Religion: Connecting Cognition and Culture*. Cambridge: Cambridge University Press.

Lyons, J. 1986. *Semantics*. Vol. 2. Cambridge: Cambridge University Press.

Radford, A. 1981. *Transformational Syntax: A Student's Guide to Chomsky's Extended Standard Theory*.

London: Cambridge University Press.

Sacks, H., E.A. Schegloff and G. Jefferson. 1974. "A Simplest Systematics for the Organization of Turn-Taking in Conversation." *Language* 50(4): 696–735.

Schutz, A. 1945. "The Paradox of the Transcendental Ego." In *Collected papers IV*, 190–192. Haag: Kluwer Academic Publishers.

———. 1957. "The Problem of Transcendental Intersubjectivity in Husserl." In *Collected Papers III*, 51–91. Haag: Kluwer Academic Publishers.

———. 1973. *Collected Papers I*. Haag: Martinus Nijhof.

———. 1975. *Collected Papers III*. Haag: Kluwer Academic Publishers.

———. 1996. Collected papers IV. Haag: Kluwer Academic Publishers.

Welton, D. 1997. "World." In *Encyclopedia of Phenomenology*, edited by L. Embre *et. al.*, 736–743. Dordrecht: Kluwer Academic Publishers.

# Index

For Product Safety Concerns and Information please contact our EU
representative GPSR@taylorandfrancis.com
Taylor & Francis Verlag GmbH, Kaufingerstraße 24, 80331 München, Germany

www.ingramcontent.com/pod-product-compliance
Lightning Source LLC
LaVergne TN
LVHW080638C11026
849160LV00014B/1250

* 9 7 8 1 8 4 5 5 3 2 9 5 6 *